ADVANCED

PLC HARDWARE & PROGRAMMING

Revision 1.0
ISBN 978-0-57-848223-1

Hardware and Software Basics, Advanced Techniques & Allen-Bradley and Siemens Platforms

By
Frank Lamb

© 2019 Frank Lamb, Automation Consulting, LLC. All rights reserved. Except as permitted under the United States Copyright Act of 1976, no part of this publication may be reproduced or distributed in any form or by any means, or stored in a data base or retrieval system, without the prior written permission of the publisher.

Information contained in this work has been obtained from sources believed to be reliable, however, the author shall not be responsible for any errors, omissions or damages arising out of use of this information. This work is published with the understanding that Automation Consulting, LLC and the author are supplying information but are not rendering engineering or other professional services. If such services are required, the assistance of an appropriate professional should be sought.

This book was expanded from the original training manual "PLC Hardware & Programming, Multi-Platform" © Automation Consulting, LLC, October 2016, Rev.1.1

www.automationllc.com

This Page Intentionally Left Blank

Table of Contents

Introduction 12

PLC Hardware & Programming - Overview 18

    What is a PLC, or Programmable Logic Controller? 18

The History and Evolution of Computers and PLCs 19

    The Babbage Analytical Engine 19

    Relay Logic 20

    Electromechanical Computers 20

    The First Electronic Computers 21

    Early Computer Memory 22

    The Evolution of Personal Computers 24

    Birth of the Programmable Controller 25

    PLC Improvements 28

        Mid-Range Modular PLCs 30

    PLC Timeline 31

    PLC and Computer History Bibliography 32

    Physical Layout of a PLC 34

    I/O (Inputs and Outputs) 34

        Digital/Discrete Devices 35

        Analog Devices 36

        Wiring - Digital 37

        Wiring – Analog 40

        Battery/Memory Back-Up 42

    Communications 43

        RS232 43

        RS485 44

        RS422 44

        A Note on Twisted Pair Communications: 44

        USB 45

        Ethernet 45

- Ethernet Terminology 46
- Ethernet Addressing: 46
- Industrial Communication and Control: 47
- Exercise 1 48

PLC Memory 49

- Numerical Data Types 50
  - Understanding How Bits Can Become Numbers: 51
  - Data Formats: 51
  - Data Structures: 55
- Exercise 2 57
- Data Memory Organization: 58
  - I/O Addressing 59
  - Program Memory Organization 59
- Hardware Configuration 60
- Exercise 3 64

Program Processing 65

- IEC 61131 65
  - Ladder Logic (LAD) 66
  - Function Block Diagram (FBD): 67
  - Instruction List (IL) 67
  - Structured Text (ST) 69
  - Sequential Function Charts (SFC) 70
- How Does it Work? 71
  - Scanning 71
  - PLC Modes 73
- Exercise 4 74
- Ladder Logic - Tutorial 75
  - Discrete Logic 75
- Exercise 5 80
- Timers 82
  - On-Delay 82

- Off-Delay 83
- Retentive On-Delay 84
- Pulse 84
- Exercise 6 85
- Counters 88
- Exercise 7 90
- Data and File Movement 92
  - Move 92
  - Masked Move and Shift 93
  - File Copy 94
- Comparisons 94
- Exercise 8 97
- Math Instructions 98
  - Conversion 98
  - Addition and Subtraction 99
  - Multiplication and Division 100
- Exercise 9 102
- Scaling 103
- Exercise 10 105
- Advanced Math 106
- Other Instructions 106
  - String Operations 106
  - PID Instructions 107
  - Motion Control Instructions 108
  - Communications Instructions 108
  - Program Control Instructions 110
  - Miscellaneous/Other Instructions 111
- Exercise 11 113
- Maintenance and Troubleshooting 114
  - Forcing 114
  - Searching and Cross-Referencing 116

- Exercise 12 120
- The Art of Programming - Overview 121
- Machine Control vs. Process Control 121
- Advanced Topics 122
- Program Organization 122
- Structure 123
- Common Routine Types 125
  - System Routines 125
  - Input Routines 132
    - Mapping 133
  - Output Routines 134
  - Fault and Alarm Routines 135
  - Auto Sequence Routines 140
  - Homing Routines 148
  - Recipes 149
  - Part Tracking 151
  - Tips and Tricks 157
    - Hold-In or Latching Circuit 157
    - Toggling Pushbutton 158
    - "Free Running" Timer 158
    - Blink Timers 159
    - Accumulators and Decumulators 160
  - Trainers and Simulators 160
    - Standard Trainers 161
    - "Suitcase" Trainers 161
    - Simulation Trainers 161
    - Software Simulation 162
- Programming Lab: Fischertechnik Color Identification and Conveyor with Bins 163
  - The Trainer 163
    - PLC and I/O 164
    - Emergency Stop and MCR 164

- Stack Light and Horn/Buzzer 164
- The HMI 164
- Software 165
- Allen-Bradley MicroLogix 165
- Color Identification and Conveyor with Bins 174
  - Objective 175

PLC Platforms - Overview 178

Allen-Bradley PLCs 178

- Allen-Bradley SLC and MicroLogix Platforms 179
  - Rockwell Software – RSLogix 500 179
  - MicroLogix 800 180
  - MicroLogix 1000 181
  - MicroLogix 1100 181
  - MicroLogix 1200 181
  - MicroLogix 1400 182
  - MicroLogix 1500 182
  - SLC500 Series 183
- SLC and MicroLogix Memory Registers 184
- SLC and MicroLogix Instructions 185
  - Basic Instructions 185
  - Comparison Instructions 186
  - Math Instructions 187
  - Data Handling Instructions 188
  - Program Flow Instructions 189
  - Application Specific Instructions 190
  - ASCII Instructions 190
  - Block Transfer and PID Instructions 191
  - Interrupt Routine Instructions 191
  - Communication Instructions 192
- Starting and Editing a Project with RSLogix 500 192
  - Creating a Project 193

- Hardware Configuration 196
- Writing the Program 197
- Allen-Bradley CompactLogix and ControlLogix Platforms 200
- Rockwell Software – RSLogix 5000 and Studio 5000 200
  - 1769 CompactLogix 5370 Controllers 201
  - 5069 CompactLogix 5380 Controllers 202
  - 5069 CompactLogix 5480 Controllers 203
  - 1768 CompactLogix L4x and L4xS Controllers 203
  - ControlLogix 5570 and 5580 Controllers 204
- CompactLogix and ControlLogix Instructions 205
  - Basic Instructions 205
  - Comparison Instructions 206
  - Math Instructions 206
  - Data Handling Instructions 207
  - Program Flow Instructions 209
  - Application Specific Instructions 209
  - ASCII Instructions 210
  - Block Transfer and PID Instructions 210
  - Interrupt Routine Instructions 211
  - Communication Instructions 211
- Starting and Editing a Project with RSLogix 5000 212
  - Creating a Project 213
  - Configuring Hardware 215
- Writing the Program 216
  - Tasks, Programs and Routines 216
  - Editing a Routine in Ladder Logic 219
  - What is a Mnemonic? 219
- ControlLogix and CompactLogix Data 220
  - Arrays 221
  - User Defined Data Type (UDT) 222
  - Global Tags 223

- Program (Local) Tags 224
  - Aliases 225
- Add-On Instructions (AOIs) 225
  - Roll the Dice! 227
- Other Languages 232
  - ASCII Mnemonics (Instruction List) 232
  - Function Block Diagram (FBD) 232
  - Structured Text (ST) 232
  - Sequential Function Charts (SFC) 232
- Communications - Allen-Bradley RSLinx 233
  - Ethernet Devices: 235
  - Ethernet/IP Driver: 236

Siemens PLCs 238

- Siemens Terminology and Abbreviations 238
- Siemens Step 7 PLC Software 240
- Siemens Step 7 PLC Hardware 240
- S7-300 Platform 240
  - S7-300 CPUs 241
- S7-400 Platform 243
  - S7-400 CPUs 243
  - Remote and Networked IO 244
- Siemens Instructions 245
  - Bit Logic Instructions 245
  - Comparison Instructions 246
  - Conversion Instructions 247
  - Counter Instructions 248
  - Logic Control Instructions 248
  - Integer Math Instructions 248
  - Floating Point Math Instructions 249
  - Shift Instructions 250
  - Rotate Instructions 250

- Status Bit Instructions 250
- Timer Instructions 251
- Word Logic Instructions 252
- Miscellaneous/Other Instructions 253

Starting and Editing a Project with Step 7 254
- Creating the Project 254
- Configuring Hardware 256

Writing the Program 262

Siemens Blocks 262
- Organization Blocks (OBs) 262
- Functions (FCs) 263
- Function Blocks (FBs) 264
- Data Blocks (DBs) 264
- System Functions (SFBs and SFCs) 270

Siemens Data 270
- Data Type 270
- Variable Table (VAT) 270
- Symbol Table 270
- Marker Memory 270
- More on Addressing 271

Statement List (STL) 273

Other Siemens Languages 277
- Function Block Diagram (FBD) 277
- Structured Control Language (SCL) 277
- S7 Multi-language Example: Node Fault 279
- S7 Graph (SFC) 283

Siemens TIA Portal PLC Software 284

Siemens TIA Portal Hardware 284

S7-1200 Platform 284
- S7-1200 CPUs 284

S7-1500 Platform 285

    S7-1500 CPUs 285

  Starting and Editing a Project with TIA Portal 286

    Platform Differences 286

    Creating a Project 287

    Configuring Hardware 289

    Writing the Program 291

  Communications- Siemens Set PG-PC Interface 298

    Communications- Siemens TIA Portal 300

Major PLC Platforms 302

  ASCII Tables 303

  PLC Hardware and Programming Exercise Solutions 304

  The Art of Programming Lab Solution 312

  About the Author 345

Introduction

This book was originally written as "PLC Hardware and Programming Multi-Platform" in 2016 and self-published through AuthorHouse. It was intended to be a training manual for use in my custom PLC classes and took a generic approach to PLC training. So many things are common to all PLC platforms that I decided to present the material being taught in brand-specific classes in a generic way.

Since 2013, I have been fortunate enough to spend much of my time traveling the U.S. and other parts of North America teaching Allen-Bradley and Siemens classes for Automation Training, a Canadian company that provides classes on PLCs, HMIs and SCADA. Most students are looking for training on a specific platform; in North America, this is mostly Allen-Bradley and Siemens. While other platforms are common in the U.S., there are not enough requests for training outside of the manufacturers' classes to justify creating training materials. While Automation Training has classes for Omron, Mitsubishi and Modicon, the requests for classes on these platforms have been few and far between. In order to justify shipping PLC trainers and laptop computers to a training site and pay for instructor travel and expenses, a minimum of 3-4 students have to sign up for a class.

Automation Training class at my facility in Lebanon, Tennessee

In 2016, I decided to move my office from a shared space in Nashville, Tennessee closer to my home in Lebanon. I found an office space for lease only five minutes away. It was bigger than I needed for a one-person company, so I set it up to hold training classes. Automation Training has been gracious enough to hold several regional classes a year there, and I have also held classes for a local engineering company, Automation NTH.

Because my facility had a great deal of extra space, I also built a tabletop factory for custom training on advanced PLC techniques. Without actual hardware, it can be difficult to program complex sequences and interface routines with each other. I spent a lot of time searching for training simulation hardware and software; most of what I found was very expensive and didn't reflect the techniques that I needed to emphasize.

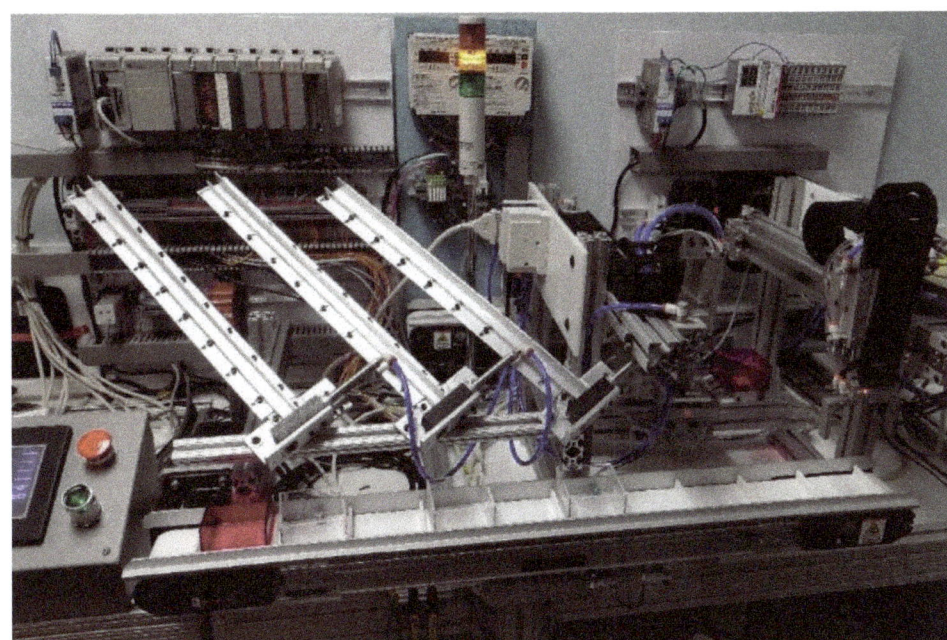

Enter the Mini-Factory. I built this trainer with an indexing cleated conveyor, escapements, dial table and pneumatic pick-and-places to teach advanced concepts such as auto sequences, part tracking and recipe management.

Automation Consulting, LLC Mini-Factory

In addition to the section shown in this picture, which is controlled by an Allen-Bradley CompactLogix PLC, there is a Siemens S7-300 on the left side and a process control area with tanks, pumps and valves. The PLCs can be connected via cables and plugs to either trainer.

Building this advanced training demo meant that I had to have written material covering advanced techniques. In addition to the Automation Training classes, which teach instruction sets from the different PLC platforms, I also teach classes to interns, engineers and customers at Automation NTH, an engineering and systems integration company near Nashville, Tennessee. Automation NTH has a training program called "NTH University", which is used internally for their employees. The program is also provided to customers who request classes. One of the standard training classes uses a conveyor and pneumatic pusher along with several sensors and a removable bin to teach a course on building a PLC application. Teaching this class required writing and refining the documentation and lab instructions. It also reinforced the value of using hands-on equipment in training, students seemed to enjoy the class very much.

Automation NTH also has a large UL panel shop and fabricates control panels for many large companies. They have also built PLC trainers for a local university and even myself. They built the equipment shown in the picture below. Teaching these classes required that I create advanced material so students could learn techniques actually used in industry. Typically, students are interested in learning a specific platform that they use in their facility; because of this, classes are usually based on teaching the instruction set for a specific brand, and the exercises reflect this. Typical trainers are built with pushbuttons, pilot lights, potentiometers and meters to interface with student programs. PLC trainers have been built this way for a long time.

The Automation NTH trainers shown in this picture have an Allen-Bradley PanelView Plus HMI built into the trainer. They also have an Emergency Stop pushbutton, cycle start and cycle stop buttons, and a cable that allows the trainer to be connected to the conveyor. A multicolored indicator is also used to simulate a stack light.

Automation NTH training class with PLC trainers and conveyor

After creating my generic approach to PLC programming manual, I also realized that most students needed information on the specific platform that they wanted to learn. I had created a hardware addition to the Automation NTH course for the Allen-Bradley ControlLogix platform. I expanded this to include the SLC and MicroLogix when I built some of my own trainers.

I have had several individual students come to my facility and take custom classes since I moved. One used the mini-factory shown previously and several have used the trainers I built myself. While I don't have nearly the fabrication capability that Automation NTH has, I have been able to build several trainers of my own on the Allen-Bradley MicroLogix 1400 platform.

I built my trainers with several important requirements in mind. I needed them to be less expensive than the trainers commercially available. Much of the cost of trainers outside of the PLC itself is in the wiring of the pushbuttons and indicators. Labor makes up a large part of most projects, and trainers are no exception.

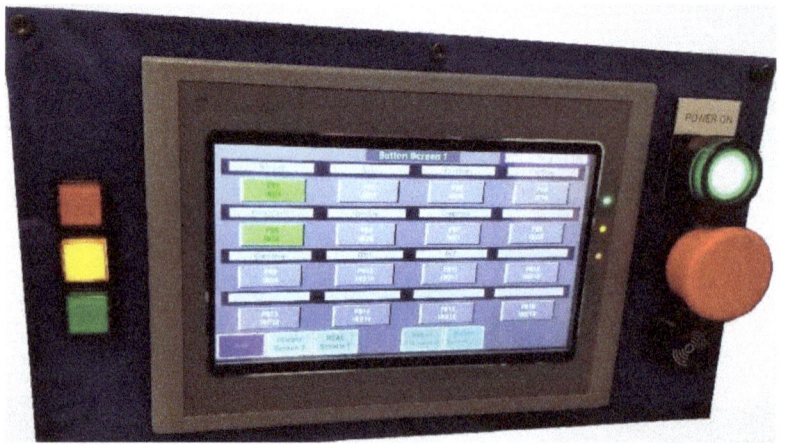

I found several inexpensive touchscreen brands and evaluated them for my trainers. Rather than wiring pushbuttons into the main panel, I decided to make them an accessory. I programmed the 7" color touchscreen with 64 pushbuttons and indicators and also made screens for displaying and modifying 64 integers and 32 real or floating-point numbers.

Trainer with Stack Light simulator, HMI, Buzzer, E-Stop and Power buttons

It was also important that my new trainers had an E-Stop and power button with MCR (Master Control Relay) and a stack light simulator. Industrial machinery uses these devices, and they are important when interfacing with PLC programs. They also provide an opportunity to teach real world applications, even in the basic classes.

I also attached a cable to the trainer to interface with external equipment, such as the pushbutton/pilot light accessory. As I mentioned previously, before building my mini-factory I had researched trainers and simulators, and found various products used in schools and factories. Most were quite expensive. I had even wandered around the Lego store at the mall wondering if they had a toy factory or something! No such luck.

I did, however, find some factory simulations that were used in Europe that were considered to be "PLC ready". A company called Fischertechnik builds a construction "toy" with a controller on a 9-volt dc system, but they also build a 24-volt system. I was not able to find anyone in the U.S. that built an interface with a PLC though, so I decided to do it myself.

Trainer with Fischertechnik High Bay Warehouse

The trainer and Fischertechnik High Bay Warehouse demo shown here was the first training system I sold outside of my own facility. I do some systems integration work for a customer in Florida and the owner expressed an interest in buying a trainer for his own employees. I wrote more on the advanced training material mentioned previously and provided it with the trainer, along with documentation for the trainer and factory demo wiring.

At this point, I had written a lot of extra material and needed to provide written instructions and a training manual to accomplish the advanced programming required for these trainers. Since I already had a generic training manual, I decided to combine the *PLC Hardware and Programming Multi-Platform* manual with my advanced material. I also modified the original PLC platform section to include in-depth information on the Allen-Bradley and Siemens platforms. Information on other brands is included also, but I concentrated on Allen-Bradley and Siemens for several reasons: One, I have been teaching these brands for several years and probably know them better than any others. Two, whether people like the brands or not, they are the most widely used. Siemens is considered to be the brand with the most installations worldwide, while Allen-Bradley has the highest market share in the United States.

I know Automation Direct and Omron PLCs fairly well and have programmed GE and Mitsubishi also. There are many other important players in the PLC market, and of course other PLC languages besides Ladder Logic, but I decided to concentrate on these brands and languages.

This book is not in a traditional format. When I published my first book, *Industrial Automation: Hands On*, (Lamb, F. (2013). *Industrial Automation: Hands On*. New York: McGraw-Hill Education) I had very little control over the formatting of the book. In the beginning, it was intended to be a reference book, and McGraw-Hill thought that it also might become a possible textbook in colleges. After it was published, I discussed the feasibility of using the book as a textbook with a local college electrical engineering professor.

I learned something important about textbooks. College professors often write their own textbooks and have them published. They are not interested in dissecting someone else's book and creating course materials such as exercises and tests, so if you don't include questions or exercises in your book, it is unlikely that it will be used as a textbook. When I wrote the original *PLC Hardware and Programming Multi-Platform* manual, I kept this in mind and wrote exercises into the book and put the answers in the back. I kept the questions in this version, even adding an exercise that relates to the trainers I build. The answers are still in the back of the book.

When it came to the advanced part of the manual, which I titled "The Art of PLC Programming", I did not create more exercise questions. This could be done, and if I receive enough pressure from customers to do this, I will do so in a future revision. Advanced PLC programming requires more than just questions; it requires actual programming. The training accessories have a completed program for every factory simulator that is made available for the instructor, so I consider actual programs to be the exercises for that section. The documentation for the Fischertechnik "Color Identification with Conveyor and Bins" demo and program for the Allen-Bradley MicroLogix 1400 is included in this section.

I also placed captions under the pictures with figure numbers; this is required by most publishers and certainly recommended. Since this section was originally a separate document, the figure numbering started at 1, I did not go back and add figure numbering to the PLC Hardware and Programming section. Unfortunately, it changes the position of many of the objects in that section, which would require extensive reformatting.

The original PLC Hardware and Programming manual also had an index. I spent a lot of time putting the index together and realized it would take possible months of work to create an index for the rest of the book. Also, as sections are added or moved, it changes the page numbers in the index. I did not include an index in this book for that reason, simply the extra time it would take. I did, however, make the Table of Contents more comprehensive and detailed.

Finally, the hardware specific part of the manual was very time-intensive when it came to acquiring technical information from the vendors' documentation. I was also not permitted to use pictures from some sources, so I had to draw the products myself in CAD. There is a point at which one simply needs to say they are done, so there is much more information available directly from manufacturers and their websites.

This book is written in three sections, which I have titled separately:

1. PLC Hardware and Programming
2. The Art of PLC Programming
3. PLC Platforms

There are also a number of appendices at the end of the book that programmers will find useful. As mentioned before, the answers to the exercises for the first section are there. The complete program for the Fischertechnik demo is there. An ASCII table for string operations are included. In the PLC platforms section, there are lists of most instructions for Allen-Bradley and Siemens, along with a list of websites for PLC manufacturers.

This book was truly self-formatted and self-published. Rather than worrying about whether it was appropriate as a textbook, I built the structure of the book around the classes that I myself teach. Because I own all of the rights to the book, I can title it, make changes to it, and release parts of it as I wish. If an instructor wishes to use sections of the book in a class or wants me to change parts of the book to make it more useful for his or her training, it can be done.

Through the magic of print on demand, this book will be able to be revised and changed as needed. While it is intended for use in my own training classes, I certainly hope it is used as a reference or even a textbook by others.

I'd like to thank the many people who have helped me with this book and my journey from being a programmer and designer to becoming an instructor. Thanks to Steve Woodhouse, owner and president of Automation Training for allowing me to teach classes across North America for the last 5 years. Thanks to Jeff Buck, VP/Operations of Automation NTH for the opportunities NTH has given me for training and integration jobs. Thanks to my daughter Mariko Hickerson and her company Huckleberry Branding for the formatting and branding help, not only on this book but also on my training products and website. And finally, thank you to my wife Mieko for the ideas, support (and panel fabrication) she has provided throughout this process!

- Frank Lamb, April 2019

PLC Hardware & Programming - Overview

The beginning of this book approaches PLC training from a generic viewpoint. Most PLC platforms have many things in common; before beginning the study of a particular brand of PLC, it is important to learn the things that are common to *all* platforms. This section does this, pointing out some of the exceptions and different ways of doing things along the way.

A history of computing devices and the birth of the PLC is also covered.

Resources used in the preparation of this manual include information from many of the major PLC manufacturers; software examples are primarily drawn from Allen-Bradley RSLogix5000 and Siemens Step 7. This section was originally released as a separate training manual titled *PLC Hardware and Programming Multi-Platform*, published by AuthorHouse in November 2016.

What is a PLC, or Programmable Logic Controller?

A *PLC*, Programmable Logic Controller or Programmable Controller, is a digital computer used to control electromechanical processes, usually in an industrial environment. It performs both discrete and continuous control functions and differs from a typical computer in several important ways:

1) It has *Physical I/O;* electrical inputs and outputs bring real world information into the system and control real world devices based on that information.
2) It is *Deterministic;* it processes information and reacts to it within defined time limits.
3) It is often *Modular;* it can have I/O modules, communication modules or other special purpose modules added to it for expansion.
4) It is programmed using several defined *Languages*. Some languages allow the program to be changed while the machine or system being controlled is still running.
5) Software and Hardware are *Platform Specific;* components and programming software usually can't be used between different manufacturers.
6) It is **Rugged** and designed for use in industrial environments.

Unlike computers, PLCs are made to run 24 hours a day, 7 days a week and are able to resist harsh physical and electrical environments.

According to a *Control Engineering* magazine poll in 2012, major applications for PLCs include machine control (87%), process control (58%), motion control (40%), batch control (26%), diagnostic (18%), and other (3%). (Results don't add up to 100% because a single control system generally has multiple applications.)

The History and Evolution of Computers and PLCs

While PLCs have only been in existence for fewer than 50 years (as of 2016), the evolution from computing machines into their current form is interesting and illuminating. This section examines their history and parallel development with computers. To understand the history of programmable controllers, it is useful to examine where many of their elements came from.

The Babbage Analytical Engine

Even before the use of electrical devices to solve mathematical and logical problems, a mathematician and inventor from England, Charles Babbage, had an idea for a mechanical device to calculate astronomical and mathematical tables. In 1823, he was awarded £1700 by the British government to begin work on his "Difference Engine", but after spending more than £17,000 over 15 years or so, no working device had been built. The government viewed this as a total failure on the mechanical side, though they did consider the economically produced tables themselves worthwhile.

In 1837, he proposed a mechanical general-purpose computer to solve arithmetic problems. The "Analytical Engine" was designed to solve general mathematical polynomial equations using a system of gears. The work Babbage did on the Analytical Engine made the original Difference Engine idea obsolete, at least in his mind. Because of conflicts with his chief engineer and what he considered inadequate funding, he was never able to build a working device.

Many of the concepts that came out of Babbage's designs laid the groundwork for future computers and processors. For instance, his designs incorporated an Arithmetic Logic Unit, Control Flow (conditional If-Then branching and Loops), and integrated memory in the form of gear positions. This structure is similar to the electronic version of the computer.

If the Analytical Engine had been built, it would have been digital and programmable. It also would have been "Turing-complete", a term that when applied to programming languages means that it has conditional branching (*e.g.*, "if" and "goto" statements, or a "branch if zero" instruction) and the ability to change an arbitrary amount of memory.

The original engine would, however, have been very slow. In *Sketch of the Analytical Engine*, Luigi Federico Menabrea reported "Mr. Babbage believes he can, by his engine, form the product of two numbers, each containing twenty figures, in three minutes".

Relay Logic

Before the development of computers, automated control of machinery was accomplished by wiring relays and other devices together. Electromechanical relays and switches could be used to control pumps, heaters and motors when connected in a "relay rack". This method was expensive, took up a lot of space and was difficult and time-consuming when it came to making changes.

Methods of designing control logic using hand-drafted "ladder diagrams" had been used since the advent of electrical systems in the 1800s. Ladder diagrams were named after their resemblance to the rungs of a ladder, with the left side being the line, or energizing side (L1) and the right side being the neutral, or L2 side.

The electromechanical relay itself, invented in the early 1800s, was not used widely until telegraphs and telephone switching circuits required them in the late 19th century. In the early 1900s, inventors realized that electrical relay circuits could be used to direct a series of mathematical calculations automatically. Simple relays were fairly inexpensive by the 1930s, but it was still less costly to build "adding machines" using mechanical cams and gears.

For more complex math functions, relay circuits were more flexible than mechanical systems. They could be arranged in rows and columns on a rack and connected together with wires, according to what one wanted the circuit to do.

Electromechanical Computers

In 1937, the concept of a general-purpose computing machine was presented to IBM by Howard Aiken. The project was approved in 1939 after a feasibility study and completed in 1944. The first version, the Mark I, was shipped to Harvard and used to determine whether implosion was a viable choice to detonate the atomic bomb that was used a year later. Interestingly, Aiken's ideas were clarified after studying Babbage's work one hundred years earlier. The Mark I was also used to compute and print the same kinds of mathematical tables that were Charles Babbage's original goal.

The Mark I was built using switches, relays, rotating shafts and clutches. It weighed about 10,000 lbs. (4500kg) and was 51 feet long. It used 765,000 components and 500 miles of wire. There were 3500 multipole relays with 35,000 sets of contacts, 2225 counters and tiers of 72 adding machines, each of which could do math with 23 significant digits. The basic calculating units were synchronized mechanically by a 5-horsepower (4kW) electric motor.

Data could be entered by using 60 sets of 24 switches but could also read instructions from a 24-channel punched paper tape.

The Harvard Mark I was followed by the Harvard Mark II in 1947, the Harvard Mark III/ADEC in 1949, and the Harvard Mark IV in 1952. While the Mark II was an improvement over the Mark I, it was still based on electromechanical relays. The Mark III used vacuum tubes and crystal diodes but still included rotating mechanical drums for storage and relays for transferring data between drums.

The Mark IV was all electronic, replacing the drums with magnetic core memory. The Mark II, III and IV were all sold to the military (U.S. Navy and Air Force). The Mark I remained at Harvard and was retired in 1959. An early picture of the computer displays the name "Aiken-IBM Automatic Sequence Controlled Calculator Mark I".

The First Electronic Computers

The first electronic general-purpose computer was designed to calculate artillery firing tables for the U.S. Army but was also used to study the feasibility of a thermonuclear weapon. It was originally announced in 1946 and called a "Giant Brain" by the press. It was designed to be about 1000 times faster than an electromechanical computer.

This computer was called ENIAC, an acronym for Electronic Numerical Integrator and Computer. It was operational from 1946-1955, but improvements were made in 1948 (a read-only stored programming mechanism was added), 1952 (high-sSpeed shifter added) and 1953 (100-word BCD Core Memory added). These improvements increased the speed and capabilities of the computer quite a bit.

ENIAC contained 17,468 vacuum tubes, 7200 crystal diodes, 1500 relays, 70,000 resistors, 10,000 capacitors and about 5,000,000 hand-soldered connections. It was 100 feet long and weighed about 27 tons (54,000 lbs.). It required 150kW of electricity; the rumor was that when it was switched on, the lights in Philadelphia dimmed.

Input was possible using an IBM card reader and output was via a card punch or by using an IBM accounting machine. It was modular, consisting of individual panels to perform different functions. Some panels were accumulators that performed math functions. Other units included: the Initiating Unit, which started and stopped the machine; the Cycling Unit, which synchronized the other units; the Master Programmer, which controlled loop sequencing, the punch card reader, the printer, the constant transmitter and three function tables, which were programmed using switches.

ENIAC used octal-base radio tubes, which were common. Several tubes burned out daily, making it non-functional about half the time. Higher reliability tubes appeared in 1948, reducing downtime.

Mapping a mathematical problem into ENIAC could take weeks; it was quite complex. Manipulations of switches and cables followed by verification and debugging was required to execute the program step by step. There were six female programmers who not only input the data but also debugged problems by crawling inside the machine to find bad solder joints and tubes.

The picture to the left shows the back of one section of ENIAC, full of vacuum tubes.

Near the end of World War II, the U.S. Navy had approached MIT about creating a flight simulator for bomber pilots. The project was funded under the name "Project Whirlwind".

After seeing a demonstration of ENIAC, an engineer at MIT suggested that a digital computer was the solution. By 1947, a high-speed stored-program had been built. Most of the previous computers at this time operated in bit-serial mode, using single-bit arithmetic and feeding in large words of 48 or 60 bits in length. This method was not fast enough for the simulation task, so the Whirlwind computer included 16 math units that operated on 16-bit words in bit-parallel mode. This made Whirlwind sixteen times faster than other machines of its era. This legacy has been handed down to the CPUs of today, which almost all use this bit-parallel system to do arithmetic.

Official Whirlwind construction began in 1948 and took three years to build. It began operation on April 20, 1951. The design used approximately 5,000 vacuum tubes.

Early Computer Memory

In the first electromechanical and electronic computers, memory took the form of mechanisms, latched relays or vacuum tubes held in a fixed state. This method was slow and not very flexible.

The original Whirlwind computer design called for 2048 words (2K) of 16 bits each for random-access storage. In 1949, when the design was being done there were only two memory technologies that could hold this much data: **mercury delay lines** and **electrostatic storage**.

A mercury delay line was a complex system that consisted of a long tube filled with mercury, a mechanical transducer on one end and a microphone on the other. Pulses were sent in at the transducer end, detected by the microphone, reshaped and sent back through the delay line, much like a spring reverb unit used in audio processing. The delay line operated at the speed of sound, which was very slow even by the standards of computers in that era. Whirlwind designers discarded the delay line as a possible memory resource due to its speed and complexity.

The alternative form of memory, electrostatic memory tubes, used a cathode ray tube similar to an early TV picture tube or oscilloscope tube. The most popular type then was known as the Williams tube, developed in England, but this design was incompatible with the Whirlwind specifications, so the designers chose a different model.

After spending many months testing the system, it was determined that the electrostatic tubes were too slow for the project requirements, and a suitable replacement was sought. The leader of the project was an engineer named Jay Forrester. He had come across an advertisement for a new magnetic material that he recognized as a possible data storage medium and set up a workbench in the corner of the lab to evaluate the material. During several months at the end of 1949, he had invented the basics of magnetic core memory and demonstrated its feasibility.

After two more years of work, the design team had built a core plane that was made up of 32 x 32, or 1024 individual cores, holding 1024 bits of data. Two additional core planes were built later, increasing the total memory of the system to 3072 bits.

Core memory used tiny magnetic rings with wires threaded through them to read and write information. The cores can be magnetized in two different ways, clockwise or counterclockwise, the bit stored is a zero or a one based on the magnetization direction.

Wires are arranged so that individual cores could be set to a one or zero by changing its magnetization, but reading the core caused it to be reset to zero, erasing it. This is known as destructive readout. This problem was solved in 1951 by An Wang, who invented a write-after-read cycle that used a one-dimensional shift register of cores, essentially using two cores to store each bit.

When not being read or written to, the core retains its value, even when power is removed. This makes them **non-volatile**.

By using smaller rings and wires, the memory density slowly increased; by the late 1960s, a density of 32 kilobits per cubic meter was typical. Costs for manufacturing these core units declined from $1 per bit to 1 cent per bit by 1970.

The first semiconductor-based memory (Semiconductor Random Access Memory or SRAM) was introduced in the late 1960s and began to erode the core market. In 1972, the first successful DRAM (Dynamic Random Access Memory) Intel 1103 was brought to market at 1 cent per bit. Improvements in semiconductor manufacturing led to rapid increases in capacity and further decreases in price, driving core from the market by 1974.

The Evolution of Personal Computers

Computer technology improved, making computers more powerful and smaller through the 1960s and 1970s, however they were still too large and expensive for individuals or even small companies to own one. Requests for computer use at universities and larger businesses had to be filtered through an operating staff or time-shared.

By 1972, electronic calculating devices began to be available to individuals and small businesses. After development of the microprocessor, individual personal computers were low enough in cost that they could be purchased, though they were often sold as a kit and of interest mostly to hobbyists and technicians.

By the 1980s, several commercially-available computers were being sold, including the TRS-80 (released 1977 by Radio Shack), Atari's 400 and 800 (late 1970s), and the Commodore 64 (1982).

IBM introduced its 5150 model in 1981, and through the 1980s, the term "PC" began to refer to desktop computers compatible with IBM's PC products. In 1970, IBM was estimated to have had 60% of the computer market; by 1980, this had declined to 32%. By 1984, it was estimated in a *Fortune* magazine survey that 56% of American companies with personal computers used IBM PCs, compared to 16% for Apple. A 1983 study of corporate customers found that two-thirds of large customers standardizing on one computer chose the IBM PC, compared to 9% for Apple. Through the 1980s, the only product that kept a significant market share without being compatible with IBM personal computers was the Apple Macintosh family.

The 5150 was based on the Intel 8088 and was IBM's first open architecture product, allowing competitive companies to create peripheral products for its computer, including software. IBM didn't market its own software until 1984, instead relying on licensing programs like Microsoft's BASIC. The PC/AT was introduced in 1984 with an Intel 80286 CPU and originally running at

6MHz clock speed. This was followed by the XT 286 in 1986, which was the first personal computer designed to support multitasking.

Almost immediately after the IBM PC was released, rumors began circulating of compatible computers created without IBM's approval. In June 1982, Columbia Data Products introduced the first IBM-PC compatible computer, followed by Compaq in November of 1982. Other manufacturers reverse engineered the BIOS to make non-patent-infringing copies of the operating system.

Early IBM PC compatibles used the same computer bus as the original PC and AT. This was later named the Industry Standard Architecture by manufacturers of compatible computers. In June 1983, PC Magazine defined a "PC Clone" as "a computer that can accommodate the user who takes a disk home from an IBM PC, walks across the room and plugs it into the 'foreign' machine".

Other manufacturers such as Tandy (Radio Shack), Hewlett-Packard, Digital Equipment Corporation and Texas Instruments introduced computers that were compatible with Microsoft's "MS-DOS", though these were not always completely hardware and software compatible with the IBM PC itself. As more companies began manufacturing these "clone" computers, IBM began to lose market share again. Into the 1990s, compatibility became more of a matter of software than hardware as Microsoft created the Windows series of operating systems and most software companies concentrated on Windows compatible products. IBM finally exited the personal computing market with the sale of its consumer PC division to Lenovo in 2005.

In the 2000s, Hewlett-Packard and Dell became the largest American PC manufacturers. Processors have become more powerful every year and the footprint has shrunk, with laptops and tablets taking up much of the market. Major foreign manufacturers including Acer, Lenovo, Sony and Toshiba are also major competitors in the market, and prices have been lowered extensively to where often software costs more than the hardware it runs on.

Tablets with detachable keyboards and laptops with touchscreens have further blurred the lines between handheld devices and full computer platforms. In addition, connectivity and remote servers (the "Cloud") have distributed software services between in-house computers and service-based software.

Birth of the Programmable Controller

In 1968, a group of engineers at General Motors presented a paper at the Westinghouse conference detailing the problems they were having with reliability and documentation of the machines at their plant. One of the engineers, Bill Stone, also presented design criteria for a "standard machine controller".

The criteria stated that the design would need to eliminate costly scrapping of assembly-line relays during model changeovers and replace unreliable electromechanical relays. It also needed to:

- Extend the advantages of static circuits to 90% of the machines in the plant.
- Reduce machine downtime related to controls problems
- Be easily maintained and programmed in line with already accepted relay ladder logic.
- Provide for future expansion. It had to be modular to allow for easy exchange of components and expandability.
- Work in an industrial environment with dirt, moisture, electromagnetism and vibration.
- Include full logic capabilities, except for data reduction functions.

These specifications, along with a proposal request to build a prototype, were given to four controls builders:

- Allen-Bradley, by way of Michigan-based Information Instruments, Inc.
- Digital Equipment Corporation (DEC)
- Century Detroit
- Bedford Associates

The DEC team brought a "mini-computer" to GM, which was rejected. A lack of static memory was one of the major reasons.

Allen-Bradley was a major manufacturer of relays, rheostats and motor controls. Even though this new idea would compete with one of its core businesses, electromechanical relays, they went from prototype to a production unit in 5 months. The first attempt was the Program Data Quantizer, or PDQ-II. This was judged to be too complex and difficult to program and was quite large. The next attempt was the Programmable Matrix Controller, or PMC. Though smaller and easier to program, this was still not sufficient for GM.

At the time of GM's design criteria, Bedford Associates was already working on a design. Its system was modular and rugged, it used no interrupts for processing and mapped directly into memory. Since this was the 84th project for the company, they named this unit the 084. The project team included Richard Morley, Mike Greenberg, Jonas Landau, George Schwenk and Tom Boissevain. After obtaining funding, the team formed a new company called Modicon, an acronym for MOdular DIgital CONtroller.

From left to right: Dick Morley, Tom Boissevain, Modicon 084, George Schwenk, and Jonas Landau

The Modicon 084 was built ruggedly, with no on-off switch, no fans, and totally enclosed. Richard Morley explained, "No fans were used, and outside air was not allowed to enter the system for fear of contamination and corrosion. Mentally, we had imagined the programmable controller being underneath a truck, in the open, and being driven around in Texas, in Alaska.

Under those circumstances, we wanted it to survive. The other requirement was that it stood on a pole, helping run a utility or a microwave station which was not climate controlled and not serviced at all".

In 1969, Bedford and Modicon demonstrated their 084 Programmable Controller to GM and won the contract. The controller consisted of three components: the processor board, the memory, and the logic solver board, which used a form of ladder logic to solve the algorithms.

According to Morley, the original machine only had 125 words of memory and did not need to run fast. In his interview with Howard Hendricks, he said:

> "You can imagine what happened! First, we immediately ran out of memory, and second, the machine was much too slow to perform any function anywhere near the relay response time. Relay response times exist on the order of 1/60th of a second, and the topology formed by many cabinets full of relays transformed to code is significantly more than 125 words. We expanded the memory to 1K and thence to 4K. At 4K, it stood the test of time for quite a while. Initially, marketing and memory sizes were sold in 1K, 2K, 3K and 4K. The 3K was obviously the 4K version with constrained address so that field expansion to 4K could easily be done."

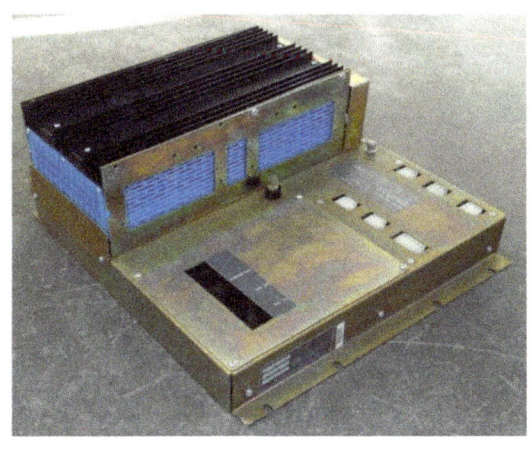

Allen-Bradley Bulletin 1774 PLC

Meanwhile, Allen-Bradley had gone back to the drawing board. By 1971, engineers Odo Struger and Ernst Dummermuth had begun to develop a new concept that improved on their PMC, Programmable Matrix Controller. This concept became the Bulletin 1774 PLC. Allen-Bradley named this the "Programmable Logic Controller"; the term later became the industrial standard when the acronym PC became associated with personal computers.

In 1972, Allen-Bradley also offered the first computer for use as a programming terminal. Other manufacturers in the 1970s and 1980s typically used dedicated programming terminals with (or without) a small screen. Instructions were entered as three- or four-letter mnemonics. As technology improved, these terminals were reduced in size to a hand-held device.

By the later years of the 1970s, several other companies had entered the PLC market, including General Electric, Square D, Omron and Siemens.

Modicon improved on the 084 in 1973 with the 184, which made it the early leader in the market. This was followed in 1975 with the 284 and 384 models. The 984 was produced in 1986 and remained a Modicon standard for many years. In a joint venture with AEG Schneider Automation, the Quantum series of controllers was released in 1994. In 1977, Modicon was bought by Gould Electronics, and later in 1997 by Schneider Electric, who still owns them today (2016).

PLC Improvements

Mitsubishi F1-20P Programmer, 1980s

The 1980s saw many new companies entering the PLC market. Japanese companies such as Mitsubishi and Omron entered the U.S. market as automotive manufacturers began using PLCs extensively in their manufacturing processes. Giants like Westinghouse, Cutler-Hammer and Eaton created products, as well as machine tool manufacturers such as Giddings & Lewis. In 1980, the market was estimated at $80 million, and it had grown to a billion dollars worldwide by 1988.

As IBM-compatible personal computers became smaller and less expensive, companies began developing DOS-based software for use in programming. This allowed users to enter the program graphically. Rather than only seeing the alphanumeric characters of text commands, ladder logic could be visualized on a CRT monitor.

With the release of the Windows 3.0 operating system in the early 1990s, software had improved with colored graphics and multitasking. PC clones made computers less expensive, and the laptop computer all but replaced handheld programmers. Many companies began producing smaller, cheaper "brick" PLCs for simple applications. Allen-Bradley's MicroLogix 1000 in 1995 competed with relatively unknown names, such as Eagle Signal's "Micro 190" and PLC Direct, which brand labeled Koyo PLCs from Japan.

Koyo had produced PLCs for Texas Instruments, Siemens, and GE since the 1980s. In 1994, they established a company in the U.S. that began marketing PLCs by mail order. Tim Hohmann founded PLC Direct in Atlanta as a joint venture, which was renamed Automation Direct in 1999.

Allen-Bradley remained the dominant brand in the U.S. during the 1990s. They had been bought by Rockwell in 1984 and spun off Rockwell Software in 1994 after purchasing ICOM, which had made a competing programming software product for the Allen-Bradley PLC. The SLC500 line, a smaller modular controller, was released in 1991, followed by the first MicroLogix product in 1995.

Siemens became the dominant player outside of the U.S. and Japan. The S5 controller, developed in 1979, had a large installed base in Europe through the 1980s. When the S7-200, S7-300 and S7-400 series were released in 1994, many companies began upgrading their existing platforms. Siemens was an early innovator in the use of User Defined Data Types (UDTs) and advanced programming using their version of Instruction List, known as STL (Statement List). They also allowed for use of re-useable code by defining local variables within subroutines, or functions.

In 1994, the International Electrotechnical Commission (IEC) began to define the languages that PLCs would be programmed in, data types, and other details pertinent to Programmable Controllers. IEC 61131-3 defined the rules manufacturers followed in order to standardize their products. Five languages were defined: Ladder (LD), Instruction List (IL), Function Block Diagram (FBD), Structured Text (ST), and Sequential Function Charts (SFC).

Mitsubishi gained the largest market share in Japan and much of Asia, while Omron saw gains worldwide.

By the 2000s, PLCs had become much more powerful and began gaining traction in process control, which had long been the domain of DCS (Distributed Control Systems). With the ability to use I/O networks such as DeviceNet, Profibus and Ethernet, these more powerful platforms became known as "Programmable Automation Controllers", or PACs. With improved memory, higher speed processors and the ability to control thousands of analog and digital points at once, PACs could control large chemical processing plants, wastewater treatment and pipelines.

Multi-axis motion control also began to be integrated into PACs in the early 2000s. Allen-Bradley, Siemens, Modicon and Mitsubishi all have integrated controllers that can operate independently of the central CPU. Multiple or redundant CPUs can also be used within the same rack. Variable Frequency Drives (VFDs) and robots often now contain microprocessors that can be programmed in ladder logic. Hybrid HMI touchscreen controllers have also become common.

Today's landscape includes more than 20 PLC manufacturers with international markets, with about 15 that have a 1% or more market share each. Open platforms have appeared allowing smaller manufacturers to offer their own PC-based or board-level controllers that program in ladder or other IEC 61131-3 languages. Companies such as Codesys now provide a platform for some of the major PLC manufacturers such as Modicon, ABB and Bosch.

With higher speed Ethernet-based communication and control networks, systems have become more distributed, with microprocessors in "smart nodes" of I/O to detect errors and perform autonomous logic and monitoring tasks. As of 2016, the PLC market seems to be converging on Ethernet/IP based control networks.

Mid-Range Modular PLCs

Allen Bradley ControlLogix

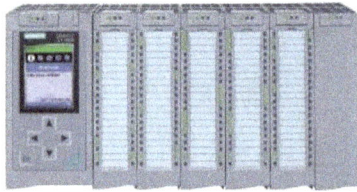
Siemens S7-1500

Mitsubishi Q Series

GE 9030

Koyo DL405

Omron CJ1

Modicon Premium

Shown above are just a few of the different brands of PLCs in the modular, mid-range size. While these are some of the major brands of PLC in use across the world, there are hundreds of companies that manufacture PLCs. Other formats include smaller "brick" styles, and large rack systems. Configurations may also combine features from rack based and self-contained types.

PLC Timeline

			PLC Timeline 1968 – 2016				
Event	Event	Network	Product		Company	Acquisition	PLC	Platform
Year	Events	Modicon	AB	GE	Omron	Siemens	Mitsubishi
1968	GM Design Spec						
1969		Modicon 084					
1970			PDQ II				
1971			PMC	PC45			
1972		Modicon 184	Prog.Computer				
1973				Logitrol	SYSMAC-M1R		
1974			1774 PLC				
1975		Modicon 284/384					
1976						Simatic S3	
1977		Gould	PLC2		Standard SYSMAC		
1978							
1979		MODBUS	Data Highway			Simatic S5	
1980	Ethernet		PLC3				
1981				Series 6			MELSEC FX
1982	Koyo SR21					Simatic TI305	
1983				Series 1	Host Link		
1984			Rockwell			S5-135U	
1985			PLC5	Series 3			A Series
1986		Modicon 984		GE-Fanuc			
1987				Series 5	C200H		
1988							
1989	Profibus						
1989	TI Series 305						
1990	MS Windows 3.0			GE 90-30			
1991			SLC500	GE 90-70	CV Series		
1992							
1993	Profibus DP		DeviceNet		CQM1		
1994	IEC 61131-3	Quantum Series	Rockwell Software			Simatic S7	
1994	PLC Direct, CodeSys						
1995			MicroLogix 1000				
1996					SYSMAC Link		
1997		Schneider	ControlLogix L1			PCS7	
1998				VersaMax			
1999	Automation Direct				CS1		
2000			ControlLogix L55				
2001			Micrologix 1500		CJ1	Profinet	
2002			ControlLogix L60	Rx7i	CJ1M		Q Series
2003				Rx3i			
2004							
2005			Micrologix 1100		CP1H		
2006		M-340					
2007					CP1L		
2008			Micrologix 1400				
2009						S7-1200, 1500	
2010			ControlLogix L70		CP1E, CJ2M		
2011							
2012							
2013							
2014							iQ-R Series
2015							
2016			ControlLogix L80				

PLC and Computer History Bibliography

Babbage's Analytical Engine –

Collier, Bruce (1970). The Little Engines That Could've: The Calculating Machines of Charles Babbage (Ph.D.). Harvard University.

Menabrea, Luigi Federico; Lovelace, Ada (1843). "Sketch of the Analytical Engine invented by Charles Babbage... with notes by the translator. Translated by Ada Lovelace". In Richard Taylor. Scientific Memoirs 3. London: Richard and John E. Taylor. pp. 666–731.

Birth of the Programmable Controller –

Segovia, Vanessa Romero; Theorin, Alfred (2012). History of Control, History of PLC and DCS.

The History of the PLC (as told to Howard Hendricks by Dick Morley) http://www.barn.org/files/historyofplc.html

W. Bolton, Programmable Logic Controllers, Fifth Edition, Newnes, 2009

Early Computer Memory –

Edwin D. Reilly, Milestones in computer science and information technology, Greenwood Press: Westport, CT

Jay W. Forrester, "Digital Information In Three Dimensions Using Magnetic Cores", Journal of Applied Physics 22, 1951

Electromechanical Computers –

Cohen, Bernard (2000). Howard Aiken, Portrait of a computer pioneer. Cambridge, Massachusetts: The MIT Press.

The First Electronic Computers –

Goldstine, H. H.; Goldstine, Adele (Jul 1946), "The Electronic Numerical Integrator and Computer (ENIAC)", Mathematical Tables and Other Aids to Computation

ENIAC specifications from Ballistic Research Laboratories Report No. 971 December 1955, (A Survey of Domestic Electronic Digital Computing Systems)

Redmond, Kent C.; Smith, Thomas M. (1980). Project Whirlwind: The History of a Pioneer Computer. Bedford, MA: Digital Press.

The Evolution of the Personal Computer –

Freiberger, Paul (1982-08-23). "Bill Gates, Microsoft and the IBM Personal Computer". InfoWorld. p. 22.

Krasnoff, Barbara (3 April 1984). "Putting PC Compatibles To the Test". PC Magazine. pp. 110-144.

Norton, Peter (1986). Inside the IBM PC. Revised and enlarged. New York. Brady.

Ward, Ronnie (November 1983). "Levels of PC Compatibility". BYTE. pp. 248-249.

PLC Improvements –

Webb, John W. (1988), "Programmable Controllers, Principles and Applications", Merrill Publishing Company

Hughes, Thomas A. (2001), "Programmable Controllers", The Instrumentation, Systems and Automation Society

Websites as listed in the Major Platforms section of this book

Relays –

"The Electromechanical Relay of Joseph Henry" http://history-computer.com/ModernComputer/Basis/relay.html

Physical Layout of a PLC

A block diagram of the physical arrangement of a PLC is shown below:

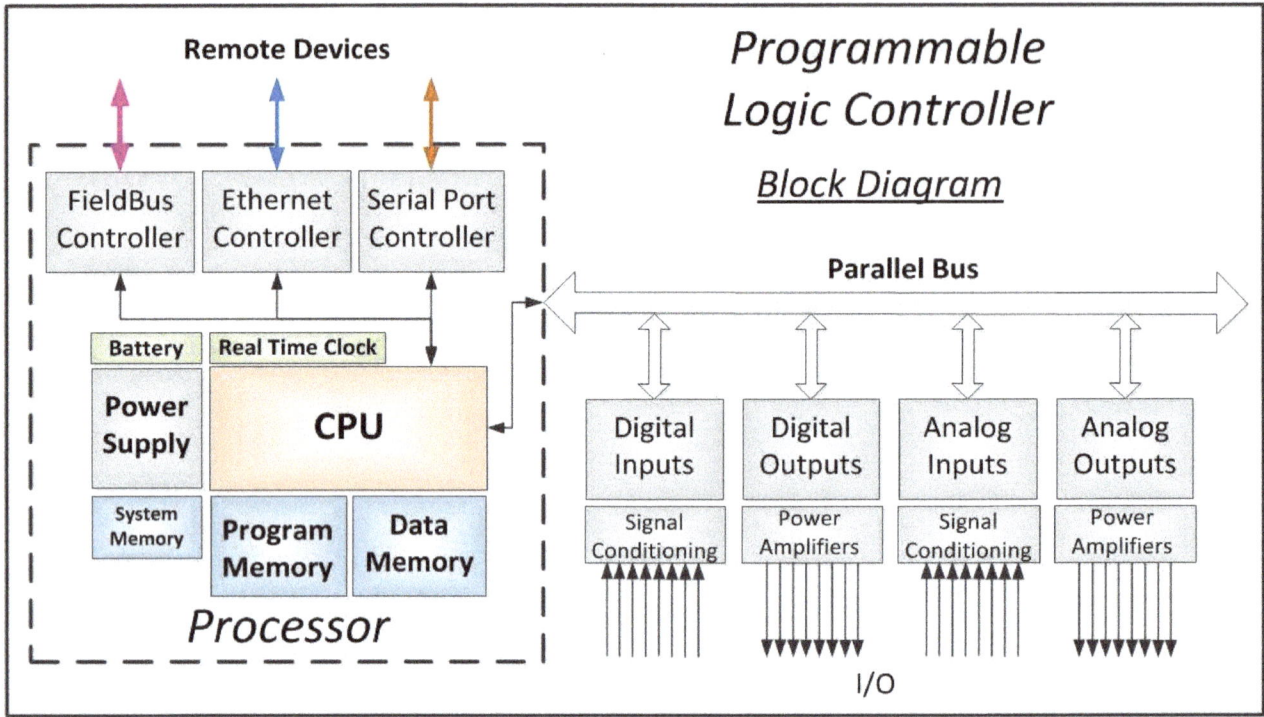

Figure 1 - PLC Layout

The CPU processes all of the logic loaded into the controller and also contains the operating system. It usually has a real-time clock built in, which is used for various functions. The system memory is also associated closely with the CPU.

Not all of the items shown in this diagram are present on every PLC, but this will give you an idea of a typical configuration.

I/O (Inputs and Outputs)

Physical I/O can be *Discrete*, that is single signals bits that are either on or off, or *Analog*, that is they can be signals that change amplitude in either voltage or current.

Digital/Discrete Devices

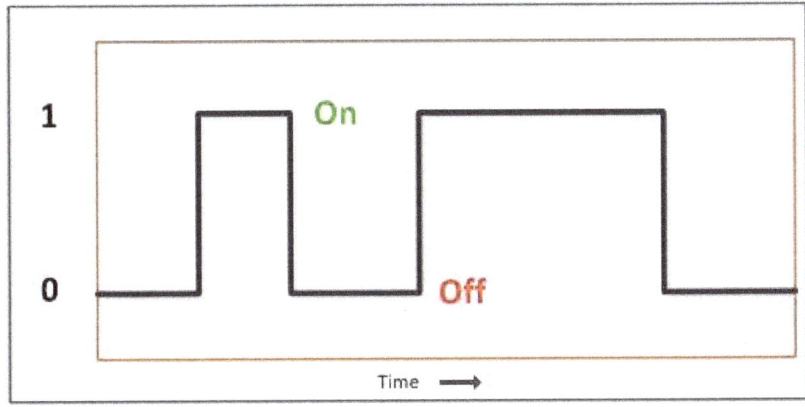

This diagram shows a discrete signal. Typical signal levels for discrete inputs and outputs are 24V DC and 120V AC, but other levels may be present depending on the type of device or input card. In addition to the designation one and zero or on and off, discrete signals may be described as being true or false.

Examples of digital input devices are shown below:

Pushbutton   Photoeye   Proximity Switch

Here are some digital output devices:

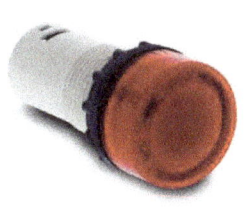  

Pilot Light   Motor Starter   Solenoid Valve

And here is a device that is both a digital input and a digital output!

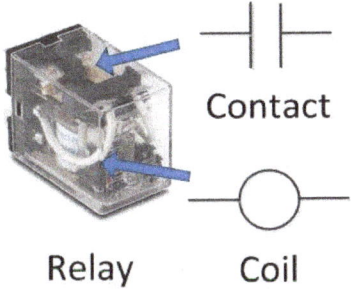

Relay   Coil

Analog Devices

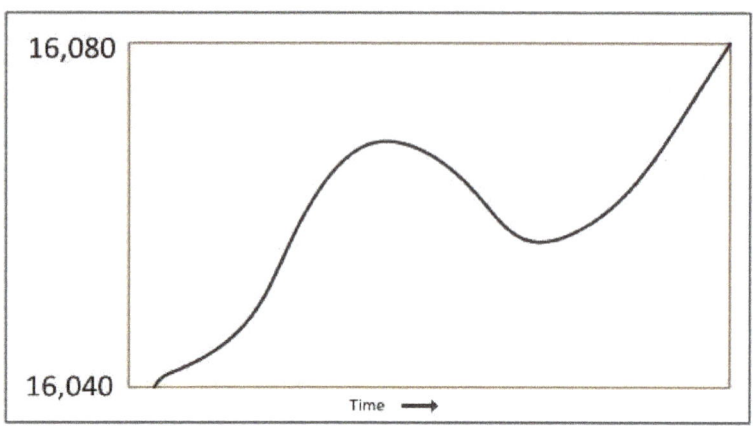

Analog signals vary in either voltage or current. Ranges are typically 0-10V DC or -10 to +10V DC (voltage type) or 0-20mA or 4-20mA (current type). The electrical signal is then converted into a number for use in the PLC program.

Here are some examples of analog input devices:

   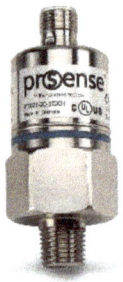   

Potentiometer   Pressure Transducer   Platinum RTD

And these devices use analog outputs:

      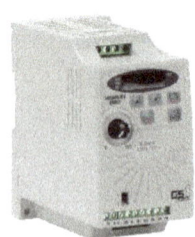

Proportional Valve   Linear Actuator   VFD Speed

Analog signals are converted by Analog to Digital Converters (ADCs) for inputs, or Digital to Analog Converters (DACs) for outputs.

ADCs capture a signal from an input device (such as a pressure or temperature sensor) and convert it into a 16-bit Signed Integer. This means that for the full range of the sensor, there are a possible 65,536 different values, ranging from -32,768 to 32,767. Since most sensors only provide positive values, this means that only 0-32,767 are possible. If the ADC has a full 16 bits of resolution, this means that every value within that range may occur; however, if the converter only has a 14-bit resolution (more common for PLC signals), the values will increment by 4 (i.e. 0-4-8-12 etc.). This means only 8192 possible values can occur!

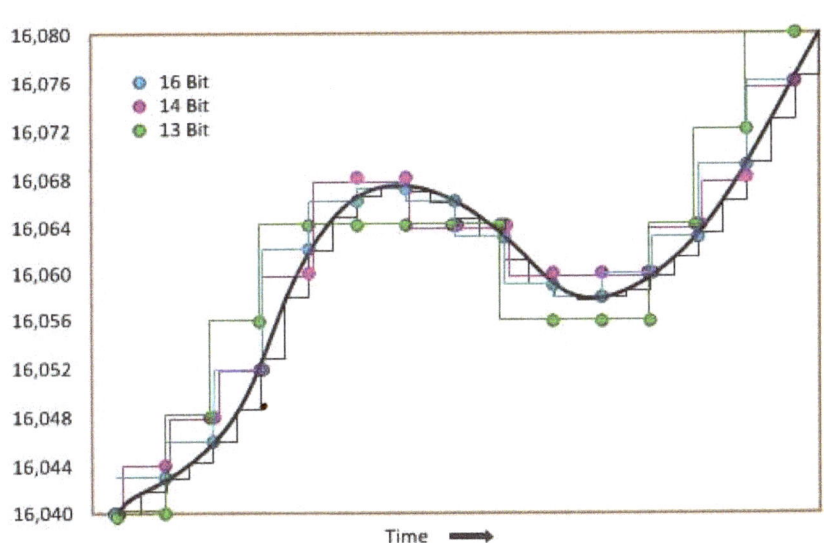

This picture shows the results of converting an analog signal to 16-, 14- and 13-bit resolution.

DACs take a number from the PLC program (memory) and convert it into a signal to control a device such as a proportional valve or a Variable Frequency Drive (VFD) speed reference. Like ADCs, the signal may be a full 16 bits, but 12-14 bits of resolution are more common for PLC analog output cards.

4-20mA signals are considered to be more immune to noise interference, while 0-10v signals are often used to control VFD speeds.

In the Programmable Logic Controller block diagram of Figure 1, ADCs are shown as "Signal Conditioning", while DACs are shown as "Power Amplifiers".

Wiring - Digital

Digital I/O wiring is based on the type of signal that is connected to the I/O point. Discrete I/O may be AC or DC, Input or Output. Relay cards can also be used to transmit different types of signals.

Discrete solid-state DC devices have two different types, Sinking (NPN), and Sourcing (PNP). NPN and PNP are transistor types. PNP devices *source* a positive DC signal, usually 24v, while NPN devices sink current from a sourcing input card.

These diagrams show typical wiring diagrams for DC devices and sensors.

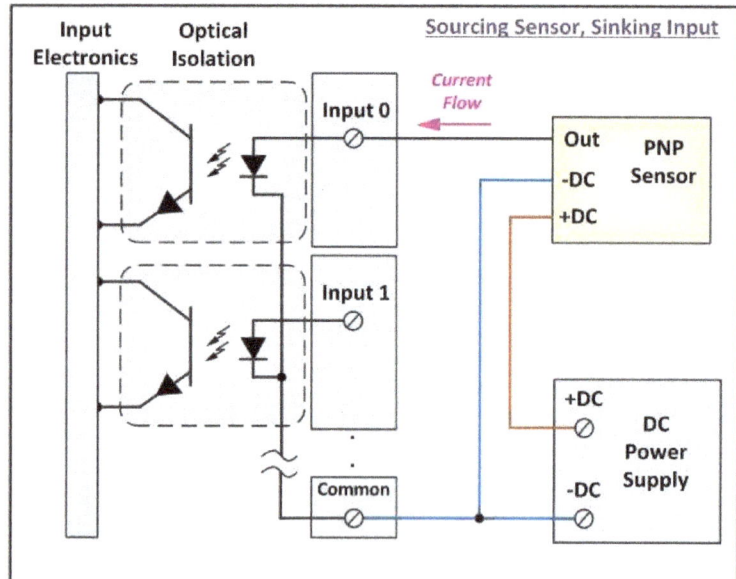

A Sourcing (PNP) sensor applies a positive voltage to a Sinking (NPN) input. The input card and the sensor need to share the –DC voltage, which is usually grounded.

The optical isolation protects the lower voltage TTL circuit in the input card. The input card receives its operating voltage (usually 5 volts DC) from the PLC's power supply. The sensor's power is supplied from an external DC power supply, typically at 24 volts DC.

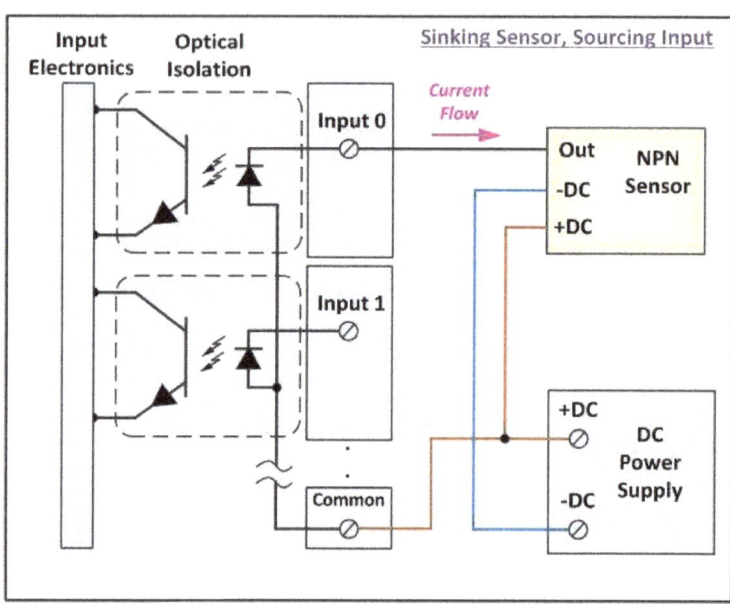

A Sinking (NPN) sensor provides a path for current to the sourcing input. As with the sinking input card, it is important that the sensor and input card share the same +DC reference.

DC input cards are often made to allow either PNP or NPN sensors or devices to be used. The card reacts based on the polarity of voltage applied to the common terminal.

Common terminals are also often grouped with inputs, allowing PNP and NPN sensors to be used on the same card if desired. For example, inputs 0-3 may have a positive DC common, while inputs 4-7 may have a negative.

Most U.S. and European machines use PNP sensors and sinking inputs, while NPN sensors are common in Japan and on Japanese equipment.

AC sensors are used with AC type input cards, in this case polarity will not matter. The common terminal will usually be neutral (0 volts, grounded) and the 120 VAC signal will be applied to the input. The main reason AC sensors would be used instead of the safer, lower voltage DC is the relative immunity from noise and the longer distances allowable for wiring. Conveyor systems and large industrial facilities often use AC inputs and outputs.

As with DC inputs, DC outputs may be sinking or sourcing. They are also optically isolated as shown in the diagram.

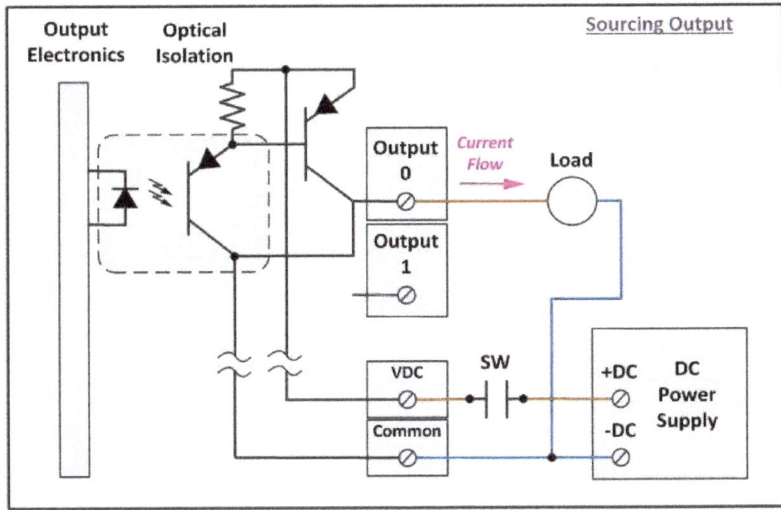

DC output cards require both a positive and negative DC signal to operate. This is because of the power requirements of the outputs; DC outputs are often capable of handling up to 2 amps each.

The VDC connection is often switched as shown by the "SW" contact. This is for safety, which must be hardwired to an Emergency Stop circuit if the load controls a potentially hazardous device. The diagram above shows a sourcing card. The output provides a positive voltage to the load. A sinking card would provide a path for current *from* the device and sink it to the common terminal.

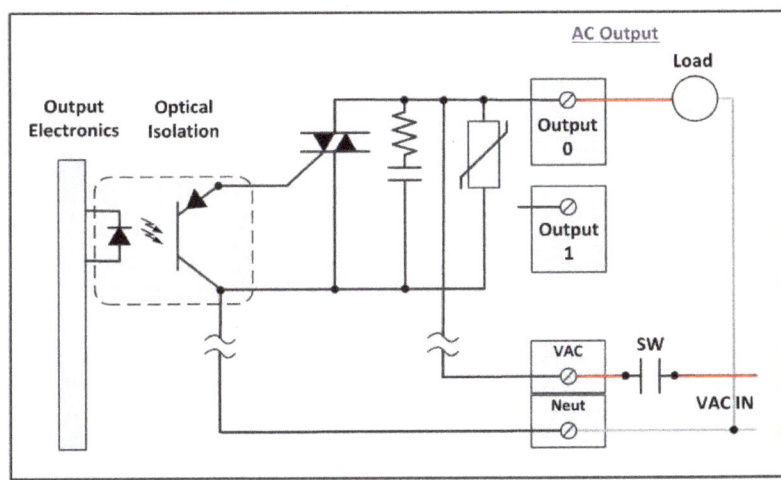

AC output cards usually use a TRIAC to provide AC voltage to the load. Outputs are usually wired through fuses for circuit protection. 120 and 240 VAC cards are available from most manufacturers. They are often used for switching motor starters.

Some AC output cards use zero voltage detection to ensure that the output only switches on when current flow is at zero, reducing surges at the load.

Relay output cards can switch either AC or DC voltages. They often group the outputs with different commons, allowing mixed use cards. The specifications for relay cards show a much lower lifetime than those of solid-state cards because of the mechanical relays.

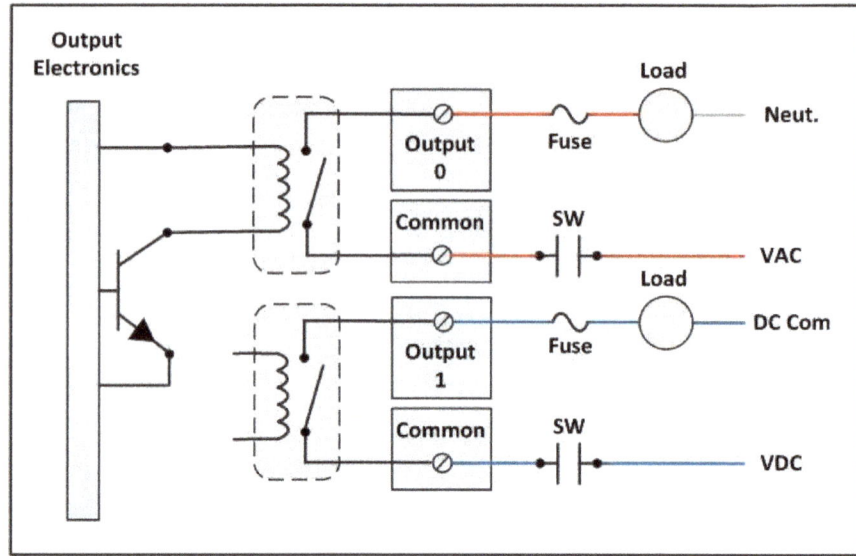

Relay cards that have a common terminal assigned to each output are known as *individually isolated*. This type of card is often used with external equipment that has its own power source.

Since the number of operations of each relay is limited compared to that of a DC output card, external relays are often used instead. Small terminal block style relays are an inexpensive alternative to a relay card.

Wiring – Analog

Analog signals are based on changes in either voltage or current. Voltages may be either 0-5, 0-10, or -10 to +10 volts. Current signals are either 0-20mA or 4-20mA.

Voltage signals are more susceptible to noise, but wiring runs can be longer than with current signals. Analog signals should be connected using shielded cable whenever possible with the shield connected on <u>one end only</u> to reduce noise.

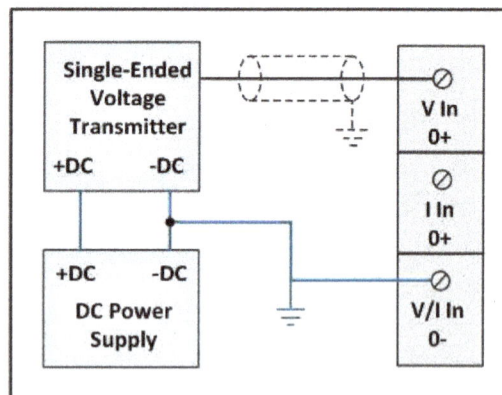

Analog card input configurations may differ on various brands, but this is typical for a universal (voltage and current) input card.

This is a single-ended transmitter, which only has one output wire. It generates the variable voltage signal from the power supply and shares the grounded –DC signal.

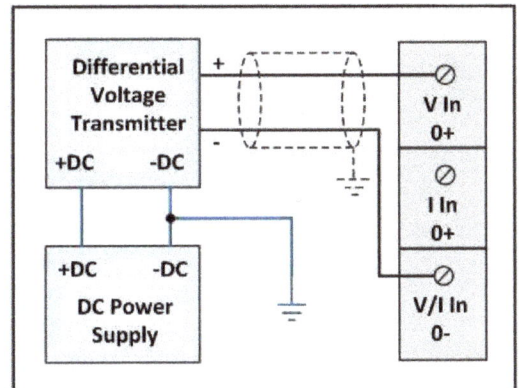

Differential transmitters have a positive and a negative terminal; these are connected to the two voltage inputs of the card. While the shield needs to be grounded, there is no need to ground the negative lead of the sensor or transmitter.

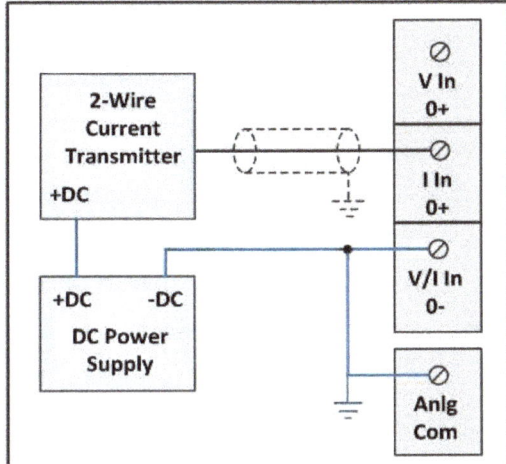

Two wire current transmitters are placed in-line with the power supply and are also known as *loop-powered* transmitters. They deliver a 4-20 mA signal to the current input of the card.

In this diagram, note the "Anlg Com" terminal; this is a group common termination that is present on many analog input cards. When in doubt, it is always a good idea to jumper the negative side of the power supply to this terminal and ground it.

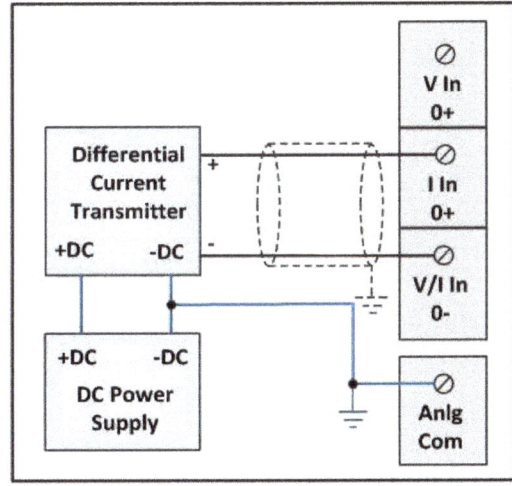

A differential current transmitter is powered by the power supply independently of the current loop. This allows these types of devices to have features like LCD indicators that display current and set-point values. These transmitters are typically more expensive than 2-wire transmitters but have more setup features such as scaling and programmability. The negative lead of the transmitter does not need to be grounded.

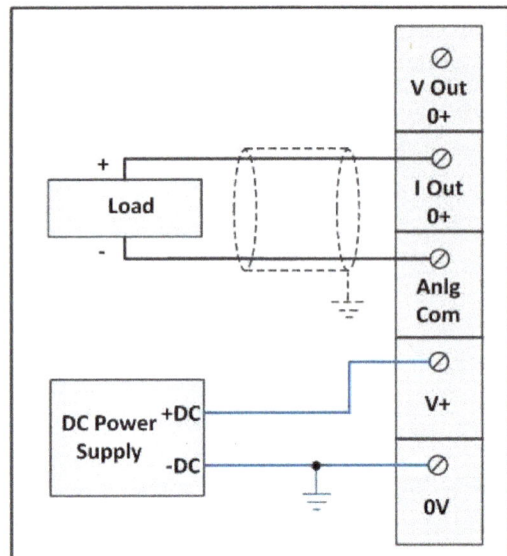

Analog output cards provide either current or voltage to a load. As with input modules, 0-20 and 4-20mA and 0-5, 0-10 and -10 to +10 VDC signals are standard.

This card shows the wiring for a typical analog output to a load. Some cards use an external power supply as shown, while others are powered from the 24v supply on the I/O bus.

As with analog input wiring, it is important to use shielded cabling and ground the shield on <u>one end only</u>.

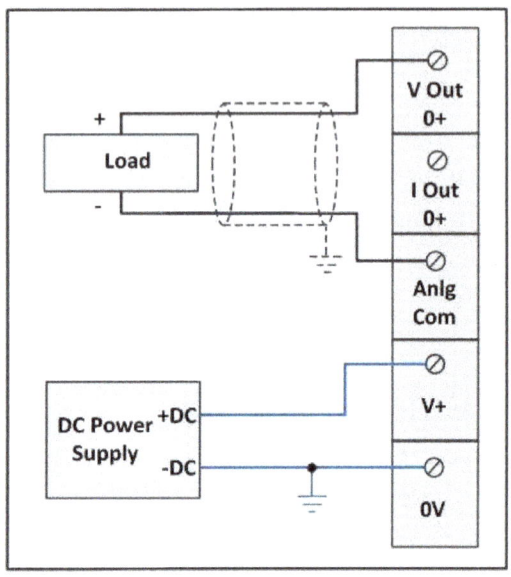

Analog voltage outputs also may be powered with an external power supply or across the bus. An example of a typical load for an analog voltage output would be the speed reference for a Variable Frequency Drive (VFD) or a position for a proportional valve.

Battery/Memory Back-Up

Program and data memory in a PLC is contained in "RAM" (Random Access Memory). This type of memory may be volatile or non-volatile, and it can (and will) be overwritten often. The program itself is in one area of RAM and must be kept in memory even when the PLC is powered off. Older PLC systems required a battery or a "super-capacitor" to back up the program when power was removed for long periods of time. Newer platforms can save the program to non-volatile memory such as CompactFlash and Secure Digital (SD) RAM. On these older battery-backed platforms, if the battery died, the program was lost.

Communications

All PLCs need some method of communicating with a programming device, and possibly even with other devices such as Operator Interface Terminals (OITs), computers or remote I/O nodes. A PLC processor will usually have at least one built-in communications port, and possibly more. In addition, modular communication devices can be added to the parallel bus, or rack.

Serial Communications means that bits are transmitted on one wire at a time in a series of high and low electrical signals, or "1's and 0's". This differs from parallel communications, such as printers, where the bits are transmitted and received on many wires in parallel with each other. Serial communication uses different physical formats, usually in the form of RS232, RS422 or RS485.

RS, or "Recommended Standard" communications include RS232, RS422 and RS485 serial communications. RS defines the wiring and format of the message, but not the language or protocol of the message.

RS232

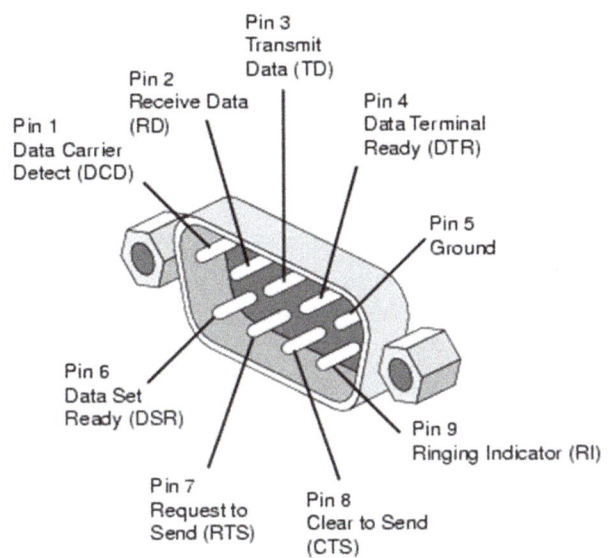
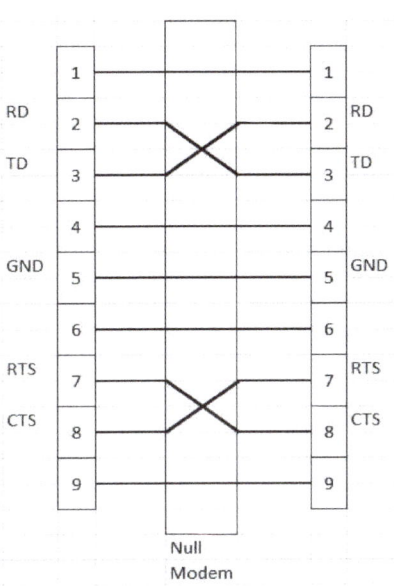

RS232 protocol is used for many of the programming interfaces between laptops and PLCs. Because serial ports are often not present on newer computers, a USB adapter may have to be used between the computer and cable. RS232 requires that parameters such as Baud rate (speed), bits and parity be set the same on all stations for communication to take place. If the transmit and receive pins (TD and RD) are the same on both devices, a null modem adaptor will be needed.

Each PLC manufacturer will have its own driver for its language; for example, in Allen-Bradley's PLC the protocol is called DF1, while in Siemens it is called MPI.

RS485

RS485 is another common standard for serial communications. It uses a single twisted pair of wires and can be "daisy-chained" between multiple controllers.

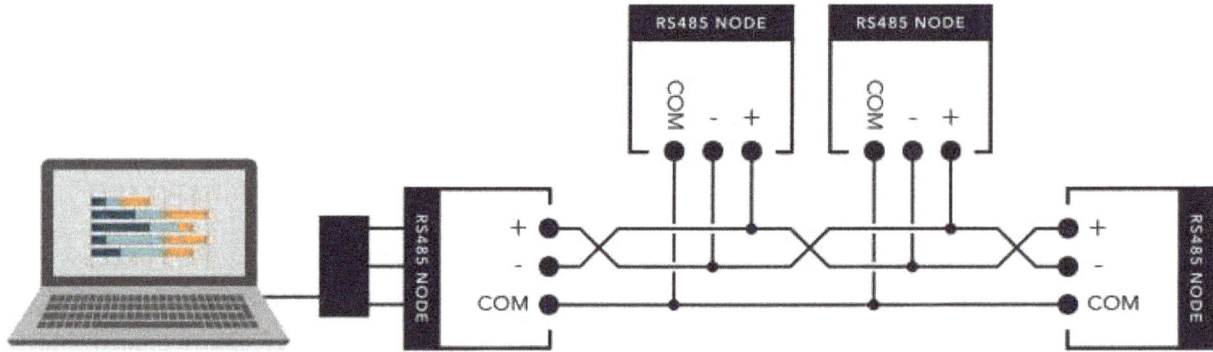

RS485 is commonly used for remote I/O and multi-drop networks. Programming devices such as computers can be used on these networks through the use on an adapter. As with RS232, the Baud Rate and protocol must be set the same on all member stations or "nodes".

Examples of RS485 networks are Profibus, DeviceNet, Modbus and Data Highway/DH+.

RS422

RS422 is often called peer-to-peer, or PtP. It also uses a twisted pair of wires, but this protocol only supports communication from a single device to the PLC.

A Note on Twisted Pair Communications:

Twisted pairs of wires reduce the amount of electrical interference from other signal wires that run in parallel. Noise can be further reduced by shielding the pairs as a whole or individually; it is important that the shields are only grounded on <u>one end</u>, otherwise even more noise may be introduced.

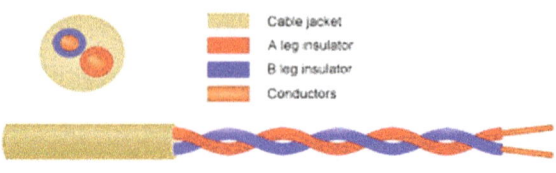

Acronyms are used for the different types of twisted pair cables used for communications.

UTP signifies Unshielded Twisted Pair as shown in the diagram to the left. STP, or Shielded Twisted Pair, means that a single shield surrounds all of the pairs in a cable, while ScTP or "Screened Twisted Pair" means that individual shields are wrapped around each pair.

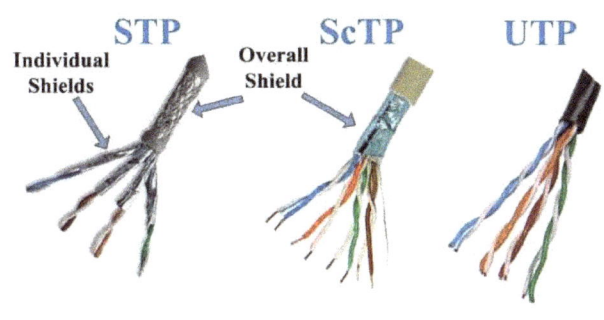

This diagram illustrates some of the types of twisted pair cables used in industrial communications. A bare (uninsulated) wire called a drain is usually run in contact with the foil shield.

USB

Universal Serial Bus, or USB is a standard developed in the 1990s to connect computer devices and peripherals such as keyboards, digital cameras and pointing devices to computers. It not only provides communications, but also can power devices. USB is much faster than standard serial connections.

Ethernet

Ethernet is a family of computer networking technologies consisting of a set of wiring and communications standards. Systems communicating over Ethernet divide a stream of data into shorter pieces called **frames**. Each frame contains source and destination addresses, and error-checking data so that damaged frames can be detected and discarded; most often, higher-layer protocols trigger retransmission of lost frames.

Cabling for Ethernet may consist of coaxial cable, several twisted pairs as shown above, or fiber-optics. Most PLC connections use standard twisted pair CAT 5 cable and RJ45 connectors.

Ethernet follows a seven-layer structure defined by the Open Systems Interconnect model. These layers describe the lowest or physical layer, as well as various methods of interconnecting and networking between different areas, or domains.

Many different protocols are included on this definition, including TCP/IP (for connecting dissimilar devices across the internet), BOOTP (for setting initial addresses), and SMTP (for e-mail).

PLC manufacturers generally define their own language and protocols for using Ethernet to both communicate and connect to inputs and outputs (I/O).

Because it is important that I/O communications are **deterministic,** PLC manufacturers follow the Common Industrial Protocol, or CIP. This ensures that signals are sent and received within a specified period of time and are therefore predictable. Examples of CIP include Ethernet/IP for Allen-Bradley and ProfiNet for Siemens.

Ethernet Terminology

Following is a list of terms that it is important to know when discussing an Ethernet network:

Client: A computer or device that initiates a request for data.

Server: A computer or device that responds to the client by providing or accepting data.

LAN (Local Area Network): a network that that connects computers or devices in one single location. Usually administered by a Server computer or a domain controller.

WAN (Wide Area Network): a network of LANs connected by Gateway or Router devices.

Workgroup: Computers in a workgroup can share files folders and printers.

Domain: A collection of host computers supervised by a domain controller computer.

Bridge: A device that interfaces between two similar networks.

Gateway: A device that allows modules on two different communication networks to interface.

Hub: A solid-state device that connects Ethernet modules in a star configuration.

Switch: A solid-state device that connects Ethernet modules that has buffering capability to reduce collisions.

Router: A solid-state device that has the capabilities of a switch, but connects Ethernet modules on different networks. Often includes a **Firewall**.

TCP/IP (Transport Control Protocol/Internet Protocol): A common protocol that allows computers with different operating systems to exchange data.

CIP (Common Industrial Protocol): A method of communicating used in industrial automation. Provides deterministic communications for control, safety, synchronization, motion, configuration and information.

Socket: A package of subroutines that provides access to TCP/IP and CIP functions.

NIC (Network Interface Card): A communication adaptor with an Ethernet port in a computer.

Ethernet Addressing:

There are several classes of Ethernet addresses allowing for different numbers of devices to be connected together in the same network. Most industrial control networks use class C, which comprises a single LAN.

The format of an Ethernet address is xxx.xxx.xxx.xxx, where values of xxx can range from 0-255. Each of these sections or "octets" therefore represents a Byte, or 8 bits. An Ethernet address then contains 32 bits.

The Ethernet address is then "Masked" with a group of 1's and 0's to allow only groups with similar numbers to communicate. A class C network will often use numbers in the "192.168.0.xxx" series with a mask of "255.255.255.0".

Industrial Communication and Control:

An example of an Industrial Control Network is shown in the diagram below:

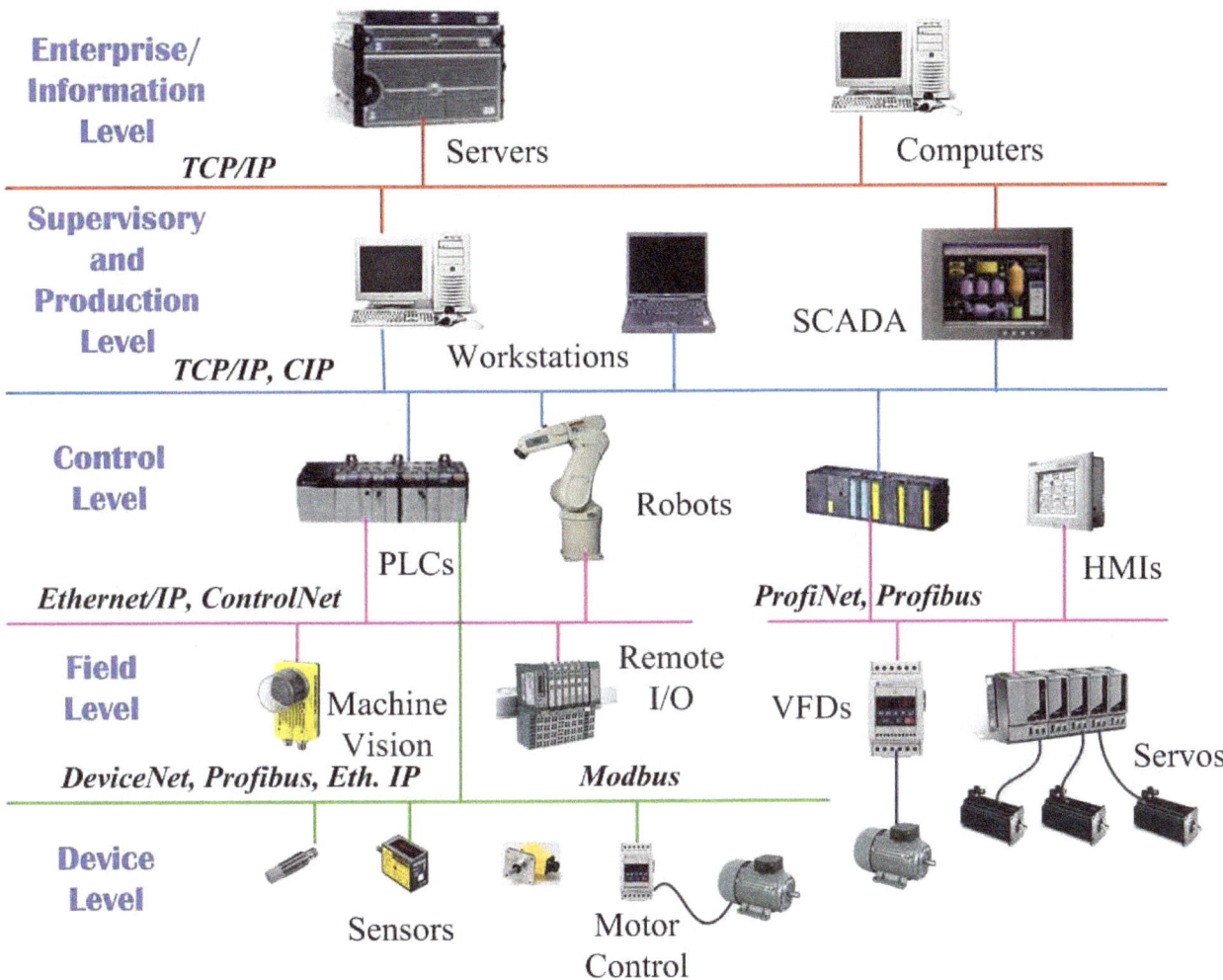

This illustrates how different controllers and elements of the system interface and bridge the various networks. All of the networks below the control level need to respond quickly to changes, while the communications at the higher levels are concerned with transferring and saving large amounts of information.

 Exercise 1

1. What kind of PLCs do you have in your plant? Brand name (platform) and type?
   _____
   _____
   _____

2. What type of I/O do they use? (Analog/Digital, voltages, etc.)
   _____
   _____

3. What types of communication networks do you have in your plant?
   _____
   _____
   _____

4. What is the Ethernet address and Subnet Mask of your computer?
   _____

5. On which level of communication network are PLCs usually found?
   _____

PLC Memory

PLC Memory consists of the operating system and firmware of the processor, sometimes called **System Memory**, the module firmware (if any), and the program and data that is used by the programmer. In the previous hardware section it was explained that there are volatile and non-volatile areas of memory, and that the volatile part of memory needs a battery, "super-capacitor" or other rechargeable energy storage module to hold its program and/or data.

Though the program can be saved on Flash or SD RAM cards without a battery, the data exchange rate is too slow to use this for the actual interfacing of the program with its data. When the PLC is powered on, the program is loaded from non-volatile RAM cards into the user memory of the controller. Not all PLC platforms back up the user memory with a battery or other energy storage device, data memory may be lost when a processor loses power. Some platforms, however, ensure that the data is kept intact even when power is lost by use of battery backed RAM. This means that the values in data registers will be retained and the program will start in its last state.

Other PLC platforms assign some parts of RAM to be "retentive" and other parts non-retentive. Omron separates its retentive bits into "holding" relays and non-retentive "CIO", and its data into the retentive DM Area and non-retentive "Work Area". Siemens allows its general "marker" memory to be assigned as retentive or non-retentive and defaults to only 16 bytes of retentive marker memory, but it can be changed. Siemens data blocks, however, are retentive unless defined not to be. Allen-Bradley's memory is all retentive.

The operating system itself on a processor is held in non-volatile System memory, called "firmware". To change the firmware on a PLC a "Flash" program or tool needs to be used to download it. This is usually included with the programming software.

I/O, communications and other modules also often have firmware built in. The firmware update tools can also update these modules and the firmware is usually available from the manufacturer's website. It is necessary to have software that is at least as up-to-date as the firmware being installed.

The RAM part of memory in a PLC can be separated into two general areas: **Program Memory** and **Data Memory**.

Program memory consists all of the lists of **Instructions and Program Code**. This is what is sent to the processor. The act of sending the program instructions to the PLC is called "downloading" on most brands of PLC, however this may differ on some platforms.

Data memory includes the **Input and Output Image Tables** as well as **Numerical and Boolean Data**. You will find that *most of the data used in the PLC program is internal memory and not directly related to I/O!*

As the program executes, it keeps track of whether bits (BOOLS) are on or off and the values of numbers in Data Memory. Different platforms have different ways of organizing data; some of these methods will be discussed in this section.

Before discussing the way in which memory is organized, it is important to discuss the types of data that will be stored there.

Numerical Data Types

BOOL or Bit: This is the simplest form of data, it only has two possible states, on and off, or one and zero. If the states are called "True" and "False", it is more properly called a BOOL, which implies a logical operation. A bit may be just one part of a larger element, such as a byte or word.

Byte or SINT: A Byte is a group of 8 bits. This is sometimes called a Single Integer, or SINT. A Byte's value ranges from 0 to 255, or 0000_0000 to 1111_1111. What is a half of a Byte, or four bits? A **Nibble** of course!

Integer (INT) or WORD: 16 bits make up an integer, or INT. While an Integer always implies a number that mathematical functions can be applied to, a WORD may sometimes only be used for logic functions, such as AND or OR. An Integer can have 65,536 possible values, from 0000_0000_0000_0000 to 1111_1111_1111_1111.

The bit on the far right of the string of 1's and 0's is known as the Least Significant Bit, or LSB, while the bit at the far left is the Most Significant Bit, or MSB. An **Unsigned Integer** ranges in value from 0 to 65,535. A **Signed Integer** uses the MSB as a "sign bit"; if it is a 1 the value is negative, if it is a 0 the value is positive. A Signed Integer ranges in value from -32,768 to +32,767. Most PLCs use Signed Integers.

Double Integer (DINT) or Double Word (DWORD): A Double Integer has 32 bits. Like an Integer, there are Signed and Unsigned versions, determined by the MSB. A DINT ranges in value from 0 to 4,294,967,295 (Unsigned) or – 2,147,483,648 to +2,147,483,647 (Signed).

REAL or Floating Point: A REAL number is a 32-bit number that can express fractional values, or decimal points. Because the decimal can be shifted to the right or left within the value to change its magnitude, the decimal point can be said to "float". Unlike Bytes, INTs and DINTs, it is not possible to look at individual bit values to determine the value of the number. A REAL number is composed of a Mantissa, an Exponent and a Significand, and the bits within the number are not used for anything else other than expression of the value. REALs have a range of 1.1754944e-38 to 3.40282347e+38.

Additional data types such as Long Integers (LINT, 64 bits), Long Reals, (LREAL, 64 Bits) and STRINGs (an array of Bytes signifying text characters) are also used in PLCs.

Understanding How Bits Can Become Numbers:

0	0
0	1
1	0
1	1

If the previous talk of ones and zeros is confusing, think of it this way: for a single bit, there are only two possible values; 0 or 1, off or on.

For two bits, there are four possible values; both off or 0,0; the first off and the second on or 0,1; The first on and the second off, or 1,0; and both on or 1,1.

4's	2's	1's	
0	0	0	0
0	0	1	1
0	1	0	2
0	1	1	3
1	0	0	4
1	0	1	5
1	1	0	6
1	1	1	7

For three bits, the possible number of combinations increases to 8 as shown: the values above the columns show the value of each "place" in the row. The string of ones and zeros is known as binary, a base 2 system. As a new position is added to this string, the value of the previous column is doubled.

For four bits, the possible number of combinations increases to 16, and the value of the next column becomes the "8's" position. The values then start at 0000, or zero decimal, and increase to 1111, or 15 decimal.

The value of each column then doubles as a placeholder (16, 32...), as does the possible number of combinations (32 @ 0-31, 64 @ 0-63); remember, computers and PLCs can only "think" or process data in terms of ones and zeros since they are really just a collection of on-off switches, albeit very tiny ones.

Data Formats:

In addition to the data types listed above, data can also be expressed in different ways. For instance, a Byte can be shown as a string of ones and zeros (Binary or Base 2): (0110_1011), as a decimal (Base 10) number: (107) or as a Hexadecimal (Base 16) number: (6B).

Integers: Computers and PLCs operate most efficiently when using numbers that are multiples of two; this is because at its heart, a microprocessor is just a collection of on-off switches. An Integer is therefore just a series of binary values signifying increasing values for each bit as shown below:

(MSB/Sign)															(LSB)
32,768	16,384	8192	4096	2048	1024	512	256	128	64	32	16	8	4	2	1
0	1	1	0	1	0	0	1	0	1	1	0	1	0	0	1

To determine the decimal value of a Signed Integer for the binary value, it is necessary to add up all of the place values where there is a 1: 16,384 + 8,192 + 2,048 + 256 + 64 + 32 + 8 + 1 = 26,985.

(MSB/Sign)															(LSB)
32,768	16,384	8192	4096	2048	1024	512	256	128	64	32	16	8	4	2	1
1	0	0	1	0	1	0	1	1	0	1	0	1	0	1	0

This Signed Integer value has a 1 in the MSB. The decimal value is determined by adding all of the other place values: 4,096 + 1,024 + 256 + 128 + 32 + 8 + 2 = 5,546; and then subtracting the result from 32,768, or -27,222. This is also known as 2's Complement.

BCD: Prior to the use of OITs (Operator Interface Terminals), digital devices such as thumbwheel switches and 7-segment displays were used for entering and displaying decimal values in the PLC.

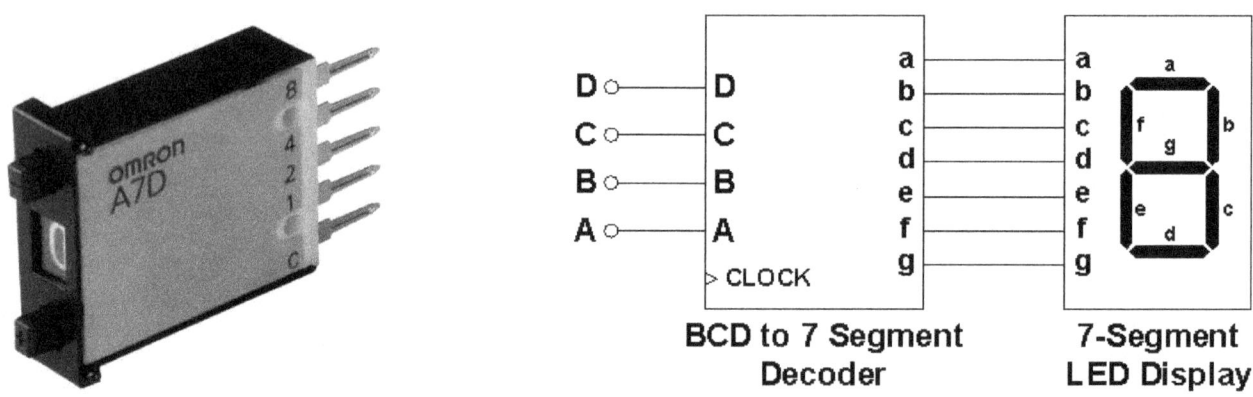

BCD to 7 Segment Decoder

7-Segment LED Display

To enter a number into a PLC's memory location, each thumbwheel switch required 4 digital inputs into the PLC. Decimal numbers from 0-9 could be set for each decimal digit; when the thumbwheel reached 9, it would then roll over to the zero position again. This meant that combinations such as 1010 (10) or 1011 (11) were not possible.

In the same way, outputs were used to illuminate the individual segments of a 7-segment display, each display requiring 4 digital outputs. It was important in this case that illegal combinations of outputs (decimal 10 and above) could not be sent to the device.

Binary Code A B C D	Decimal Number	BCD Code B_5 B_4 B_3 B_2 B_1
0 0 0 0	0	0 | 0 0 0 0
0 0 0 1	1	0 | 0 0 0 1
0 0 1 0	2	0 | 0 0 1 0
0 0 1 1	3	0 | 0 0 1 1
0 1 0 0	4	0 | 0 1 0 0
0 1 0 1	5	0 | 0 1 0 1
0 1 1 0	6	0 | 0 1 1 0
0 1 1 1	7	0 | 0 1 1 1
1 0 0 0	8	0 | 1 0 0 0
1 0 0 1	9	0 | 1 0 0 1
1 0 1 0	10	1 | 0 0 0 0
1 0 1 1	11	1 | 0 0 0 1
1 1 0 0	12	1 | 0 0 1 0
1 1 0 1	13	1 | 0 0 1 1
1 1 1 0	14	1 | 0 1 0 0
1 1 1 1	15	1 | 0 1 0 1

This coding of bits into decimal equivalents is known as **Binary Coded Decimal**, or BCD. Some operator interfaces still require BCD for display of values, so conversion of integers into BCD format is sometimes needed. In addition, there are PLC platforms that use BCD for Timer and Counter values.

As the table to the left shows, BCD codes above 1001 require that another 4 digit value be used for the next decimal number. To display the number 9,999, the 16-bit pattern would read 1001_1001_1001_1001

The equivalent 9,999 expressed as a Signed Integer would be 0010_0111_0000_1111.

To express a Signed BCD number, the most significant four bits are used as a "sign" character, as with the Signed Integer. A negative number is signified by a 0001, while a positive number uses 0000.

The range of a 16 bit Signed BCD number is then only -999 to +999!

Binary	Decimal	HEX
0000	0	0
0001	1	1
0010	2	2
0011	3	3
0100	4	4
0101	5	5
0110	6	6
0111	7	7
1000	8	8
1001	9	9
1010	10	A
1011	11	B
1100	12	C
1101	13	D
1110	14	E
1111	15	F

Hexadecimal: In order to display the full 65,536 possible values in sixteen bits in only four characters, a base 16 numbering system called **Hexadecimal** is used. The only reason the base 10 Decimal system is used is that humans calculate best in this format; all because we have 10 fingers and 10 toes. As mentioned before, computers are most efficient when they calculate in multiples of two.

After a group of four binary digits (a "Nibble", or half of a Byte) reaches 1001, the next value of 1010 can't be expressed as a numeral without using something outside of the values 0-9. In order to describe a base 16 number using 16 different symbols, after the number 9 the symbols use the letters A-F as shown to the left.

Because Hexadecimal is base 16 and a multiple of 2, it is very easy to convert from Binary to Hexadecimal. Simply separate the binary into groups of four and convert each group.

Octal: Base 8 numbers are also commonly seen in PLC systems. For instance, Siemens I/O addressing is octal; this means that only the numbers 0-7 are used. The numbers after 7 would be 10, 11 ...17, 20, 21 and so on.

Like Hexadecimal, because Octal is a multiple of 2 it is very easy to convert to and from Binary. Separate the binary into groups of three and convert each group; the highest value in any group of three is 111, or 7.

Viewing Data Types: PLC software allows the user to view a number in many of the formats listed here, removing the need for a calculator. It is not always clear in what format the data is being viewed, but there are often designators. In Siemens, a signed integer (decimal or Base 10) will have no designator, however a Hex number will have a W#16 prefix, indicating that it is base 16. A REAL will have a decimal point or be expressed with an exponent, while a binary representation may have a prefix or appear as a string of ones and zeros.

Dot Fields and Separators: If a single bit of an integer is designated, it may be shown with a separator such as a slash or a dot; for example, N7:5/3 (Allen-Bradley, the fourth bit of the sixth word; numbering starts at zero) or Q3.2 (Siemens, the third bit of the fourth output byte).

Dot fields are also often used to designate an element of a complex data type, such as a timer. For example, Timer1.ACC designates the accumulated value (integer or double integer) of Timer 1. It is important to understand how memory is addressed for your particular PLC before beginning a program.

Tags: Many modern PLC platforms don't use numerical data registers at all. Instead, they allow users to create memory objects as required in the form of text strings. Allen-Bradley's ControlLogix and Siemens' TIA Portal platforms are examples of this. Most major PLC manufacturers make a PLC with tag-based data. Tags are also called Symbols on some platforms, but a symbol is not necessarily a tag; it may simply be a mnemonic address or shortcut to a register address. Tagnames are downloaded into the PLC and used instead of an address.

Tags are usually created in a data table as required. Instead of numeric addresses such as "B3:6/4" or "DB2.DBW14", symbolic names such as "InfeedConv_Start_PB" or "Drive1402.ActualSpeed" are created as memory locations. As tags are created, details such as the data type (BOOL, Timer, REAL) and the display style (Hex, Decimal) need to be selected.

Tags have the advantage of being more descriptive than numerical register numbers. In addition, descriptions and symbols from register addresses were only present in the computer and were not downloaded into the PLC. With tags, since the address is the actual register location, a tag-based program can usually be uploaded straight from the PLC.

Also, the same tags from the PLC program can be used in the HMI or SCADA program directly. This saves time rather than having to map PLC addresses to HMI tags.

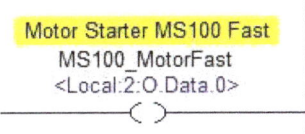

Of course, I/O addresses are still created from the hardware configuration of the PLC, but manufacturers have created various ways to connect I/O addresses with tags. One of the most useful of these is Allen-Bradley's ControlLogix platform, where any tag or address can be "aliased" to any other and both show up in the ladder logic, as shown in the figure to the left. Modicon's newer platforms also allow "Symbols" or tags to be connected to I/O points in a similar way.

Data Structures:

Array: An Array is a group of *similar* data types. For instance, an array can be defined that contains ten integers, or 50 REALs, or 32 BOOLs. Data types can't be mixed in an array.

Complex data types such as Timers, Counters or User Defined Types (UDTs) can also be placed into an array. Typically, an array will be shown with square brackets, such as Delay_Tmr[6]. This designates element number 6 of the array.

Some platforms allow multidimensional arrays to be defined, such as Integer [2,4,5]. This means Integer number five of the fourth group of the second group.

Elements that are composed of more than one data type are known as **Structures**. A structure may be defined by the programming software, as with instructions, or by the programmer.

User Defined Type (UDT): A UDT is a group of *different types* of data, or a structure. Later in this course, data types that consist of more than one type of data will be discussed; for instance, Timers and Counters consist of two integers or double integers and several bits, all combined into a structured data type called "Timer" or "Counter".

A UDT can only be used with Symbols or Tags; this is because a UDT is not data. Instead, it is a *definition* of data. After defining a UDT, a tag or symbol must be created using the new data type.

A common reason to build a UDT is to describe an object more complex than a simple element of data. As an example, a Variable Frequency Drive has many pieces of data that may be associated with it. For instance, a motor needs to start and stop. It has various numerical parameters to describe its movement such as commanded speed, actual speed, acceleration and deceleration. We may also want to know its status; whether it has faulted and what kind of fault has occurred.

UDT Name:	"Drive"	
Name	Type	Description
Run	BOOL	Run Command
Stop	BOOL	Stop Command
Alarm	BOOL	Drive in Alarm
Running	BOOL	Run Status
CMD_Speed	REAL	Commanded Speed %
Act_Speed	REAL	Actual Speed %
Accel	INT	Acceleration ms
Decel	INT	Deceleration ms
AlarmStatus	SINT	Alarm Number

Name	DataType	Status	Description
Drive_5207	"Drive"		Spindle Drive VFD 5207
Drive_5207.Run	BOOL	0	Run Command
Drive_5207.Stop	BOOL	1	Stop Command
Drive_5207.Alarm	BOOL	1	Drive in Alarm
Drive_5207.Running	BOOL	0	Run Status
Drive_5207.CMD_Speed	REAL	27.34	Commanded Speed %
Drive_5207.Act_Speed	REAL	0.00	Actual Speed %
Drive_5207.Accel	INT	40	Acceleration ms
Drive_5207.Decel	INT	50	Deceleration ms
Drive_5207.AlarmStatus	SINT	4	Alarm Number

On the left is a UDT named "Drive" defined in the software, and on the right is a tag made from the UDT. The definition does not get downloaded to the processor; it can only be modified on the programming device.

The sub-elements of the tag are an example of the dot fields described earlier. By making a UDT, many drives can be added to a program without a lot of extra typing. UDTs are in important element in Rapid Code Development.

TIP: On non-tag-based systems, UDTs can cause problems if the commented program is not available. Remember, descriptions and non-tag symbols are not kept in the processor. This is why it is difficult to reconstruct a Siemens S7 program if you don't have the original code, the downloaded data blocks do not contain the names of the elements.

 Exercise 2

1. Convert the Binary number "0110_1100_1011_0111" into...
   Decimal _____ Hexadecimal _____ Octal _____
   Can this number be converted to BCD? Why or why not?
   _____
   _____

2. How would you write the Binary number "1001_0101_1000_1001" as a Signed Integer?
   _____

3. Convert the Decimal number 417 into:
   BCD _____ Binary _____ Hexadecimal _____

4. Write the number 2A9E in Binary : _____
   What is this number in Decimal? _____

5. How many Bytes are in a Double Integer? _____

6. What is a "Tag"? Is it the same as a Symbol? _____
   _____

Work Area:

Data Memory Organization:

Data memory is organized in different ways depending on the type or brand of PLC. Some PLCs have registers assigned to specific data types, that is, Bit, Integer or Real, (Allen-Bradley SLC and Micro), whereas other brands may separate data by whether it is retentive (Holding Relays, Omron) or place all data together ("V Memory", Koyo/Automation Direct).

It is important when learning a new PLC's programming platform to first understand how its memory is organized. In the older **GE** PLCs for instance, data memory and I/O share the same space. It could be quite embarrassing if you were to cause actuators to move when intending to simply save an integer to a register!

A-B SLC	
O	Outputs
I	Inputs
S	System
B	Bits
T	Timers
C	Counters
R	Control
N	Integers
F	REALs

Siemens S7	
I	Digital In
Q	Digital Out
PIW	Analog In
PQW	Analog Out
M	Memory (M)
DB	Memory Blocks

Omron	
CIO	Basic I/O (Discrete)
CIO	Special I/O (Analog)
CIO	CPU Bus I/O
W	Work Area
H	Holding Area
A	Aux Relay Area
TR	Temp. Relay Area
D	Data Memory Area
E	Ext. Memory Area
T	Timers
C	Counters
TK	Task Flags
IR	Index Registers
DR	Data Registers

Koyo		
T	Timer Curr. Val	V0-377
	Data Words	V400-777
CT	Counter Curr. Val	V1000-1377
	Data Words	V1400-7377
	System	V7400-7777
	Data Words	V10000-35777
	System	V36000-37777
GX	Remote Inputs	V40000-40177
GY	Remote Outputs	V40200-40377
X	Disc. Inputs	V40400-40477
Y	Disc. Outputs	V40500-40577
C	Ctrl. Relays	V40600-40777
S	Stages	V41000-41077
T	Timer Status	V41100-41117
CT	Counter Status	V41140-41157
SP	Special Relays	V41200-V41237

This table shows the layout of several PLCs' memory areas. The first list, **Allen-Bradley's SLC** and **MicroLogix** family, shows that data is segregated into numbered files, O0, I1, S2...F8. Each data file is expandable up to 255 words, but after that, new file numbers have to be added, for instance N9, B10 and so on.

The next table shows **Siemens' Step 7**. I/O is assigned during hardware configuration rather than by slot number as with Allen-Bradley. The general memory area "M" is of a fixed size, whereas Memory Blocks or Data Blocks (DBs) contain a mix of data types and can be up to 64KB in size!

Omron's memory sizes are fixed in dimension for each data type, memory is not allocated dynamically as in the previous two examples. It is unique in that it separates retentive memory (Holding Area) from non-retentive (Work Area).

Koyo (Automation Direct) uses a large data area much as the GE system described earlier, each type of data is fixed in size and can't be expanded. All data can be accessed by direct addressing, the "V" addresses.

I/O Addressing

I/O addressing varies from brand to brand. Inputs may be addressed as I or X, outputs as O, Q or Y, and analog I/O designations may use a completely different format than digital ones.

Some brands, such as Allen-Bradley, designate I/O based on the slot number where the card is assigned while configuring hardware; this can't be changed. Other platforms, like Siemens, have a default location where I/O is assigned during configuration, but this can be overridden by the programmer. Addressing may also be Octal, Decimal or even Hexadecimal!

	Allen-Bradley SLC-500	Allen-Bradley ControlLogix	Siemens S7	GE 311	Omron CP1E	Mitsubishi FX2N	Codesys
Digital Input	I:1/3	Local:1:I.Data.3	I0.3	I0103	I0.03	X003	%IX4000.3
Digital Output	I:2/5	Local:2:O.Data.5	Q3.5	Q0205	Q1.05	Y025	%QX4002.5
Analog Input	I:3.2	Local:3:I.Ch1Data	PIW272	AI06	CIO200	D302	%IW2022
Analog Output	O:4.3	Local:4:O.Ch2Data	PQW800	AQ007	CIO210	D403	%QW2036
Address Base	10	10	8	10	10	8	

The table above shows some examples of I/O addressing from several platforms.

Program Memory Organization

All PLC platforms have a routine that is designated to run first. It is important when learning a new platform to identify which routine this is. This is sometimes known as the "Main Routine".

As with data memory, the program itself can be organized in different ways. Major PLC platforms all have some form of subroutine, though they may be called by different names. Allen-Bradley's PLC5 and SLC500 platforms organize their subroutines by file number, where file 2 or "Ladder 2" is the routine that runs first and calls the other routines. As new routines are created, they are called Ladder 3, Ladder 4, etc.

Siemens' routines can take different forms; OBs or Organization Blocks each have a special purpose based on their number. For instance, OB1 is the continuous routine that runs first, OB86 runs if there is a network fault, and OB35 runs on a periodic basis set by the programmer, such as every 100 milliseconds. Functions, or FCs, are much like standard subroutines, however they have local temporary memory. Function Blocks, or FBs, are like FCs but have retentive memory in the form of the data blocks mentioned earlier.

Koyo or Automation Direct's PLCs have subroutines, but they also have special routines called "stages" that automatically deactivate the stage from which it was called. Omron has subroutines called "sections" and also periodic routines called "tasks".

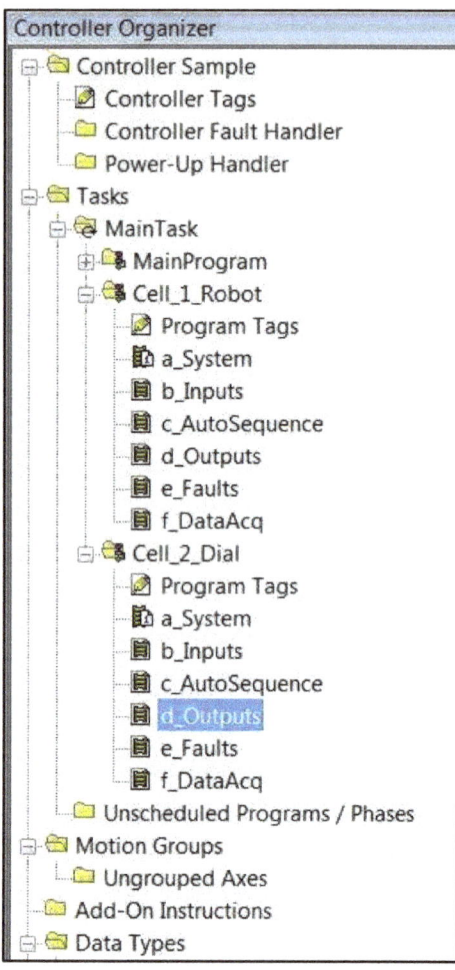

TIP: It is very important to consider whether memory will be Global (available to all programs and routines) or Local (only available to part of the program) before beginning a program. Think about whether you will have multiple instances of the same code.

More powerful PLCs may also allow multiple programs to be placed into a task, as shown in this Allen-Bradley ControlLogix Controller Organizer figure. While the programs are still scanned one at a time, this allows data tables or tag lists to be assigned to one program, rather than being global. Programs are then scheduled to run in a specific order under the task.

This also allows programs to be duplicated under different names, but with the same tag names. This allows for rapid code development, since a program can be written and tested, then copied, addresses and all.

Tasks can also be assigned to run on a periodic basis as with the OB35 block mentioned earlier in the Siemens description. Analog I/O processing is often done this way to accommodate PID instruction cycle frequency.

Hardware Configuration

PLCs are manufactured in a wide variety of configurations. The simplest "brick" types may only have a few digital inputs and outputs, with no analog and no expansion capability. Sometimes these are called "Smart Relays".

The picture to the left shows a small "brick" PLC from Panasonic.

Other PLCs have a very large footprint and are placed into a rack with many slots for expansion. Remote racks can be added to extend the parallel bus and may allow for thousands of local I/O points. Communication modules can further extend the I/O capability as shown in the network diagram in the Communications section.

 The picture to the left shows an Allen-Bradley PLC-5 rack, while the one on the right is a Siemens S7-400.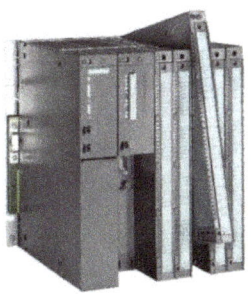

Both of these allow for multiple (additional) rack configurations. These are examples of "large rack" PLCs.

In addition to digital and analog I/O, specialty modules for motion control, high speed counting, PID or process control and various other purposes can be added to the rack. Multiple processors or communications modules can also be configured.

 There are also many configurations that fall between these very small and very large systems. Some of the most common PLCs are mid-range rack or extendible bus versions. The PLC at the left is a Koyo (Automation Direct) DL405.

Processors for these mid-range modular PLCs may have I/O built in or be stand-alone; large rack systems' CPUs do not have I/O built in. Also, rather than having a physical rack, PLC modules may connect together end to end. While this means modules in the middle can't be easily removed, these types of PLC are typically less expensive, and spare slots don't have to be allocated for expansion.

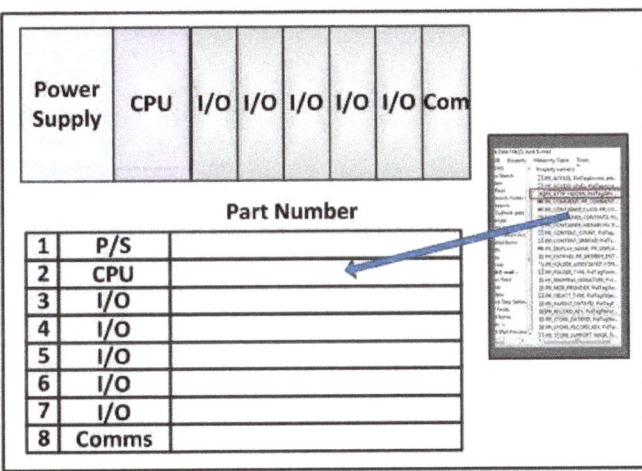 The first step in beginning a new PLC program is configuring the hardware. This is because different processors have different amounts of memory, and because the addresses for I/O are determined by the configuration. As cards are added, new addresses or tags are generated and available for selection in the program.

You can't write a program until you know the I/O addresses and memory configuration!

Some platforms assign I/O addresses by location in the rack (Slot Number) while others allow the programmer to assign addresses. Usually there will be a default address that can be modified in the properties of the card. As explained later in this course, some brand's memory allocation can overlap the I/O area, so careful configuration and planning is important. Selecting hardware will often also include entering a hardware or firmware number for each card. If a rack is used, selecting the rack size will be necessary before inserting cards.

After selecting the hardware for a PLC, the CPU, I/O Cards and communication ports will have additional configuration that will need to be done. Assigning memory locations for status and clock bits, scaling analog channels, and determining failure states of I/O are examples of this. A common method of additional configuration is right clicking the element and selecting "Properties", or simply double-clicking on the card.

Slot	Module	Order number	Firmware	MPI address	I address	Q address
1	PS 307 10A	6ES7 307-1KA00-0AA0				
2	CPU 313C-2 DP	6ES7 313-6CE01-0AB0	V2.0	10		
X2	DP				1023*	
2.2	DI16/DO16				124...125	124...125
2.4	Count				768...783	768...783
3						
4	DI8/DO8x24V/0.5A	6ES7 323-1BH00-0AA0			0	0
5	AI2x12Bit	6ES7 331-7KB02-0AB0			272...275	
6	CP 343-1 Lean	6GK7 343-1CX10-0XE0	V1.0	11	288...303	288...303
7						
8						

This is an example of a Siemens Step7 configuration screen in the Hardware Configuration. Notice the I and Q addresses, these can be changed by double-clicking the card and de-selecting the "default" box. Cards are selected from a catalog with folders for each type of device.

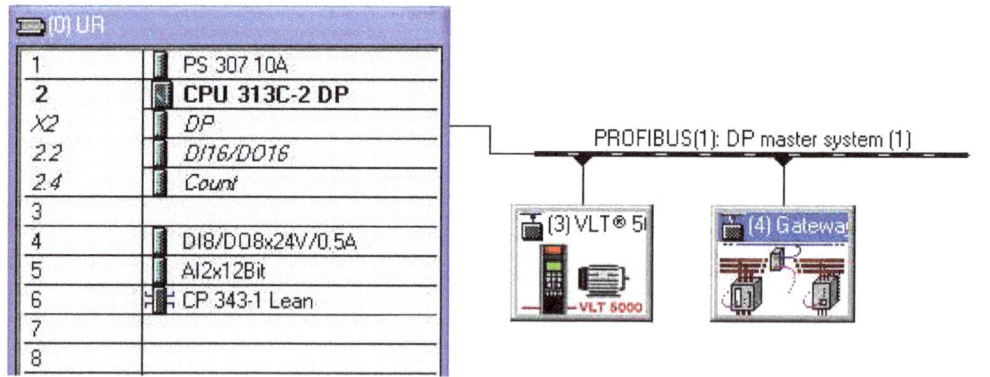

This is from the same Step7 Hardware Configuration screen; notice that the Profibus I/O network can also be configured from here.

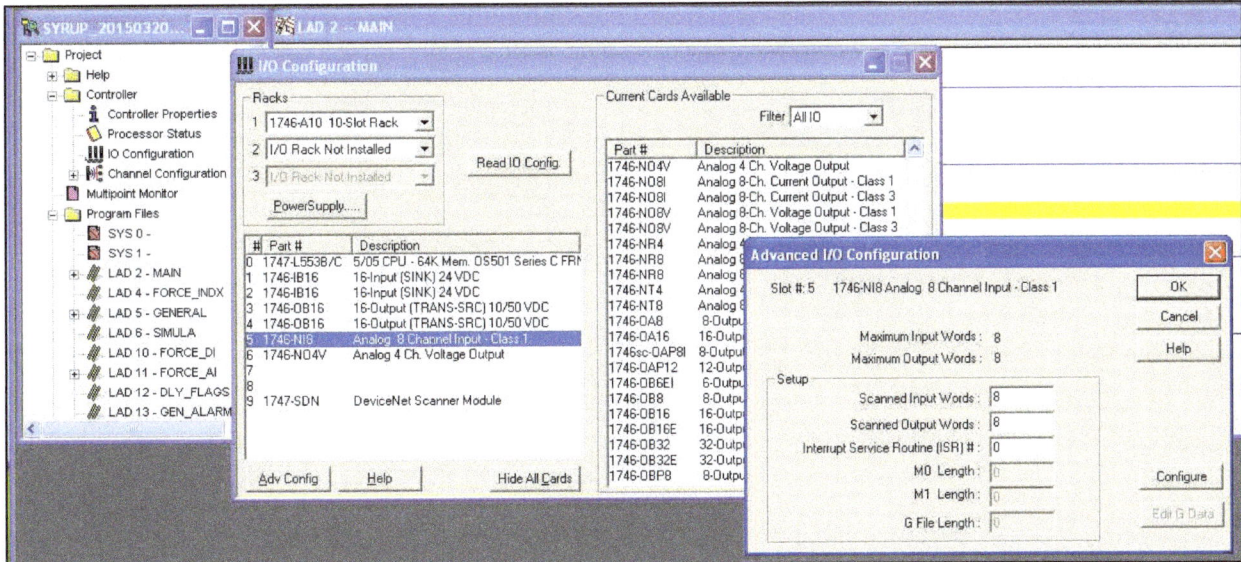

This is an example of Hardware Configuration from Allen-Bradley's RSLogix500, which is used to program the SLC500 series and MicroLogix. Double-clicking the analog card brings up the advanced configuration for the card as shown. Unlike the Siemens configuration, the addresses can't be changed, they are assigned by slot number.

Also, configuring the DeviceNet card shown in slot 9 requires separate programming software called RSNetworx. Because of this the network does not show here as it does in the Step7 software.

 Exercise 3

1. List 3 different numbering systems that PLC manufacturers may use for addressing data registers:

   _____

   _____

   _____

2. Where does a PLC program begin its scan? (In which routine?) _____
   _____

3. Before writing code in a PLC, the _____ must be configured.

4. Besides selecting cards in a PLC, what other kinds of things can be configured?
   _____
   _____
   _____
   _____

Program Processing

IEC 61131

PLCs have evolved in different ways depending on the manufacturer. Programming software and methods of handling data can differ immensely from platform to platform. Because of this, in 1982 the International Electrotechnical Commission (IEC) created an open standard that defines what equipment, software, communications, safety and other aspects of programmable controllers should look like. After the national committees had reviewed the first draft, they decided it was too complex to treat as a single document. They originally split it into five sections as follows:

Part 1 - General Information

Part 2 – Equipment and Testing Requirements

Part 3 – Programming Languages

Part 4 – User Guidelines

Part 5 - Communications

Currently, the standard is divided into 9 different parts, and a tenth is being worked on.

The third part, IEC 61131-3, defines the languages that are used in programming. It describes two graphical languages and two text languages, along with another graphical method of organizing programs for sequential or parallel processing. It also describes many of the data types described previously in this document.

The first two graphical languages described are Ladder Logic (LAD) and Function Block Diagram (FBD). The text languages are Structured Text (ST) and Instruction List (IL), while the organizational method described above is Sequential Function Charts (SFC), which is also graphical. An additional extension language is Continuous Function Charts, (CFC), which allows graphic elements to be positioned freely; it can be considered as an extension of SFC.

The following examples illustrate the five IEC programming languages; the addresses used are generic and the logic shows selection of Auto and Manual modes, along with a timer enabling "Cycle". These examples do not come from an actual programming language or brand but are meant to illustrate uses of the languages.

The following examples perform the same function written in all 5 IEC languages:

Ladder Logic (LAD)

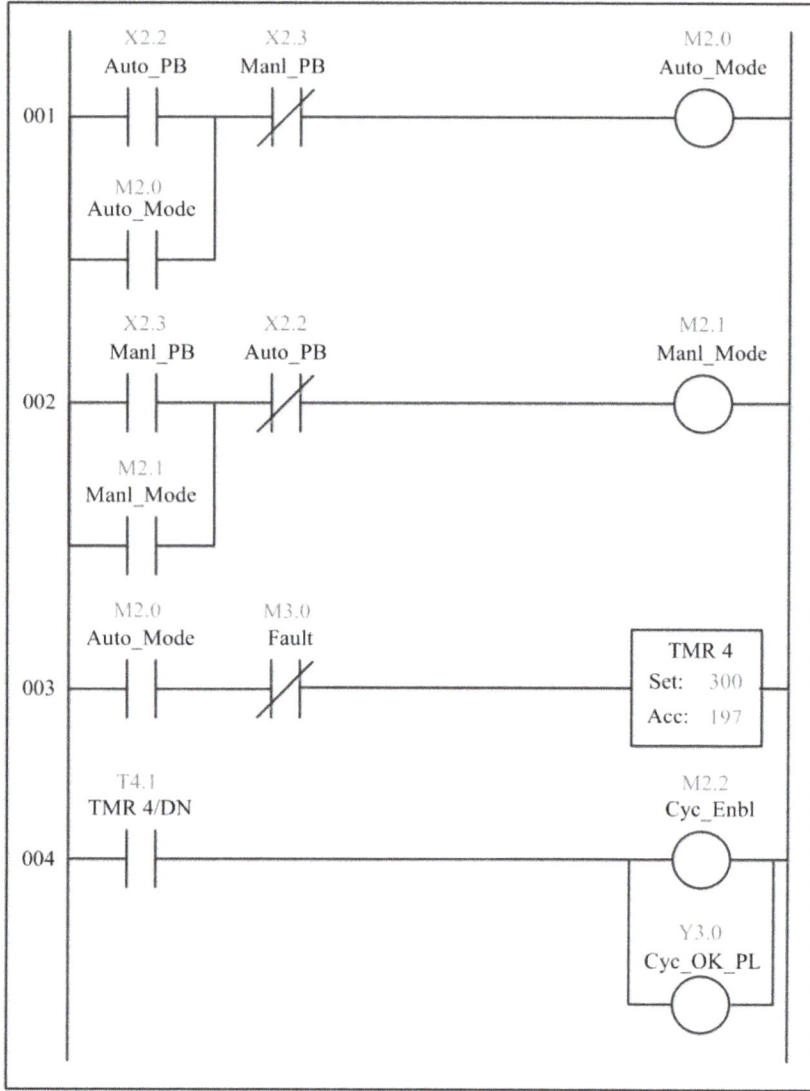

Ladder Logic evolved from electrical circuit drawings, which resemble the shape of a ladder when drawn. As a graphical language, the instructions represent electrical contacts and coils; the vertical sides of the ladder diagram are known as "rails" and the horizontal circuits are often called "rungs". In Siemens software, the rungs are known as "networks".

The "X" addresses represent physical inputs, while the "Y" address is a physical output. The "M"s are internal memory bits.

Because of the variety of addressing schemes as described previously, register representations can mean different things on different platforms.

When monitoring Ladder Logic in real time, usually contacts and coils change color to indicate their state in the logic. If a path of continuity exists from the left rail to the coil, the address will be said to be "On" or "True".

The timer shown in the diagram may also show a time base. If the above timer's preset is three seconds, the time base would then be 10 milliseconds.

More on ladder logic will be discussed in a later section.

Function Block Diagram (FBD):

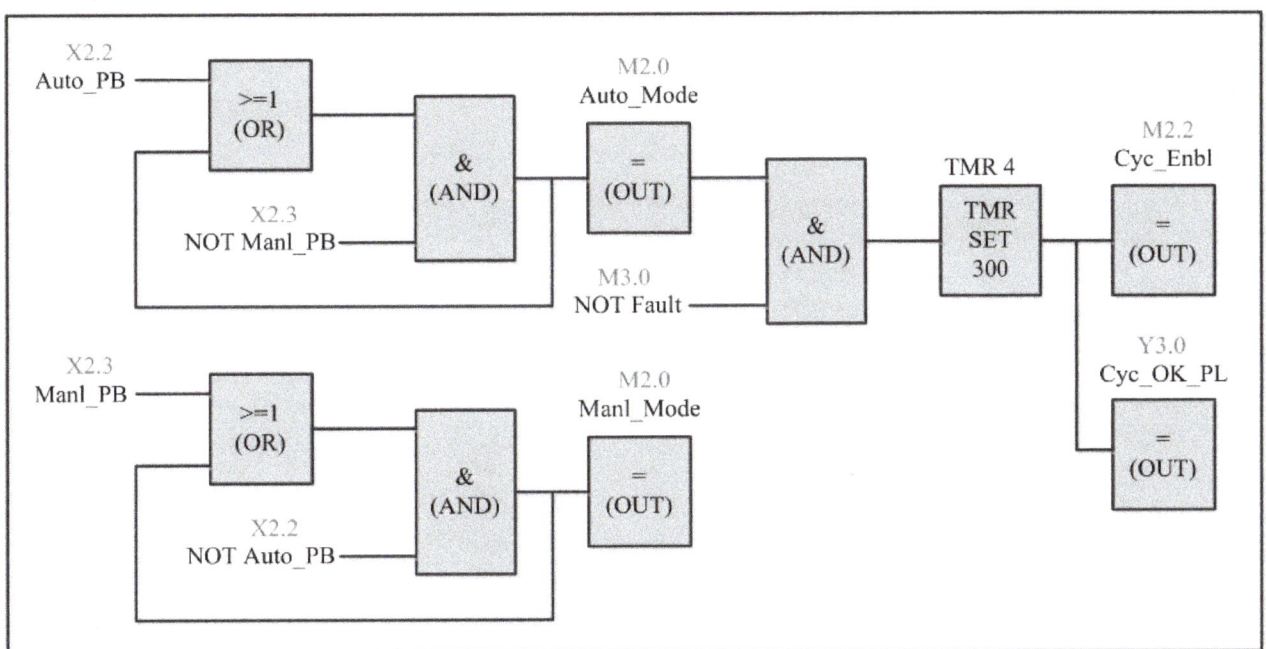

The diagram above shows the same functionality as that in the Ladder Logic diagram. The function blocks evolved from Boolean algebra, the AND and OR representing basic logic. More complex blocks are used for math, loading, comparing and transferring data, timing and counting. As with the previous example, this does not represent any particular brand of PLC.

There are some functions, such as XOR (Exclusive OR) that cannot be easily represented in Ladder Logic. Also, because of the complex nature of some FBD drawings, logic can often extend across many pages. Off-page connector symbols are used to show these connections.

Instruction List (IL)

```
LD X2.2 Auto_PB
O  M2.0 Auto_Mode
AN X2.3 Manl_PB
=  M2.0 Auto_Mode
LD X2.3 Manl_PB
O  M2.1 Manl_Mode
AN X2.2 Auto_PB
=  M2.1 Manl_Mode
LD M2.0 Auto_Mode
AN M3.0 Fault
=  TMR 4 Set 300
LD T4.1 TMR 4/DN
=  M2.2 Cyc_Enbl
=  Y3.0 Cyc_OK_PL
```

Graphical languages are usually converted into a text language called Instruction List before being compiled into another low-level code called Machine Language. Before the advent of personal computers, handheld programmers were used to type instructions into the PLC before compilation. These devices often had pictures of Ladder Logic contacts on the keys.

Some platforms, such as Siemens, make extensive use of IL in programming; Siemens version of Instruction List is called Statement List, or STL. Statement List has many commands that are not possible in Ladder Logic or FBD. Other platforms simply use Instruction List as a "stepping stone" to machine language; Allen-Bradley is one of these.

Because everything can be converted to Instruction List/STL, yet the opposite case is not true, it is considered to be more efficient than FBD or Ladder.

The table below shows an example of **Assembly Language**, a close relation of machine code. Note the list of addresses and Hexadecimal equivalents to the assembly instructions; this is very similar to the conversion of Instruction List to machine language.

Address	Assembly	Hex	Comment
6050	SEI	78	Set Interrupt Disable Bit
6051	LDA #$80	A9 80	Load Accumulator HEX 80 (128 Decimal)
6053	STA $0315	8D 15 03	Store Accumulator to Address 03 15
6056	LDA #$2D	A9 2D	Load Accumulator HEX 2D (45 Decimal)
6058	STA $0314	8D 14 03	Store Accumulator to Address 03 14
605B	CLI	58	Clear Interrupt Disable Bit
605C	RTS	60	Return from Subroutine
605D	INC $D020	EE 20 D0	Increment Memory Address D0 20
6060	JMP $EA31	4C 31 EA	Jump to Memory Address EA 31

TIP: Since IL is text based, it is easy to manipulate in third party text or spreadsheet editors such as Microsoft Excel. Instruction List can usually be imported or exported to and from PLC software in the form of .csv (comma-delimited) files or XML (eXtensible Markup Language). This makes it easy to create tables of addresses or tags with a common structure and then convert it into many repetitive rungs or blocks with different addresses. When writing large amounts of repetitive code, this can be a big time-saver!

Structured Text (ST)

```
// PLC Configuration
CONFIGURATION DefaultCfg

VAR_GLOBAL
        Auto_PB    :IN @ %X2.2       // Auto Pushbutton
        Manl_PB    :IN @ %X2.3       // Manual Pushbutton
        Cyc_OK_PL  :OUT @ %Y3.0      // Cycle OK Pilot Light
        Auto_Mode  :BOOL @ M2.0      // Automatic Mode
        Manl_Mode  :BOOL @ M2.1      // Manual Mode
        Cyc_Enbl   :BOOL @ M2.2      // Cycle Enable
        Fault      :BOOL @ M3.0      // Machine Fault
        TMR 4      :TIMER @ T4       // 10ms Base Timer
END_VAR

END_CONFIGURATION

PROGRAM Main

STRT  IF (Auto_PB=1 OR Auto_Mode=1) AND Manl_PB=0 THEN Auto_Mode=1
      ELSE IF (Manl_PB=1 OR Manl_Mode=1) AND Auto_PB=0 THEN Manl_Mode=1
      End IF

      IF Auto_Mode=1 AND Fault=0 THEN
      START TMR 4
      END IF

      IF TMR 4.ACC GEQ 300 THEN
      Cyc_Enbl=1
      Cyc_OK_PL=1
      END IF

      JMP STRT

END_PROGRAM
```

Structured Text resembles high-level programming languages such as Pascal or C. Variables are declared as to data type at the beginning of routines as well as configuration of other parameters. Comments are shown in this program as starting with "//"; this may differ depending on the brand.

Linear programming languages such as Structured Text use constructs like "If-Then-Else", "Do-While" and "Jump" to control program flow. In these languages, syntax is very important and it can be difficult to find errors in programming. Debug tools allowing for partial execution of the code one section at a time are common.

While writing PLC code in Structured Text can be difficult, it is also a much more powerful language than Ladder Logic or Function Block. Libraries can be developed to perform complex tasks such as searching for data using SQL or building complicated mathematical algorithms. At the same time, since the program proceeds step by step, it is more difficult to respond to multiple inputs at the same time; program control can be complex with many loops.

Sequential Function Charts (SFC)

SFC makes use of blocks containing code that typically activates outputs or performs specific functions. In many platforms, the blocks or "steps" can contain code written in other IEC programming languages such as Ladder or FBD. The program moves from block to block by means of "transitions", which often take the form of inputs.

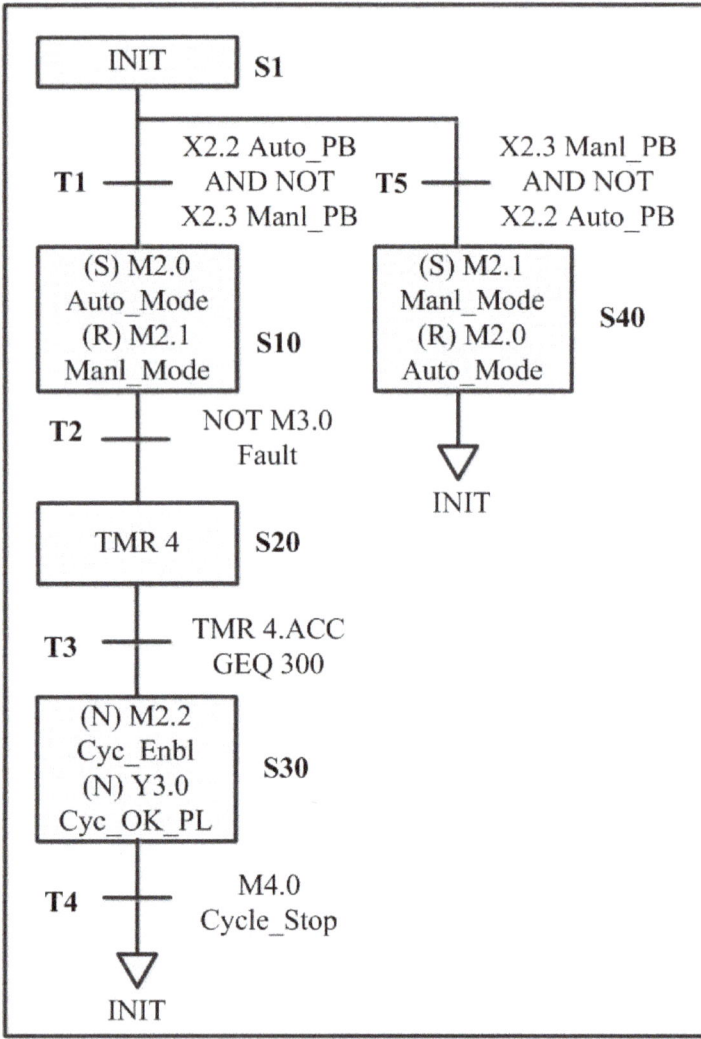

SFC is based on Grafcet, a model for sequential control developed by researchers in France in 1975. Much of Grafcet is, in turn, based on binary Petri Nets, also called Place/Transition nets. Petri Nets were developed in 1939 to describe chemical processes.

Steps in an SFC diagram can be active or inactive, and actions are only executed for active steps. Steps can be active for one of two reasons; either it is defined as an initial step, or in was activated during a scan cycle and not deactivated since. When a transition is activated, it activates the step(s) immediately after it and deactivates the preceding step.

Actions associated with steps can be of various types, the most common being Set (S), Reset (R) and Continuous (N). N actions are active for as long as the step is, while Set and Reset operate as in the other PLC languages.

Actions within the steps and the logic transitions between them can be written in other PLC languages. Structured Text is common in the action blocks, while ladder is often used for transitions. Steps and transitions are labeled as S# and T#. The top of the program will always contain an initial step; the program starts here and this is also where it returns after completion. A program will scan the logic in a step continuously until its associated transition logic becomes true; after this the step is deactivated and the next step is activated.

How Does it Work?

Scanning

The PLC processor or CPU controls the operating cycle of the program. The operating cycle, or Scan, consists of a sequence of operations performed sequentially and continuously.

There are four parts to the scan cycle as listed below:

1. Read physical inputs to the Input Image Table.

2. Scan the logic sequentially, reading from and writing to the memory and I/O tables.

3. Write the resulting Output Image Table to the physical outputs.

4. Perform various "housekeeping" functions such as checking the system for faults, servicing communications and updating internal timer and counter values.

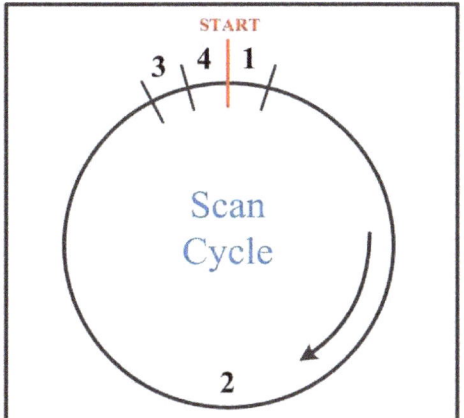

The scan cycle can be visualized as shown in the diagram on the left. When the processor is placed into "run" mode, the state of the analog and digital inputs is captured and saved into memory registers dedicated to the configured inputs. In the second part of the scan, the logic is evaluated sequentially. If the program is written in ladder logic, the rungs are evaluated one rung at a time, left rail to right, top to bottom. As the logic is processed, output and memory coils are energized or de-energized and their status is saved into their respective registers. In the case of the outputs, they are saved into the Output Image Table, which is generated during hardware configuration.

The time that it takes to execute a scan depends on the number and types of instructions in the program. Scans may be as short as 3-5 milliseconds (a very short program or a very fast processor) or as long as 60-70 milliseconds (a longer program). If the scan time exceeds this period by very much, physical reactions of actuators start to become noticeable; it may be time to evaluate a change to a more powerful PLC processor or even use multiple processors.

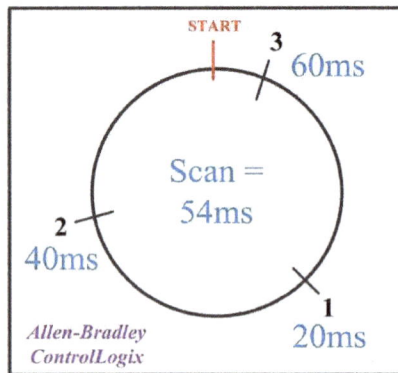

An exception to the scanning method described previously is that of Allen-Bradley's ControlLogix platform. Instead of accessing the Input and Output Image Tables at the beginning and end of the scan, each I/O card is configured with a "Requested Packet Interval", or RPI. I/O tables are updated at this rate, as shown at the left.

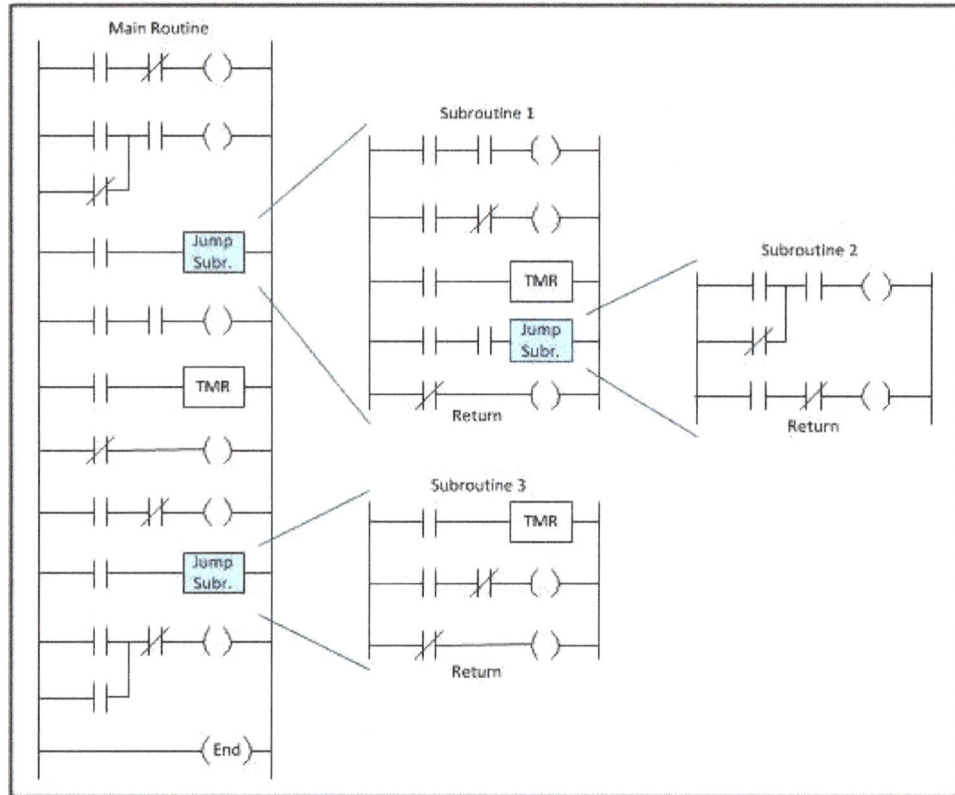

Scanning begins with the first rung of the routine designated as the main routine; all PLCs have a routine with this designation. As logic is processed, calls to other routines will occur as shown. After a subroutine is scanned, the scan returns to where the jump or call was made. Eventually, the scan always ends at the end of the main routine.

In the diagram above, the scan begins at the top of the main routine and is evaluated left to right, branch by branch, top to bottom. When the scan reaches a Jump Subroutine (also known as a "call"), the scan continues in the new routine, in this case Subroutine 1. Subroutine 1 then calls Subroutine 2, when the end of Subroutine 2 is reached, the scan continues in Subroutine 1. When the end of Subroutine 1 is reached, the scan continues back in the main routine.

Subroutine 3 is called, is scanned to the end and returns to the Main Routine. When the end of the Main Routine is reached, the resulting Output Image Table is written to the physical outputs and the housekeeping functions (Step 4) is completed. The scan then starts over at the top of the Main Routine.

Scan times will vary from cycle to cycle, depending on the number of instructions and routines active at any given time, the load on the processor from "housekeeping" activities, and communications' connections. Execution times for each type of instruction can be found in each manufacturer's documentation.

Hardware also affects scan times; newer models of processor are much faster than older ones due to advances in technology. For example, Allen-Bradley's new ControlLogix processors are nearly 150 times as fast as their older PLC5, which was the most powerful platform they made in the 1990s!

PLC Modes

PLCs can be placed into a number of operational states. When the processor is executing its program and scanning in the normal way it is said to be in "Run" mode. When initially downloading a program, the CPU is in "Stop" or Program mode, it is not executing the program and I/O is not changing state.

The mode of a PLC can often be changed by means of a switch on the front of the processor, for security this may be a key. There is often a third position on the keyswitch labeled "Remote"; this allows a computer to be used to change the state or mode. This adds an additional layer of safety; when the switch is in Program or Run the state cannot be changed from the computer.

While the computer is connected, the program can sometimes be changed while the PLC is scanning. Not every PLC allows this, and it is not always done in the same way. Some platforms such as Allen-Bradley allow each line or rung of code to be changed while online followed by accepting, testing and assembling (compiling) the program, while others such as Siemens allow each block to be compiled and downloaded without interrupting program execution.

Additionally, many platforms allow the processor to be placed in "test" or "debug" mode. This allows breakpoints to be placed in the code to stop execution at that point. This can be useful if monitoring "looped" code.

I/O can also be "forced" on many platforms. While this is not a mode as such, it does change the operation of the program. There is more information on forcing in the Maintenance and Troubleshooting section of this book.

 Exercise 4

1. List the five languages defined in IEC 61131-3

   _____
   _____
   _____
   _____
   _____

2. List the four parts of the scan cycle in a PLC

   _____
   _____
   _____
   _____

3. Can a PLC program be changed while it is running? _____

Ladder Logic - Tutorial

The most common and first PLC programming language is Ladder. This is because it is based on physical wiring diagrams, which most electrical maintenance personnel are familiar with. This makes this graphical type of programming easy to read and troubleshoot.

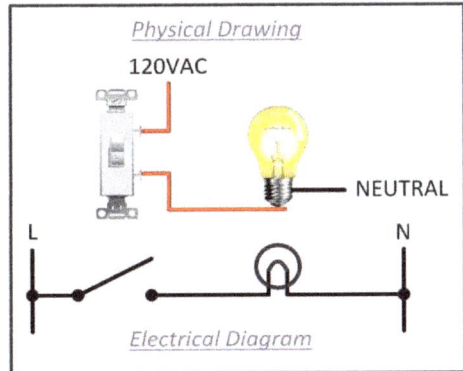

The simplest circuit to understand in an electrical system is a switch turning on a light. The top picture shows a physical drawing of a light switch passing 120 Volts AC to an incandescent light bulb; the bulb also needs a neutral wire to complete the circuit.

The diagram on the bottom shows the electrical equivalent of the physical drawing. The L symbolizes **Line** voltage and the N signifies **Neutral**.

There are many symbols used in schematic diagrams to signify the type of device used in an electrical circuit. The device on the left is generally known as a *Switching Device*, while the device on the right is called the *Load*.

Discrete Logic

Just like electrical circuit drawings, there are symbols or figures signifying the type of device used in the circuit, however they are classified by their function rather than by the type of device (i.e a pushbutton, switch or sensor uses the same contact icon).

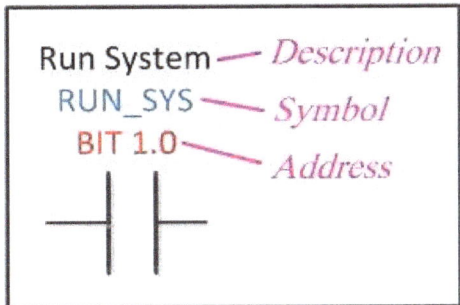

For the logic diagrams in this section, the conventions to the left are used. The **Address** is the register that is updated as the logic is processed. This applies to non-tag-based systems, however even they use addresses for I/O.

The **Symbol** is used as a short cut for the address; in most PLC platforms, typing the symbol will automatically call up the address. In the case of a tag-based platform, there may be no address, in which case the symbol is the only address that is used. For tag-based systems, the symbol is actually downloaded into the PLC as part of the program. For address-based systems it usually is not; it is a part of the program saved on the computer, but not present in the CPU. Symbols or tags are usually limited in the number and choice of characters allowed. For example, underscores (_) may have to be substituted for spaces.

A **Description** is purely for the convenience of the programmer. The description is only present in the programming computer and is not downloaded to the PLC. Typically a description can be several lines in length, unlike a symbol. It can usually contain any text character and is used to fully describe aspects of the device or contact, such as its physical location, numerical designation or purpose.

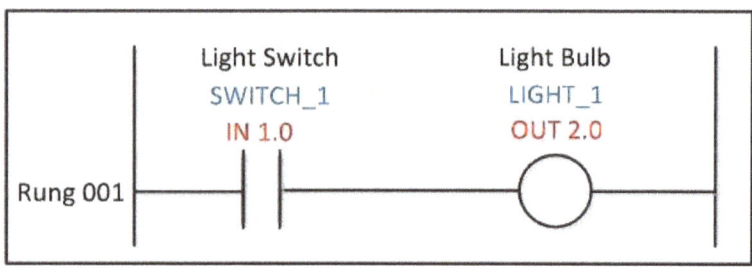

This diagram shows the same circuit as in the electrical diagram on the previous page, but in Ladder Logic. Like the electrical diagram, if the switch contacts on the left are closed, the light will be energized; however, the only information you can get from the contact while offline is that it is a *Normally Open* contact.

The device on the left is called a **contact**, and the device on the right is a **coil**. These names come from parts of a relay, which is the basis of Relay Ladder Logic. This type of contact is known as Normally Open, or NO; it is assumed to be not energized or activated in its normal condition. There are also contacts that are in the energized or "on" condition when in its normal condition, these are known as Normally Closed, or NC contacts.

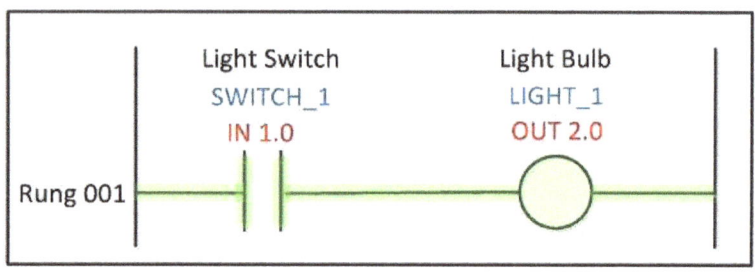

When a contact is viewed in its online condition, it is often highlighted in some way to indicate continuity. In the case of the light switch logic shown before, this indicates that while the switch is Normally Open, in this case something has activated it and the contacts are closed. The green highlight around the coil indicates that it is energized.

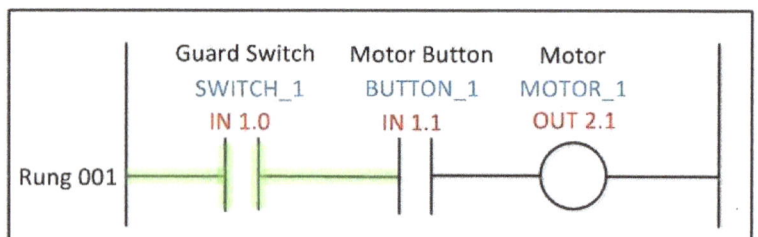

This circuit shows a normally open Guard Switch in series with a normally open motor pushbutton. If the guard switch is a switch on a motor cover, it would be assumed that if the cover is closed, the switch would allow the button to jog the motor. Therefore the switch is actually wired to be closed with the cover on; if the cover is removed, the switch opens, turning off the motor even if the button is being pushed. In other words, if the Guard Switch is closed AND the button is pressed, the motor will run. This is a classic example of an AND circuit, two contacts in series. In this case, if the circuit was being monitored through the software, the cover is on the motor (Guard Switch), but the button is not being pressed, so the motor is not running.

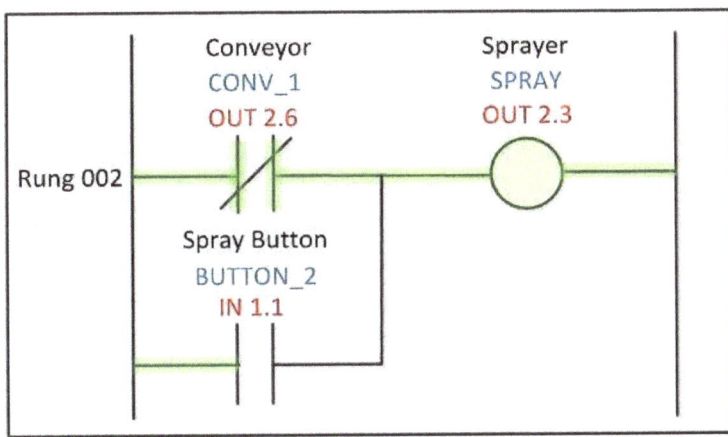

This circuit shows a Normally Closed contact from a conveyor in parallel with a Normally Open button. If the conveyor is on, (that is, if the output for running the conveyor is turned on elsewhere in the program), the sprayer will not be energized, UNLESS the button is pushed. This illustrates a new concept, the OR circuit. If the conveyor is off, OR the button is pushed, the sprayer output will activate. In this case, since the coil is activated and the highlighting extends through the conveyor contacts, the conveyor is not running! Note also that it is possible to use output addresses as contacts.

In the previous examples, the coil has indicated the status of the preceding logic. In other words, if there is a continuous path from the left rail to the coil, the coil is on. If not, the coil is off. It is also possible however to SET or RESET the coils address, making the logic Retentive. This is also sometimes called "Latch" and "Unlatch".

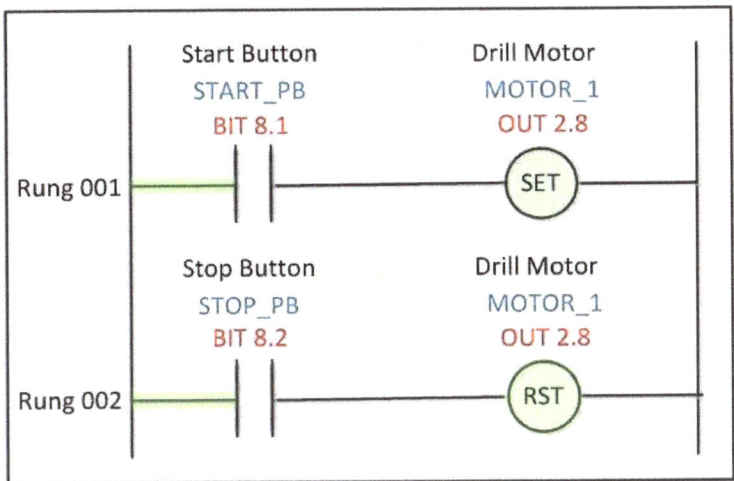

This diagram shows an output that energizes a motor. It can be "latched" on or off by pressing the start or stop button. Notice that the coil is energized, but there is no path or continuity from the contacts to the coils. This means that the output will be maintained until the Stop Button is pressed.

Notice also that in this case a "bit" address is used to control the motor. Since these aren't input register addresses as in the previous examples, that means that they must come from somewhere other than physical inputs. It is hard to tell without doing a cross reference to see where the bit's coil might be, but if a coil for that address can't be found in the program, it likely comes from an HMI, or operator interface.

Contrary to the drawing, it is not advisable to directly latch output addresses. Instead, a memory bit should be latched and unlatched that controls the output.

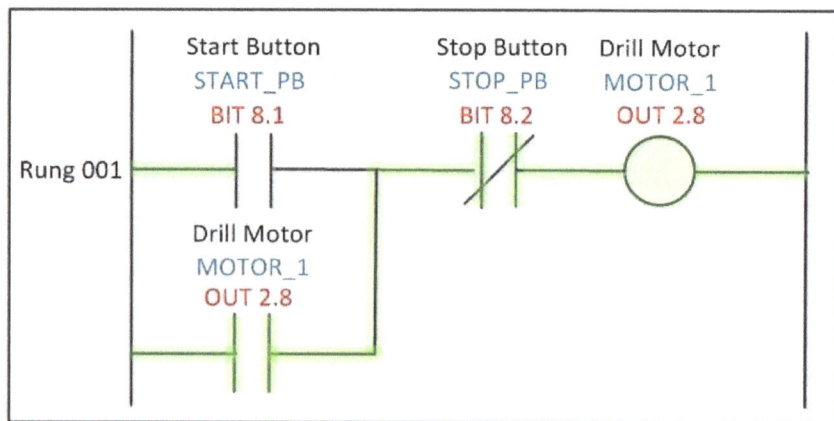

This logic performs exactly the same function as the latching circuit, but in a different way. Sometimes called a "hold-in" or "seal-in" circuit, if the start button is pressed and the stop button is not being pressed, the motor coil will energize. Even if the start button is released, the "hold-in" contact of the motor coil will ensure that the motor stays on until the stop button is pressed. Notice that in this case, the stop button needs to be a normally closed contact, even though both buttons are physically wired normally open in both diagrams.

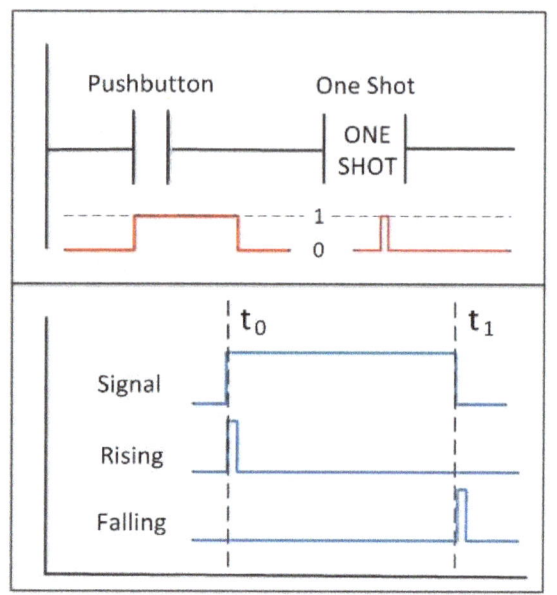

A One-Shot, single shot or differential pulse is used for a variety of reasons in ladder logic. This diagram shows the result of using a one-shot; when a signal changes state the one-shot creates a pulse exactly one scan in length. For example, when the Pushbutton is pressed, no matter how long it is held down, the logic will generate one single-scan pulse.

One-shot pulses can be generated from the rising or falling edge of a signal. They are known by different names on different platforms: Allen-Bradley uses OSR and OSF contacts called One-Shot Rising and One-Shot Falling; ONS one-shots are also sometimes used on the rising edge. Siemens calls its' one-shots Positive and Negative differentials, they look like –(P)– and –(N)– in ladder logic.

Omron's one-shots are called DIFU (Differential Up) and DIFD (Differential Down) for the rising and falling edge versions. They look similar to this: -|DIFU|- and -|DIFD|-

Mitsubishi calls them PLS or Pulse. -|PLS|-

Some types of one-shot are placed at the end of the rung as a "box" type instruction. These have two addresses, one for storage and one to be used as the output. These output contacts can be used at multiple places in a program. In-line one-shots must each have their own memory bit address; *do not use the same address for multiple one-shots!*

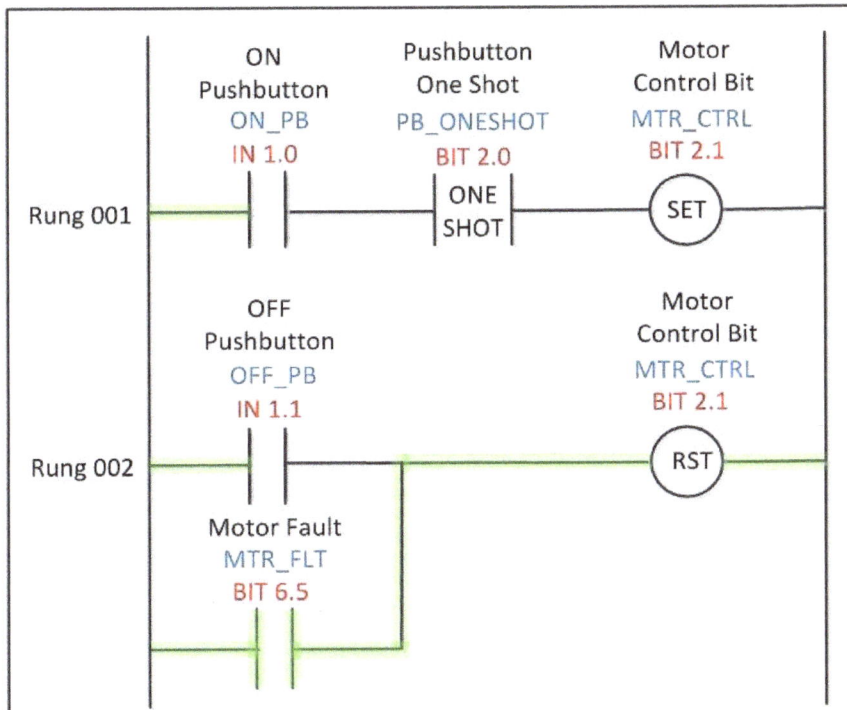

One-shots are often used with latching circuits to ensure that logic defaults to the off state as shown. If the one shot is of the "in-line" type described previously, each must have its own address.

As long as the Motor Fault is present, the motor control bit will be held in the off state. To turn the bit back on, the fault will have to be cleared and the ON pushbutton will need to be pressed again.

Notice also that even though the rung is highlighted all the way to the RST coil, the coil itself is not highlighted, indicating that the address is off. Contrast this with the latched coil picture on the previous page; this is typical when monitoring PLC logic with the software.

# Exercise 5

1. Draw ladder logic to place a machine into either Auto or Manual mode by pressing physical pushbuttons. Ensure that the machine is placed into Manual if a Fault occurs.

   Work Area:

2. Draw ladder logic to accomplish the following:
    a. Turn on a motor using a start button and a stop button using a hold-in contact. Both buttons should be <u>wired</u> Normally Open (NO). Include a NC Fault contact that will stop the motor.
    b. Latch the Fault bit if the motor's overload trips or the guard door is opened while the motor is running.
    c. Turn on a Red Light if there is a fault.
    d. Turn on a Green Light if the motor is running.
    e. Reset the Fault if there is no fault and a Reset Button is pushed.

Assign addresses to all of the contacts and coils using the generic method shown in this document; i.e. BIT, IN, OUT

Work Area:

Timers

The purpose of a timer in ladder logic is to delay the on or off state of a signal. Timers can be used to track the accumulation of time in a process, create a pulse of a fixed length, or determine whether a fault has occurred.

Before looking at how a timer operates, it is important to view its data structure.

Preset	
Accumulated	
Status Bits	...

The Preset and Accumulated values are usually either an integer or double integer value; however, Siemens uses a BCD value that incorporates the time base as part of a word (16 bits). The status bits always include a "**Done**" bit but may also include bits for Timer Enabled or Timer Timing. If the status bits don't include these, they can easily be generated using logic structures.

On-Delay

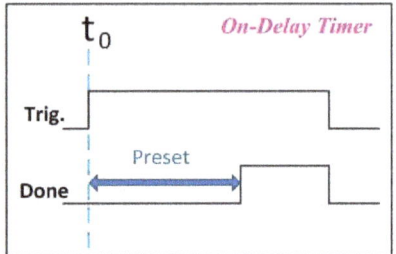

An On-Delay Timer is used to delay the ON state of its done bit as shown in this diagram.

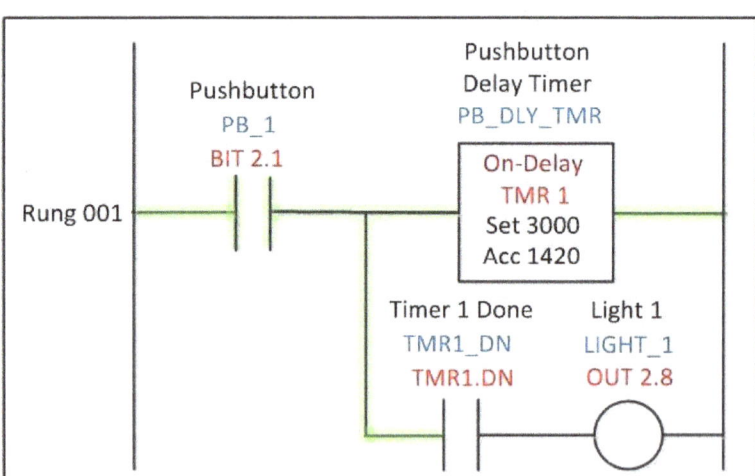

When the button is pushed, the timer begins timing. After the preset time (Set) has expired, the done bit will energize and remain on until the pushbutton has been released. If the Pushbutton is released before the Accumulated time (Acc) reaches the preset, the timer will reset to zero. When the done bit comes on it will energize the output, Light 1.

TIP: Notice that the preset is in thousands. According to IEC 61131-3, a timer counts time in milliseconds, so 3000 is three seconds. Also, the timer will count up as it times. Not all platforms' timers perform this way; some time down, some have a time base that allow them to count in 10ms, 100ms, 1 second or even 10 second increments. Most major manufacturers now have a timer available that meets the IEC definition.

Off-Delay

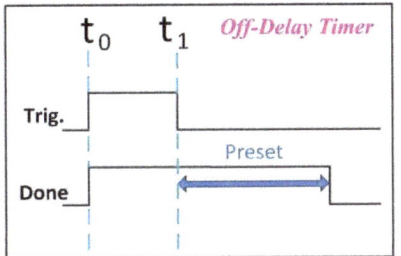

An Off-Delay Timer delays the OFF state of its done bit. Unlike the On-Delay, the done bit turns on immediately. It does not begin timing until the timer is deenergized.

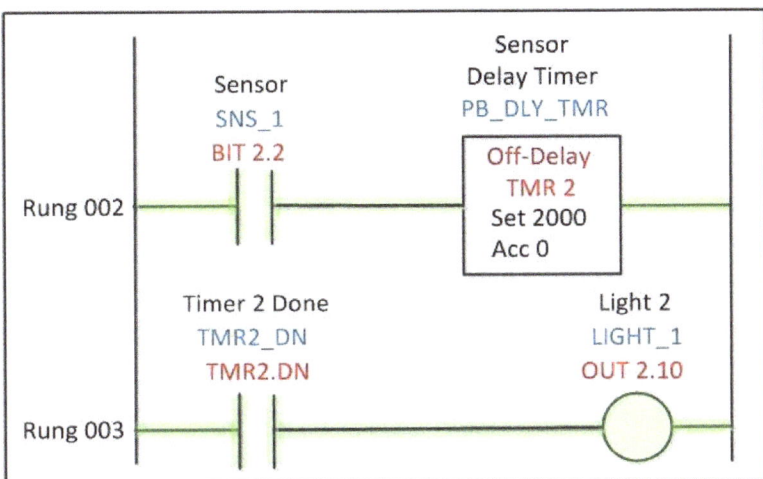

When the sensor is activated, the done bit comes on immediately, therefore the light comes on. When the sensor turns off, the timer begins timing and runs until the preset (Set) value is reached, then the done bit turns off. During this time, the accumulated value is counting up in milliseconds.

Notice that with this type of timer the done contact can't be placed in a branch after the trigger contact as with the on-delay timer. If it was, turning off the sensor would also not allow the light to come on, though some platforms have the . The off-delay timer can be thought of as a "pulse stretcher".

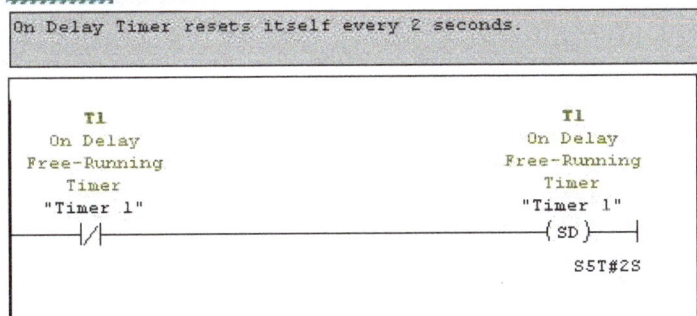

TIP: With many PLCs, the done bit is shown having the same address as the timer. This is a rung or "Network" from a Siemens Step 7 program. Note that the Symbol T1 is on both the SD (On-Delay) coil and the NC contact; the contact is the "done" bit. This rung creates a one scan-length pulse every 2 seconds.

Another aspect of a timer that is not part of its data type is the RESET coil. This coil is used to set the accumulated value to zero. All timers have this capability, though the reset is most commonly used in retentive timers.

Retentive On-Delay

A retentive timer keeps its accumulated value even when the energizing contact is not made. This is useful for accumulating run-time on a device or product. This means it must be reset.

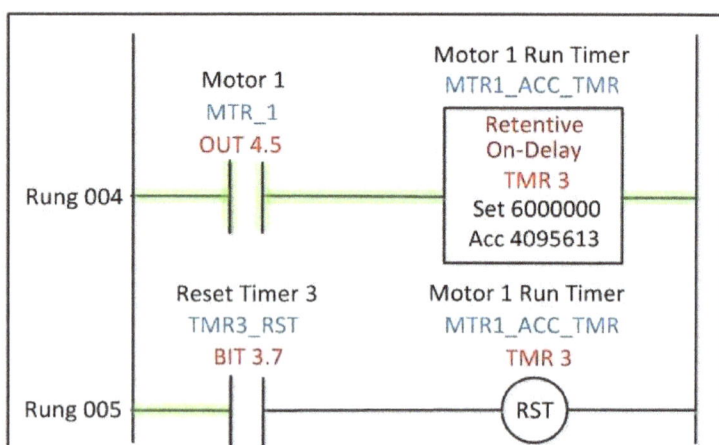

When the motor runs, the timer accumulates time. When it stops, the timer retains its accumulated value and does not reset. When the accumulator reaches the preset value, the timer stops timing, the done bit is energized, signifying that it is time to service the motor.

Usually, rather than placing a high number in the preset, a number such as 60000 (60 seconds) is placed there. The done bit is then used to increment a "Minutes" counter, which resets the timer. When the Minutes counter reaches 60, it increments an "Hours" counter; service on devices is often specified in hours. After the maintenance has been performed, the timer and counters can be reset. *A Siemens Retentive On-Delay Timer acts differently: when the trigger signal is removed, the timer continues timing until done. Siemens timers also time down from their preset.*

Pulse

Pulse Timers create a pulse of a fixed length when energized; the done bit energizes immediately and stays on until the preset is reached. Some pulse timers' done bits will de-energize if the trigger is removed from the timer early.

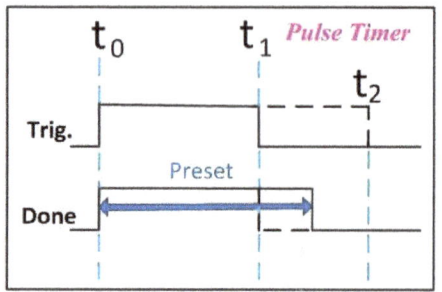

If the trigger signal stays on either shorter (t1) or longer (t2) than the Preset, the pulse output ("done" bit) stays on for the preset time. (Siemens S_PEXT). If the trigger signal is released before the preset time, some pulse timers will turn off the done bit as shown at t1 (Siemens SE). For PLC brands without a Pulse Timer, the pulse can be created using two On-Delay timers.

Exercise 6

1. Draw ladder logic to delay the start of a motor for 3 seconds. Use a start and stop button to control the motor. Ensure that if the button is released before the motor starts, it will start anyway.

   Work Area:

2. Draw ladder logic to ensure that a sprayer stays on for at least 2 seconds whenever an object passes in front of a photoeye on a conveyor. Ensure that if the conveyor is not running, the sprayer stops.

   Work Area:

3. Draw ladder logic that creates a one second on, two second off train of pulses. Use a pushbutton to sound a buzzer with this signal.

   Work Area:

4. A common component in an automated system is an Auto Cycle Start circuit. When a machine is in Auto Mode and ready to run, the operator presses a button and must hold it for three seconds before the machine will start. If the button is released early the timer restarts. Once the timer is done, the system is considered to be in "Auto Cycle". While the machine is starting, a buzzer will often sound or pulse warning that the machine is starting.

If the machine needs to be stopped, a "Cycle Stop" button is used. If the machine is in the middle of an operation, it needs to wait until it is finished to stop, this requires a memory bit, often called a "Cycle Stop Request". If a fault occurs, the machine may still stop immediately.

Draw logic to accomplish these tasks.

Work Area:

Counters

A **Counter** is used to add or subtract counts in a register. As with the Timer, a counter also has a data type associated with it.

Preset							
Accumulated							
Status Bits						...	

The Preset is the value at which the done bit will come on in most counters, again with the exception of Siemens. The Accumulated value is the current number of counts in register. Unlike a timer, which stops accumulating when the done bit activates, a counter will continue incrementing. Preset and accumulated values may be an integer or double integer depending on the platform.

Other Status Bits that MAY be present besides the "Done" bit:

a. Count Up (CU): active while up count trigger is active
b. Count Down (CD): active while down count trigger is active
c. Overflow (OV): active when Accumulated value exceeds maximum value for an Integer or Double Integer, depending on brand/platform
d. Underflow (UN): active when Accumulated value exceeds minimum value for an Integer or Double Integer, depending on brand/platform

TIP: The IEC61131-3 definition of a counter states that a counter's done bit will change state when the accumulator reaches the preset. However, Siemens counters, once again, act differently; the done bit is on if the accumulator contains a value higher than zero. Because of this, Siemens programmers often use a different method to count, Accumulators and Decumulators.

There are three types of counters: Up Counters, Down Counters, and Up-Down Counters. While every brand needs an Up and a Down counter, the Up-Down type is not available on all PLCs.

Separate up and down counters can be used with the same effect. As with retentive timers, counters need a Reset bit. Some counters will also use a "Set" bit to place the preset value into the accumulator.

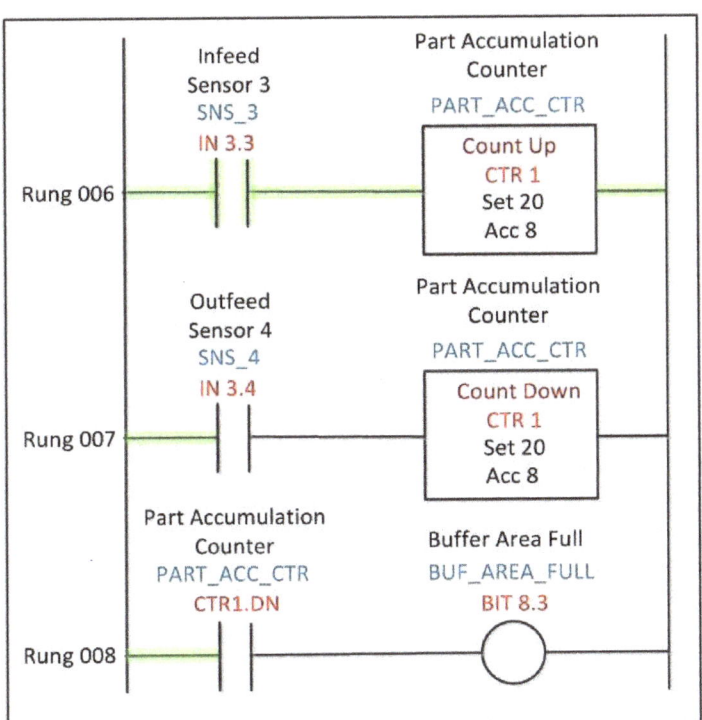

This logic tracks parts into and out of a "Buffer" area on a conveyor system. A sensor is located at the entry and exit of the area, so parts entering make the counter count up, while parts exiting make it count down. When the "Done" bit is active, a gate could be closed preventing new parts from entering the buffer area until parts have exited.

Notice that the same address was used for both counters. This is important to ensure that the same address is counting up and down.

So, what happens when a counter reaches the overflow or underflow value? It rolls over to the negative range if the positive value is exceeded and to the positive side if counting down in the negative direction. This is the purpose of the overflow and underflow status bits; they indicate that this boundary has been crossed. They can be reset (unlatched) after determining what you wish to do about the over or underflow.

If the counter is a Double Integer based device, the limits are -2,147,483,648 to +2,147,483,647.

The reset for the counter works the same as the one shown in the previous timer diagrams.

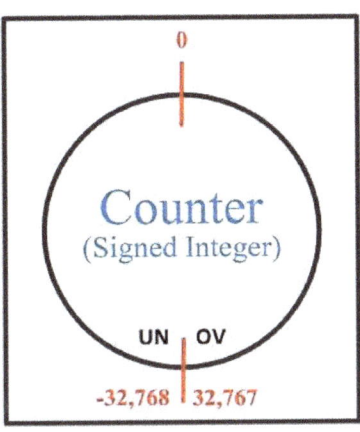

In addition to the Reset bit, a Set coil may be used to load the preset value into the accumulator. This is used for down counters, as mentioned previously, Siemens counters operated differently in the older Step 5 and Step 7 platforms. The Done bit was on when the accumulated value was NOT zero.

Exercise 7

1. Draw ladder logic that counts parts into a box. When the box is full (10 parts), latch a memory bit that controls a light, the light tells an operator to remove the box. Use a sensor to detect that the box has been removed. When the box has been removed, reset the counter and the memory bit that controls the light.

   Work Area:

2. Draw ladder logic that uses a Retentive On-Delay timer to accumulate run time on a motor. When the timer reaches one minute, use the done bit to increment a "Minutes" counter and reset the timer. When the counter reaches 60 minutes, use its done bit to increment an "Hours" counter and reset the Minute counter. When the Hour counter reaches 10,000, it is time to perform maintenance on the motor.

Do not reset the Hours counter automatically; use a pushbutton that is pressed by the maintenance technician, confirming that maintenance has been done. How can you incorporate logic to ensure that the technician really did do the maintenance?

Work Area:

Don't forget to reset the timer and other counters when the operator presses the button!

Data and File Movement

An important part of programming is the manipulation, modification and movement of data. The simplest method of doing this is to simply move a number from one location or register to another.

Move

Despite the common term "Move" that is used for this operation, it is actually a copy, since the value also remains where it was moved from.

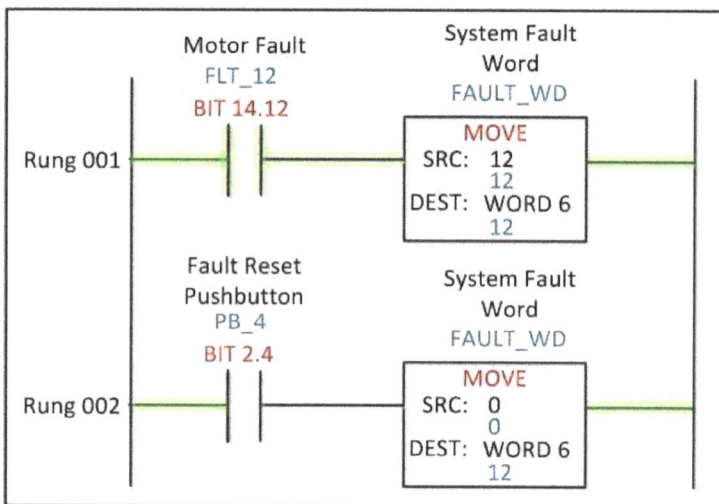

This logic shows an integer being moved into a register based on a numbered fault. The number can be used to display a message on a touchscreen or in a comparison to set the fault status by latching a bit. The numbers in blue show the actual number in the register (Word 6) if the logic was being monitored online.

In this example, constants are being moved, but numbers can also be moved from one register to another.

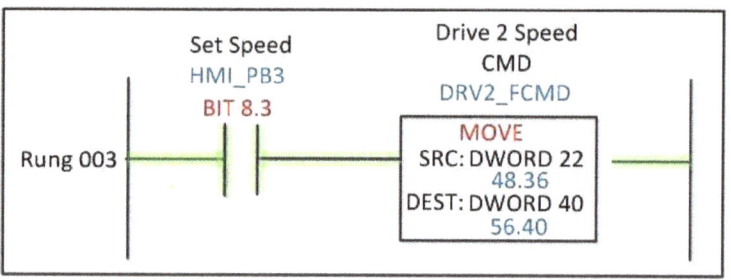

This logic shows a speed being moved into a VFD speed command. The numbers are floating point or REALs, so they require a double word, or 32 bits for the register size.

For larger data structures, a different instruction is usually used. For Allen-Bradley, this instruction is a COP, or "Copy". This instruction can also move multiple files in an array. In Siemens, a pointer must be used in Statement List (STL), Siemens version of Instruction List.

Masked Move and Shift

Parts of numbers can also be moved. If it is necessary to extract a specific part of a 16- or 32-bit number, a **Mask** can be used to move only that part, as shown in the diagram.

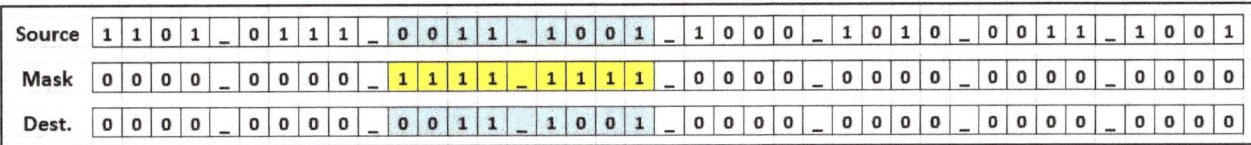

The Source is a 32-bit number, which is four bytes. If one wanted to move only the second byte into a register, the mask would contain ones wherever the data to be moved, and zeroes elsewhere. Masks are usually entered as a Hexadecimal number, which in this case would be 00FF0000, or just FF0000.

Of course, then you have the problem that the byte in the destination is in the wrong location within the double integer, this is easily remedied by using a **Shift** instruction. The result is shown below. In this case the shift is to the right sixteen spaces.

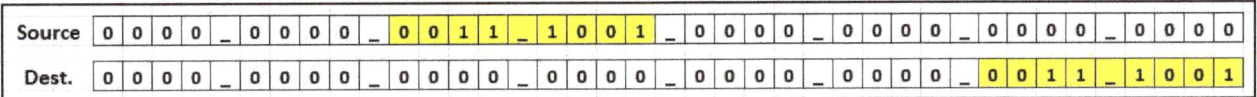

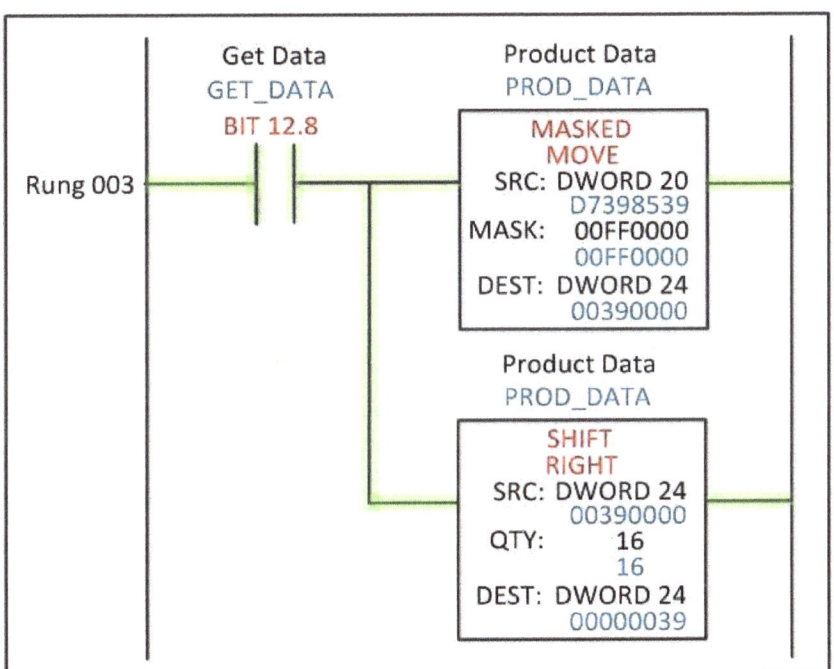

The ladder logic for a Masked Move and a Shift are shown here. The shift instruction may not have a destination as in this diagram, however it is shown here to indicate the number before and after the shift. Shift Left instructions can also be used.

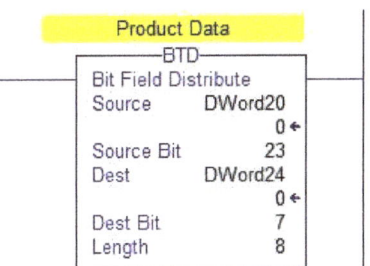

TIP: The Allen-Bradley ControlLogix 5000 platform has an instruction that does both of these functions at once (Move and Shift), called "Bit Field Distribute" or BTD.

File Copy

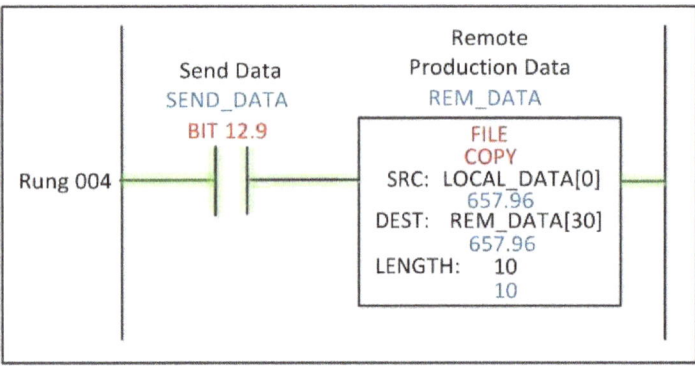

File instructions are used to move data structures that are larger than a single element or 32 bits. If a file is a single structure, such as a UDT or an individual data type such as an Array, a simple "Copy" type command can be used. If it is necessary to move specific overlapping sections, more complex pointer-based commands may be used.

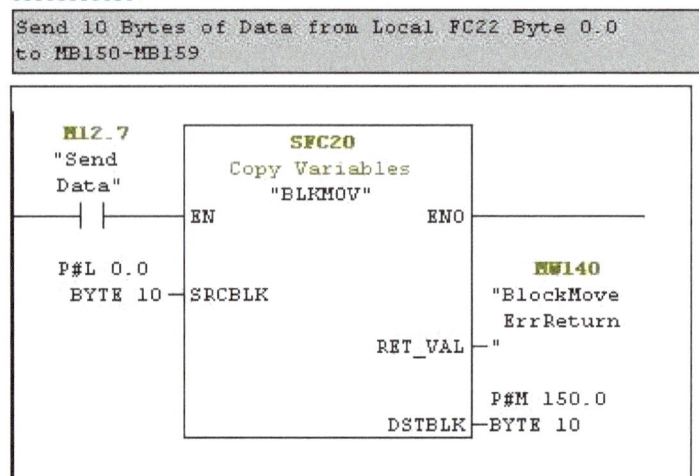

TIP: This logic illustrates the use of pointers in Siemens S7 platform. SFC20 is a Block Move command that allows the use of pointers to designate the type of data (Local and Marker Memory Bytes), the size of the data (10 Bytes) and the location within the structure (0.0 to 150.0). This command allows movements to be specified to the bit level.

The RET_VAL output is present on many Siemens instructions to allow for error return values.

Comparisons

Data comparison is an important element of PLC programming. Though comparisons deal with numerical values, they are input type instructions. That is, they evaluate to either true or false.

Standard comparison instructions that are found in any PLC include Equal (=), Not Equal (<>), Greater Than (>), Less Than (<), Greater Than or Equal (>=), and Less Than or Equal (<=).

Some PLC platforms require that the data type being compared is the same, while others allow comparisons between different types. Where data types must be the same, a conversion instruction may be required.

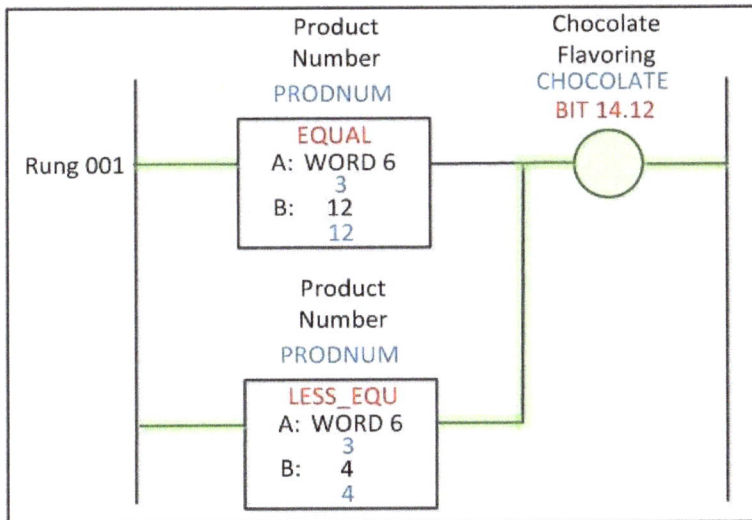

This illustrates the use of two comparison instructions used to activate a bit that is used to add chocolate flavoring to a recipe. If the selected product number is between 0-4 (inclusive) or equal to 12, the bit will be true. In this case the selected product number, in Word 6, is 3.

Comparisons are also often used in automatic sequences to move from one step to another or to determine when an output will be activated.

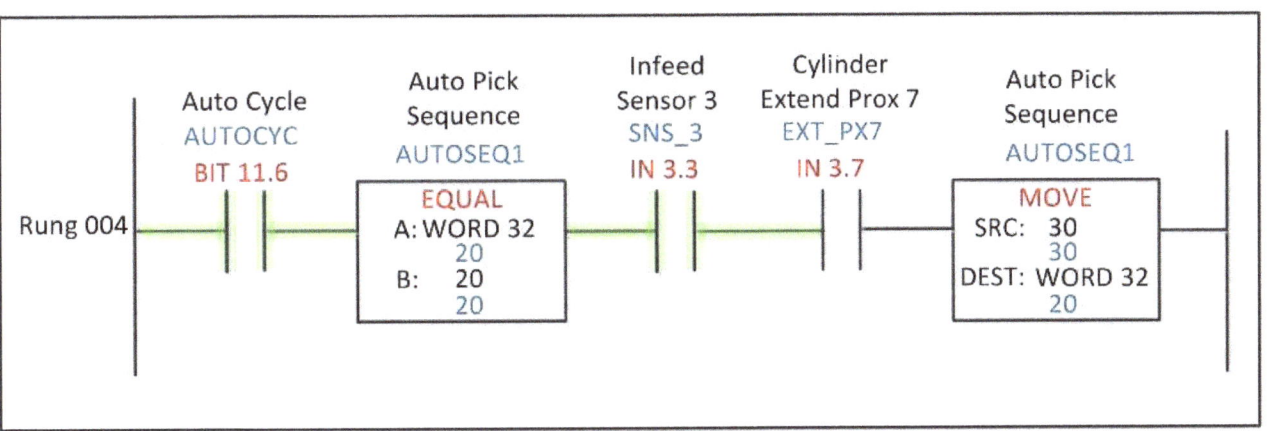

This logic increments a sequence to the next step if the machine is in Auto Cycle mode and a couple of sensors are activated. This only happens if the sequence is in Step 20.

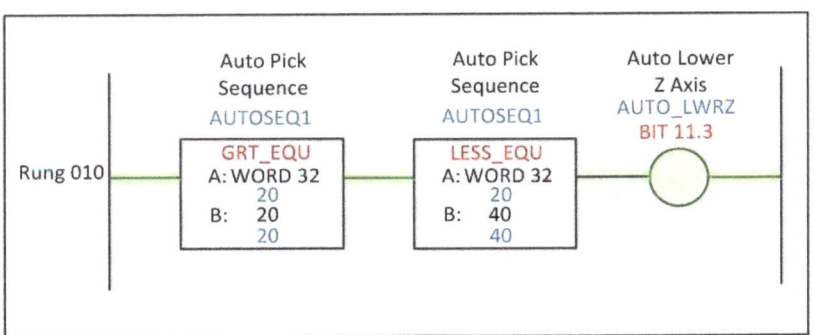

Comparison instructions can be placed in series to form a "window" during which a statement is true. In this case, the command to lower the Z axis of a pick-and-place mechanism will be on if the sequence is in steps 20-40, inclusive.

Some PLC platforms have instructions that form this range window in one instruction.

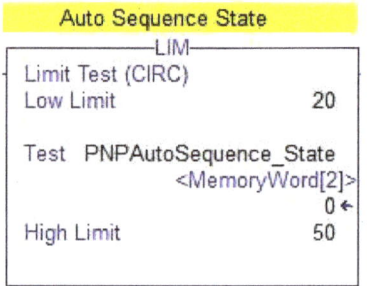

TIP: The Allen-Bradley "Limit" instruction can be used to form a window where a sequence state is tested between a low and high value. If the Low Limit value is greater than the High Limit value, the instruction operates in reverse, where if the tested value is outside of the range, the instruction is true! Siemens TIA software uses "IN_RANGE" and "OUT_RANGE" instructions for this.

A Mask can also be used with an equal instruction on some platforms. As with the Masked Move instruction, wherever there are ones in the mask, the values will be compared.

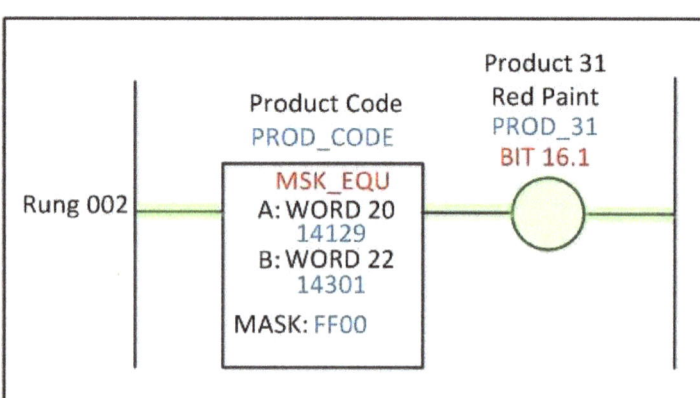

In this logic, it appears that the two values in Word 20 and Word 22 are different, yet the coil is energized. This is because only the 8 most significant bits are being looked at, while the lowest 8 bits are ignored.

The numbers are equal!

This is what the numbers 14,129 and 14,301 look like in Binary.

Page | 96

Exercise 8

1. If the Mask in a Masked Equal Instruction is 00FF, are 31,290 and 4410 Equal? _____

   Work Area:

2. Write Ladder Logic that increments an Auto Sequence through three different steps based on I/O and/or internal bits. On the last step, reset the sequence to zero.

   Work Area:

   Why do you think the Auto Sequence examples in this chapter increment by 10, rather than 1?

   _____
   _____
   _____

Math Instructions

Processing of data in a PLC often involves performing mathematical operations on data. Not all processors allow math to be performed between different data types, so it may be necessary to convert one data type to another.

Conversion

Common conversions include the following:

1. **Integer to Double Integer, Double Integer to Integer.** The inherent problem in this type of conversion is that the number in the DINT won't fit into the INT. Remember that these are usually signed values, so the largest value you can put into an integer is -32,768 to 32,767.

2. **Double Integer to Real, Integer to Real.** Some platforms will have DINT to REAL but not INT to REAL. In this case the integer will have to be converted to a double integer first.

3. **Real to Double Integer, Real to Integer.** In this case you will lose the value after the decimal point. Again, not all platforms have Real to Integer conversion.

4. **Integer to BCD, BCD to Integer.** Remember that your BCD value will contain more bits than your integer.

Platforms in which data conversions are necessary include Siemens and Koyo. BCD data conversions are possible on most platforms.

Addition and Subtraction

Addition and subtraction are commonly used for all data types. A typical use might be to increment or decrement a register by some amount:

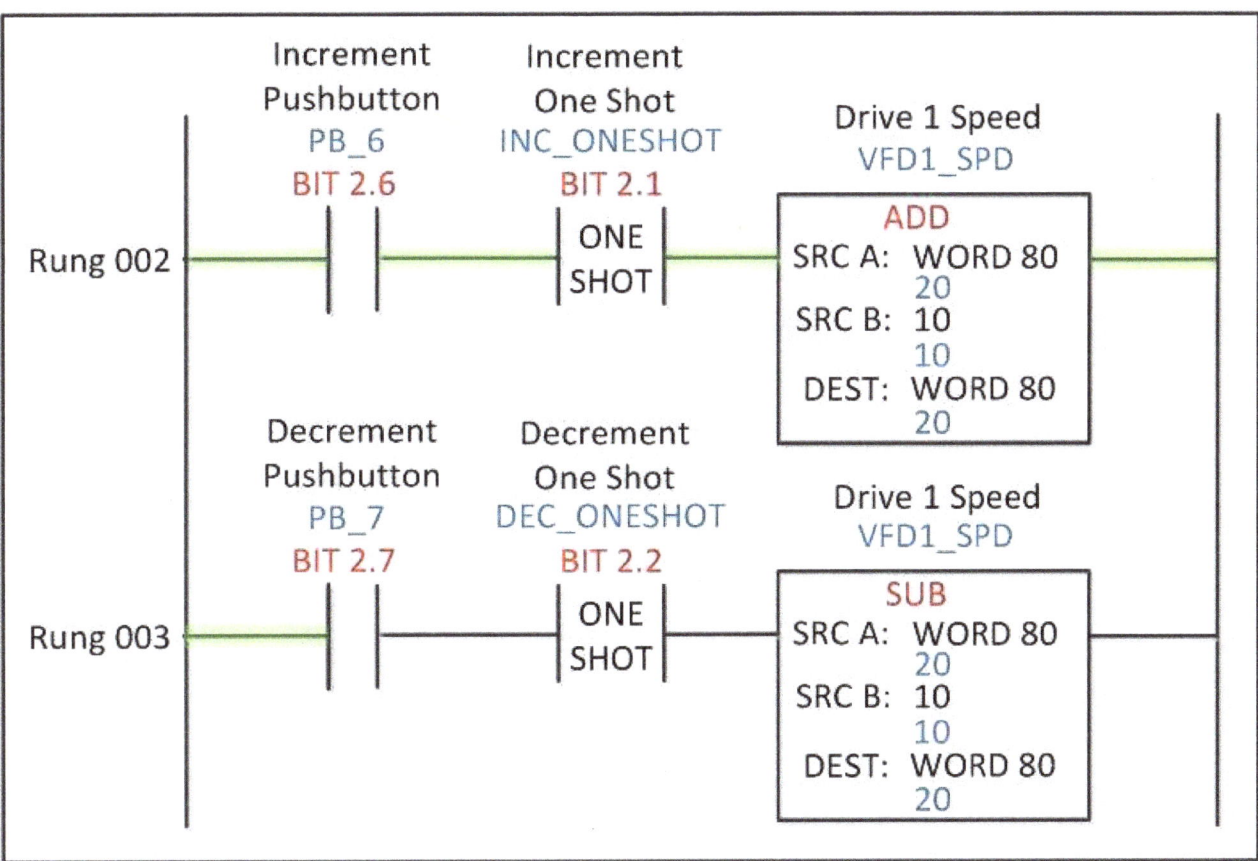

In this logic, pushbuttons are used to either add or subtract 10 from a register each time the button is pushed. Notice that the number is subtracted from a register and then placed back into the same register; this is the normal way to accomplish this. Also notice that a one-shot is used on the pushbuttons. If the one shot is not used, the instruction will not add or subtract every time the button is pushed, instead it will add or subtract *every scan* while the button is being pressed! You could end up with a very large number in Word 80 if you aren't careful!

A common name for Rung 002 is an **Accumulator**, while the following rung is sometimes known as a **Decumulator**.

TIP: A Siemens counter has quite a few limitations. The "Done" bit is on if the counter is not at zero, and the preset and accumulated values are signed BCD numbers; they only count from -999 to +999.

Because of this, Siemens programmers will often use Accumulation/Decumulation logic to count. In this case the "Done" bit will be created by using a Greater Than or Equal instruction, the Reset command moves zero into the accumulator value, and the registers would increment and decrement by 1.

Multiplication and Division

As with addition and subtraction, multiplication and division can be done with any data type, however greater care must be taken for several reasons.

If you multiply two integers, it is important to ensure that the result will fit into the destination address. For instance, if 20,000 is multiplied by 20,000, the answer of 400,000,000 will not fit into an integer register or tag.

When dividing, if the denominator is too small (or zero!) the same thing will happen.

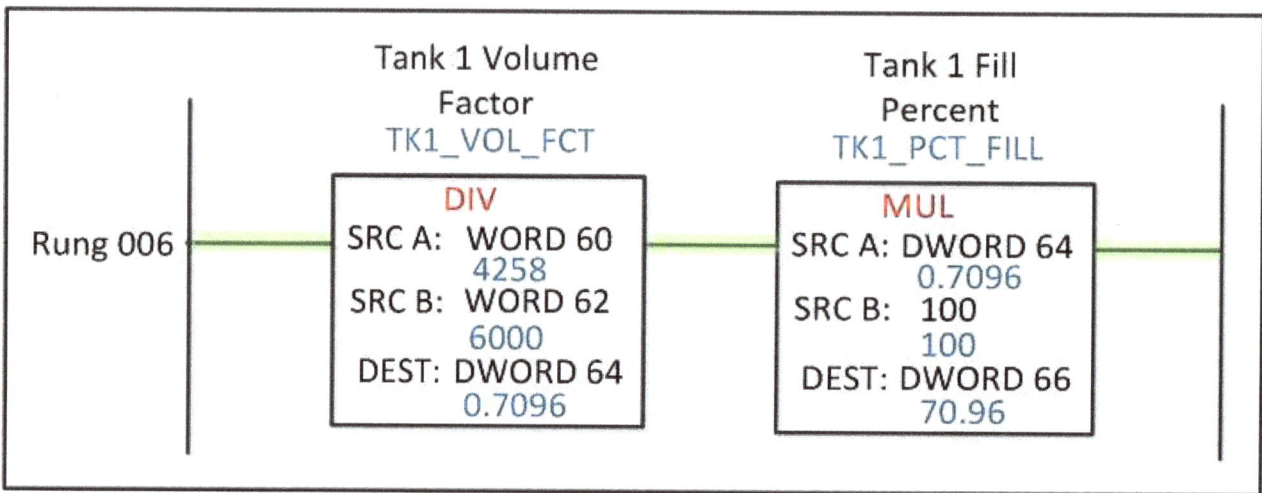

This logic calculates the percentage of fill of a 6,000-gallon tank. The measured volume (WORD 60) is divided by the total volume (Word 62), and then multiplied by 100. In this example, data types are mixed; an integer is divided by an integer and the result is placed in a REAL. The REAL is then multiplied by an integer and the result placed in another REAL. If the PLC being used did not support this functionality, data conversions from INT to REAL would be necessary.

On some brands of PLC, the data types must be the same to perform math. In this case the numbers must be converted as shown in the Siemens S7 example below.

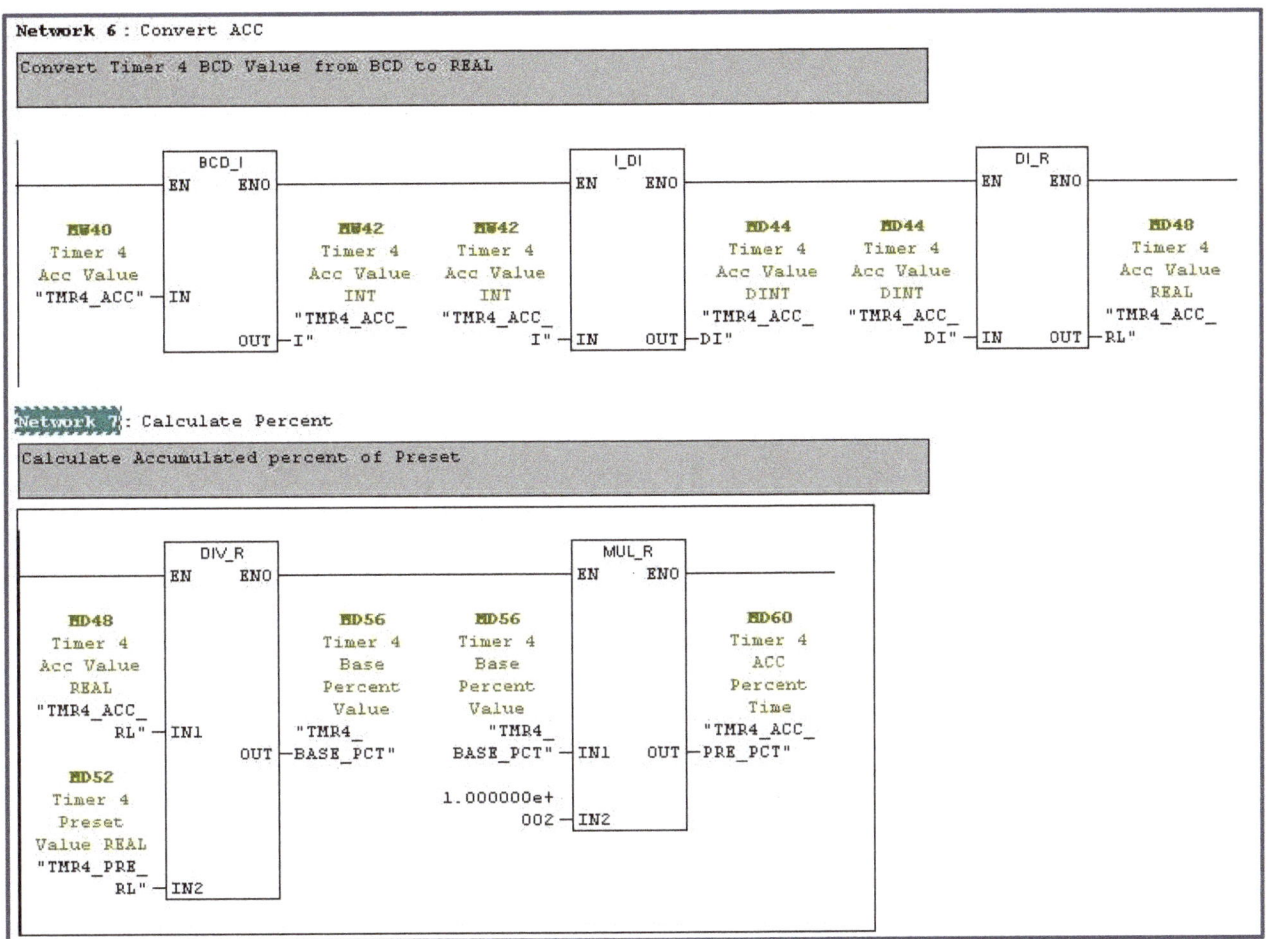

As in the previous example, this logic calculates a percentage, this time a timer's percent of completion. Siemens' timer accumulated values are in BCD, so the number is converted from BCD to Integer to Double Integer to REAL. Siemens' timers also time from the setpoint down to zero, so the final value would need to be subtracted from 100% to present the remaining time to completion (100.0 – TMR4_ACC_PRE_PCT, or 100 - MD60).

 Exercise 9

1. A Variable Frequency Drive (VFD) is used to control a conveyor. Its maximum speed is 1750 RPM, but the number sent from the drive is in integer form. At full speed, the integer reads 31,760, while it reads zero when stopped. Write ladder logic to calculate the percentage of a drive's actual speed to its total speed, also providing a REAL speed in RPM.

   Work Area:

2. Production and reject data from a manufacturing line is entered manually at the end of each shift. After completion of the third shift, there are three registers named Shift1_Prd, Shift2_Prd and Shift3_Prd containing the total parts made that shift, and 3 more registers named Shift1_Rej, Shift2_Rej and Shift3_Rej that contain the number of failed parts. Write ladder logic to calculate Total Parts, Total Rejects and Total Good Parts for the day.

   Work Area:

Scaling

An important mathematical function is that of converting raw analog values into usable units of measure, or converting one unit into another. This is known as **Scaling**, and it follows a standard formula, y =mx+b. Y is the units of the Y axis, and X is the units of the X axis. B is known as the **Offset**, while M is the **Scalar**, determined by dividing the "rise", or increase of the Y axis, by the "run", or increase of the X axis.

As an example, let's look at the graph below:

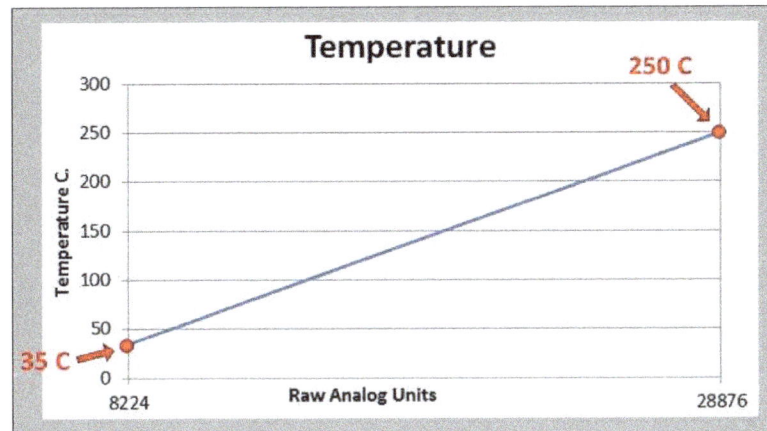

A temperature sensor produces a 0-10v signal. This is wired into an analog card which produces a signal that ranges from 0-32,767, a signed integer.

A thermometer is used to measure the actual temperature at two different points and the raw value is recorded for each measurement; the first point P1 is recorded as 8,224 on the analog card at 35 degrees C, while the second (P2) is recorded as 28,876 at 250 degrees C.

The first step in scaling the raw measurements into degrees is to calculate M, the Scalar. The "rise" or difference in Y values is Y2-Y1, 250-35, or 215. The "run" or difference in X values is X2-X1, 28,876-8224, or 20,652. Dividing the rise by the run produces a Scalar M of 0.01041061.

The next step is to calculate the offset B. Since y=mx+b, the B factor can be calculated as B=Y-MX. Substituting the values for P1, which were Y1 and X1, the calculation becomes B = (35-(0.01041061*8,224)), which yields an offset of -50.6168894. These two constants, M and B can be used to calculate Y for any inserted value of X.

As an example of how to use this formula in calculating a temperature, assume that the temperature sensor reads a value of 10,512 into the analog card. If the formula y=mx+b is used, the temperature is (0.01041061*16,512) – 50.6168894, or 121.28 degrees C.

This formula can also be used to calculate all of these variables in one group of calculations. Allen-Bradley's Scale with Parameters (SCP) instruction allows the programmer to enter Raw High and Low and Engineering High and Low units into the instruction along with the measured value from the card. It then outputs the scaled value to another variable.

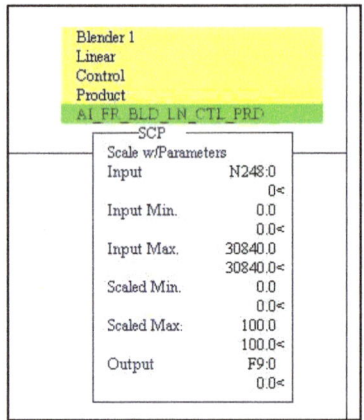

The input value is N248:0, a signed integer value from an analog card. The Minimum and Maximum Input values are taken from observed values coming from the card, while the Minimum and Maximum Scaled values of 0.0 and 100.0 represent 0 to 100 percent of the Input values. The result is moved into F9:0, a Floating Point or REAL register.

Unfortunately, many PLCs do not have this instruction, including A-B's ControlLogix, but the math can easily be reproduced as shown in the following logic:

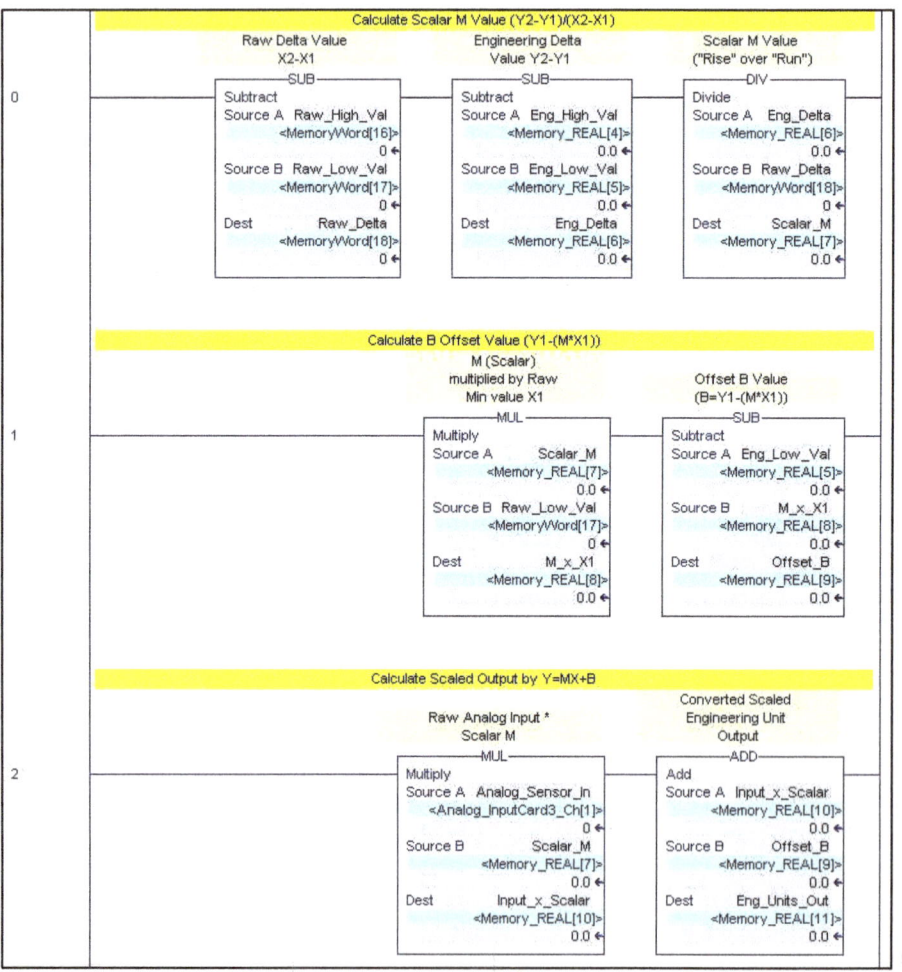

This logic can also be packaged inside of a subroutine, function or "Add-On Instruction" (AOI) where parameters can be passed into it and the results passed out.

The first two rungs use High and Low variables passed "by reference", (as shown in the SCP instruction), to calculate the internal (local) variables M and B. These are then used in the third rung to scale the Analog Sensor Input to Engineering Units Out.

Exercise 10

1. A tank used for blending juice holds approximately 8,000 gallons of liquid. There is a pressure transducer that produces a 4-20mA signal, it is wired into channel 1 of an analog card.

   The tank is filled with 6,000 gallons of juice; the reading of the analog card is recorded as 24,780. The tank is then drained completely and the value from the transducer is recorded as 96.

   Draw ladder logic that scales the raw reading from the transducer into gallons. There are 3.78541 liters in one gallon. Also calculate the number of liters.

   Work Area:

Advanced Math

In addition to the addition, subtraction, multiplication and division instructions mentioned previously, here is a list of more advanced math instructions along with their purpose:

Exponent – An exponent signifies the number of times that a number is to be multiplied by itself. The exponent is usually indicated by using the "^" sign, so 3^4 is 3 x 3 x 3 x 3.

Logarithm, Natural Log (LOG, LN) – A logarithm (LOG) is the inverse of an exponent. For instance, if 2^3 is 2 x 2 x 2 = 8 where 3 is the exponent, then the LOG of 8 in Base 2 is 3. Where a LOG is typically base 2, a Natural Log (LN) has a base of 2.718. Natural Logs are often used in math and physics (calculating decibels and pH), while LOG is often used for computer calculations.

Sine, Cosine, Tangent (SIN, COS, TAN) – Also known as Trigonometric functions, these are used to calculate geometrical coordinates. These functions -- along with their reciprocals Cosecant, Secant and Cotangent and their inverses Arcsine, Arccosine and Arctangent -- are often used in motion control applications.

Modulo (MOD) – This function calculates the remainder after a division operation.

Absolute Value (ABS) – Returns the positive version of a number even if it is negative. The Absolute Value of both -15 and 15 is 15.

Other Instructions

There are a wide variety of other instructions available on different PLC platforms in addition to those listed previously. These are just a few that are common to some of the major brands of PLC.

String Operations

As mentioned in the data section of this manual, strings are arrays of SINTs, or Single Integers (Bytes). The array elements contain ASCII characters, which can be thought of as printable characters with a few non-printable commands included. Values contained in strings can be displayed as decimal or hexadecimal numbers, or as text characters. If in text, they are often displayed with a "$" sign before the character, such as Text = $T, $e, $x, $t characters. These equate to the decimal numbers 84, 101, 120, 116 or the hex numbers 54, 65, 78, 74. These can be found in a standard ASCII table; there is one in the appendix of this manual.

Strings may also contain a length (LEN) field that contains the number of characters that exist in the string. For instance, if a string has space for 80 characters, but is filled with the characters "Today is Tuesday, September 13" then LEN = 30.

Concatenate (CONCAT) – Connect two strings together, one after another.

Middle (MID) – Copies a specified String into the middle of another String at a specified location.

Find (FND) – Locate the starting position of a specified String within another String. Usually returns the position of the found String.

Delete (SDEL) – Removes characters from a String at a specified position.

Insert (INS) – Adds characters to a String at a specified position.

Length (LEN) – Finds the number of characters in a String if length is not part of the string definition.

PID Instructions

PID, or Proportional-Integral-Derivative instructions control a process variable such as flow, pressure, temperature or level.

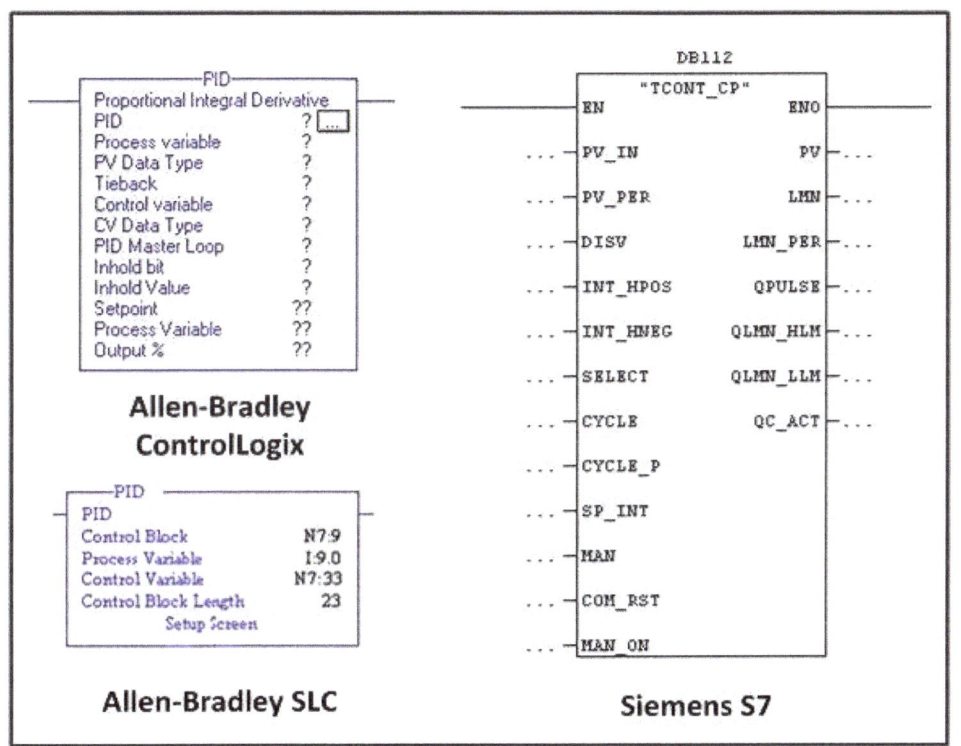

There are a wide variety of parameters that can be set for control. These may be passed in as variables as shown in the ControlLogix and S7 diagrams or set up on a special screen as shown for the SLC.

Motion Control Instructions

Newer PLC platforms may use multi-axis controller cards to coordinate movement. There are many possible commands associated with controlling axes: start, stop, jog, and direction are just a few. Numerical values such as speed, acceleration and deceleration and "go to position" commands are also common.

In addition to these individual axis commands, many coordinated movement commands are also included.

Allen-Bradley's RSLogix5000 software (ControlLogix and CompactLogix) includes 6 folders with 43 different instructions relating to motion control. These are available in Ladder and sometimes Structured Text or Function Block:

Motion Configuration – 4 Instructions. Related to tuning and parameter assignment for axes.

Motion Event – 6 Instructions. Related to arming and disarming events in the motion control card.

Motion Group – 4 Instructions. Related to issuing simultaneous command to multiple axes.

Motion Move – 12 Instructions. Related to individual issuing of motion commands to an axis.

Motion State – 8 Instructions. Related to directly controlling the operating state of an individual axis.

Multi-Axis Coordinated Motion – 9 Instructions. Related to controlling multiple axes in coordinate movement as with robotics. XYZ, XY, Articulated and SCARA configurations are all addressed.

Communications Instructions

Communications instructions are used to access a port in order to send or receive data.

Siemens Step7 software contains a range of System Functions and System Function Blocks that are used to send and receive data in a variety of ways. These are chosen based on the type of data, type of communications and, in some cases, the protocol used.

Many of these system blocks are available in the CPU; they just need to be called as a routine. Different parameters need to be filled in on the block. Most of these system blocks are categorized as COM_FUNC (Comm. Function), DP (Profibus), PROFIne2, (Profinet) and occasionally TEC_FUNC for ptp, a Siemens point to point or RS422 protocol.

Here is a list of system blocks and abbreviations for Siemens that relate to communications, please look at your help files for details:

BLOCK#	NAME	TYPE	BLOCK#	NAME	TYPE
SFB8	USEND	COM_FUNC	SFB104	IP_CONF	COM_FUNC
SFB9	URCV	COM_FUNC	SFC7	DP_PRAL	DP
SFB12	BSEND	COM_FUNC	SFC9	EN_MSG	COM_FUNC
SFB13	BRCV	COM_FUNC	SFC10	DIS_MSG	COM_FUNC
SFB14	GET	COM_FUNC	SFC11	DPSYC_FR	DP
SFB15	PUT	COM_FUNC	SFC12	D_ACT_DP	DP
SFB16	PRINT	COM_FUNC	SFC14	DPRD_DAT	DP
SFB19	START	COM_FUNC	SFC15	DPWR_DAT	DP
SFB20	STOP	COM_FUNC	SFC60	GD_SND	COM_FUNC
SFB21	RESUME	COM_FUNC	SFC61	GD_RCV	COM_FUNC
SFB22	STATUS	COM_FUNC	SFC62	CONTROL	COM_FUNC
SFB23	USTATUS	COM_FUNC	SFC65	X_SEND	COM_FUNC
SFB31	NOTIFY8P	COM_FUNC	SFC66	X_RCV	COM_FUNC
SFB33	ALARM	COM_FUNC	SFC67	X_GET	COM_FUNC
SFB34	ALARM8	COM_FUNC	SFC68	X_PUT	COM_FUNC
SFB35	ALARM8P	COM_FUNC	SFC69	X_ABORT	COM_FUNC
SFB36	NOTIFY	COM_FUNC	SFC72	I_GET	COM_FUNC
SFB37	AR_SEND	COM_FUNC	SFC73	I_PUT	COM_FUNC
SFB52	RDREC	DP	SFC74	I_ABORT	COM_FUNC
SFB53	WRREC	DP	SFC87	C_DIAG	COM_FUNC
SFB54	RALRM	DP	SFC99	WWW	COM_FUNC
SFB60	SEND_PTP	TEC_FUNC	SFC103	DP_TOPOL	DP
SFB61	RCV_PTP	TEC_FUNC	SFC112	PN_IN	PROFInet
SFB62	RES_RCVB	TEC_FUNC	SFC113	PN_OUT	PROFInet
SFB73	RCVREC	DP	SFC114	PN_DP	PROFInet
SFB74	PRVREC	DP			
SFB75	SALRM	DP			

Allen-Bradley's messaging is generally handled by the MSG instruction:

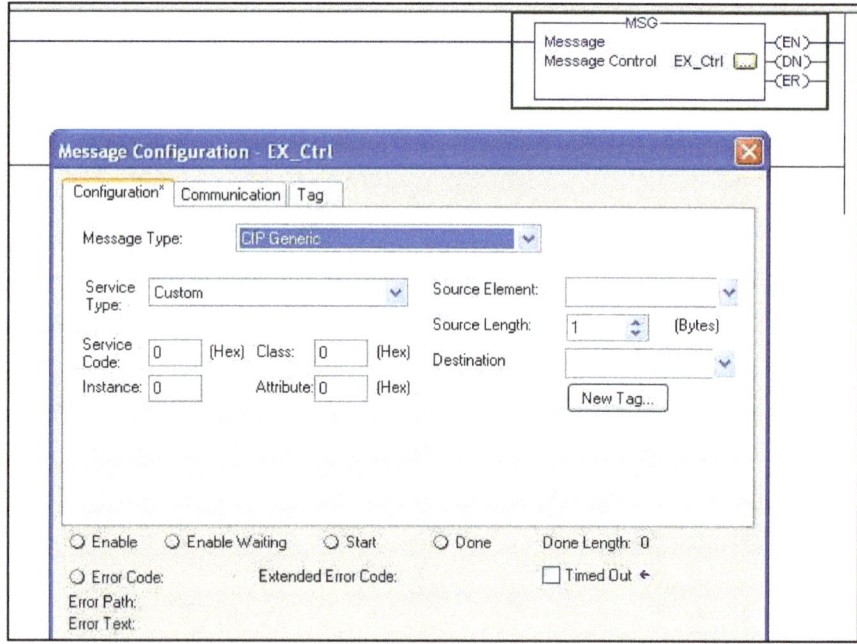

This instruction requires a message control tag specified at the controller level, shown as EX_Ctrl in this image.

After defining the control tag, a message configuration screen is accessed and the type of communications and, ideally, the path to the remote device are specified.

Protocols for Ethernet, DH485, DH+ and Serial DF1 along with SERCOS to motion control devices are available. The target node can be specified as PLC2, PLC3, PLC5, SLC and Generic CIP devices.

There are also several ASCII serial port instructions available for reading, writing, handshaking and buffer control on a basic level.

Allen-Bradley's ControlLogix platform also has a method of directly linking tags in one controller with those in another controller. These are called "Produced-Consumed Tags".

Program Control Instructions

Program control includes jumping to or calling subroutines, disabling parts of a program by jumps or "MCR" commands, looping by jumping or by using "For/Next", or redirecting the program flow in other ways. Following are some of the more common program control instructions:

Jump Subroutine/CALL – These instructions redirect the program scan to the start of a subroutine or function. At the end of the called subroutine, the flow is redirected to just after the jump or call statement.

Jump/Label – These instructions redirect the scan to a labeled point in the same routine. If jumping forward, some code will not be executed. If jumping backward, code within the zone will be executed over and over (looping) until redirected by another jump, usually associated with a counter which is preset with the number of loops to be executed.

End/Temporary End - This ends the scan of the routine and does not scan code past that point. This is often conditional, controlled by a BOOL or other logic.

For/Next, Do While – Similar to a loop as described above, a For/Next instruction is usually set to operate a specific number of times. A Do/While statement executes at least once, and remains active until a defined condition is met. Both of these instructions are seen most in Structured Text. If used in Ladder, care must be taken not to exceed the Watchdog Timer.

Master Control Relay (MCR) – This instruction is used in pairs. If the first instruction is true, the program proceeds normally. If false (not activated), the physical outputs within the MCR zone will be de-activated. This is not true for outputs that are latched on. *The MCR instruction should not be used to replace hardware MCRs.*

Miscellaneous/Other Instructions

There are a wide variety of other instructions available, far too many to list here. Every manufacturer has its own instruction set and different names for the instructions.

Here a few general categories of instructions with their uses:

LIFO and FIFO Instructions – LIFO is an acronym for "Last In, First Out", while FIFO stands for "First In/First Out". These instructions operate on a "Stack", which can be configured two different ways.

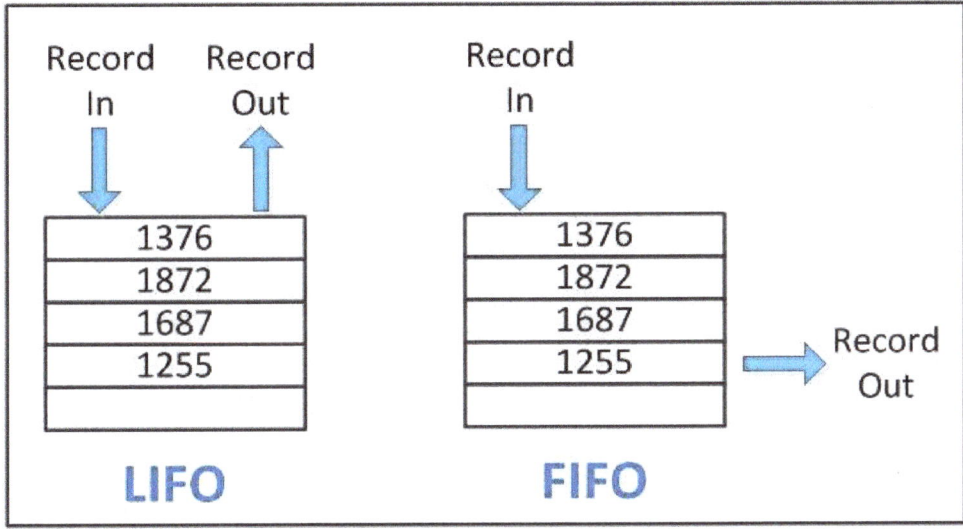

The first stack is similar to a plate dispenser at a restaurant; imagine a spring at the bottom of the stack that pushes items up as they are removed. Values are entered and removed from the top.

The second stack or FIFO allows values to be entered from one end and removed from the bottom of the stack. Each of these stacks have multiple instructions that may be used to manipulate the values. The major instructions are **Load**, which places a new value or record on the stack, and **Unload**, which removes a record or value.

Sequencer Instructions – A sequencer, sometimes called a "drum sequencer", monitors and controls repeatable operations. These instructions also use a stack, but the numbers in the stack are treated as binary values that represent conditions or drive outputs.

A Sequencer Input (SQI) instruction is used to detect when conditions are correct to index the sequencer. If the bit pattern in the designated register matches that of the next position in the sequencer, the sequencer's position value register will increment by one. The bit pattern often represents physical input states.

The Sequencer Output (SQO) instruction is used to set output conditions. These are also represented by a bit pattern, often mapped to physical outputs. The SQI and SQO instructions are usually used in pairs, with the SQI dictating the conditions that index the SQO.

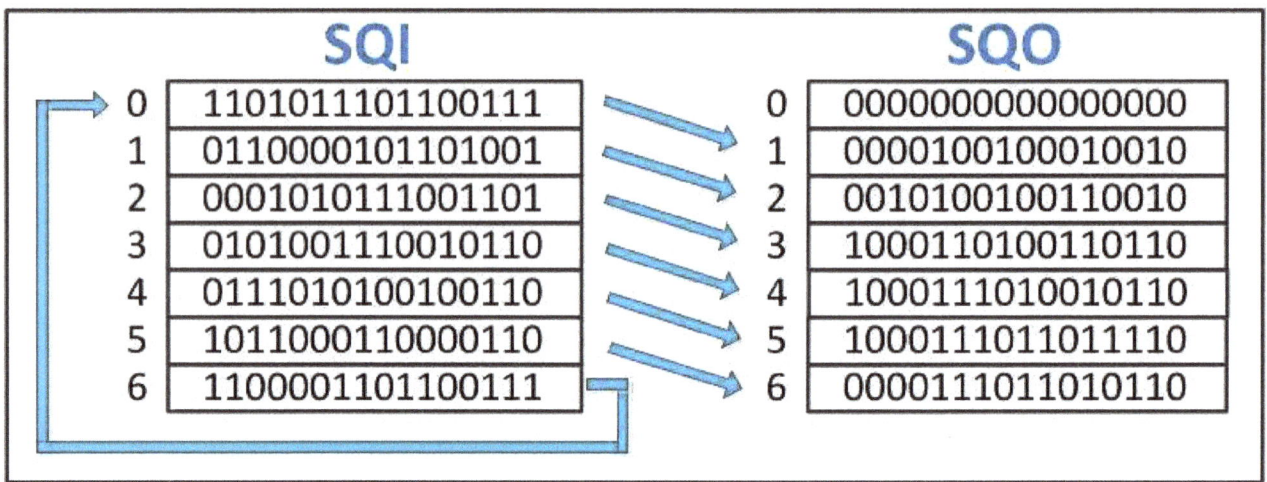

The Sequencer Load (SQL) instruction can be used to place values (bit patterns) into the sequencer's register. This is much like a "teach" function; the instruction looks at a register, often representing inputs. When the SQL is executed, the pattern is loaded into the next location in the stack.

Sequencer Compare (SQC) is another instruction that is sometimes used in order to index the position number of a sequence.

Statistical Instructions – There are various instructions available on some platforms to perform statistical math, such as standard deviation, moving averages and finding minimum and maximum signals in a specified period of time.

Other Instructions are available for Safety Functions, Signal Filtering, VFD (Drive) Control, Equipment Phasing (State Programming) and many others. For full knowledge of a specific platform, it is a good idea to read the programming manual and help files for instructions. Don't forget that many of these require the use of languages other than ladder! It is also possible to build these instructions yourself by using Add-On Instructions or Functions.

Exercise 11

1. What kind of applications are Trigonometric functions used for? _____
   _____

2. Decode the following hexadecimal ASCII characters using the table in the appendix:
   47 _____ 6F _____ 6F _____ 64 _____ 20 _____ 4A _____ 6F _____ 62 _____ 21 _____
   _____

3. Can a JUMP instruction be used to move backwards in a program? _____

4. What do the acronyms "FIFO" and "LIFO" stand for? _____

5. What is the purpose of a Sequencer instruction? _____
   _____
   _____

Maintenance and Troubleshooting

There are a number of tools and techniques common to all PLC platforms that can aid a technician in isolating the causes of problems. An important thing to remember when using these tools is that *The PLC's program cannot change without someone changing it!* Programs can't change themselves, they either run or they don't.

Forcing

One method of determining whether an input or output is working properly is using a **Force**. In the case of an input, it is obviously not possible to physically force a point on a card. You would have to place a voltage on the point in order to energize it. So what are you forcing when you force an input? *Only the input table*. This means that when you apply the force, the contacts or values related to that point will change only in the program, not on the card itself.

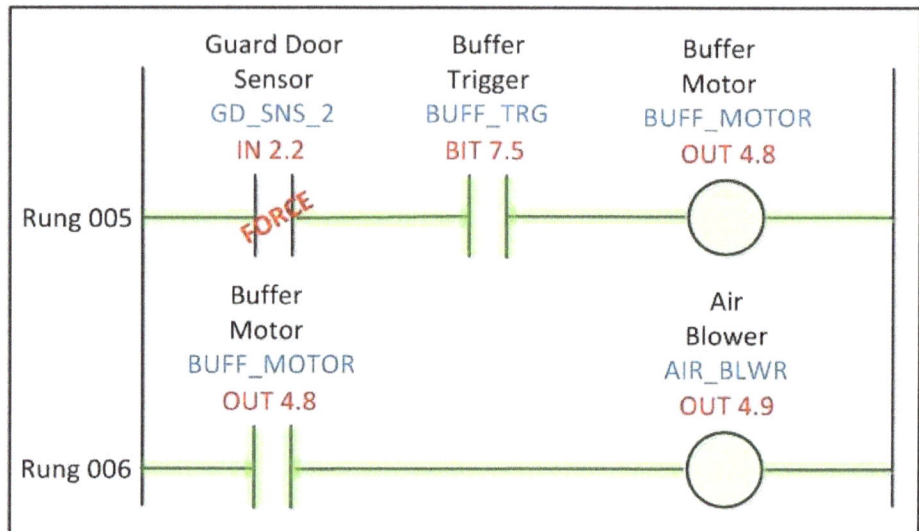

In this case, the Guard Door Sensor is not allowing the Buffer Motor to run even though the trigger signal is shown to be active. A force is applied to input 2.2 and the motor runs. This is a verification that the electrical signal to the input needs to be checked; maybe it is a bad sensor, maybe the wire is disconnected, or maybe the input point on the card itself is bad. In this case, the forcing of the input has helped to determine the problem. After the problem has been isolated, the force can be removed and the problem fixed. *The force should not be used to hide the problem!*

On most PLC platforms, there will be some kind of indication on the contact showing that the input or output is forced. There is also usually a light on the processor itself indicating that a force is present.

Forcing an output is the *exact opposite* of forcing an input. If a force is placed on the output point, the physical output will be energized, but *the output image table will not be affected*.

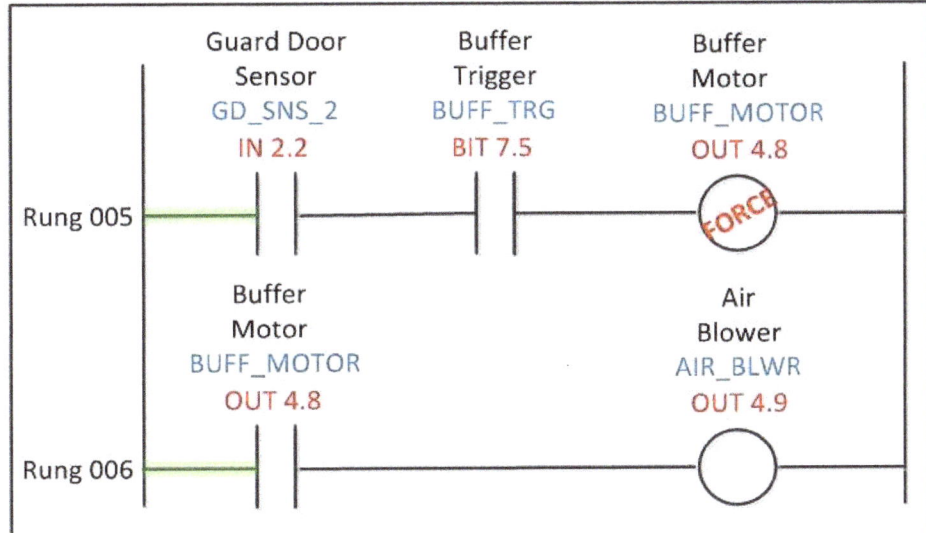

The logic energizing the Buffer Motor is not true, but a force is applied to the output. The physical output comes on, and the Buffer motor runs. Note, however, that the Air Blower that usually comes on whenever the motor runs is not energized. This is because the image table is not affected by the force. The forced coil is actually transparent to the logic; if conditions are true up to the coil, the image table will be updated and the Air Blower coil will be energized.

In most PLCs only inputs and outputs can be forced, but some platforms allow the forcing of memory also.

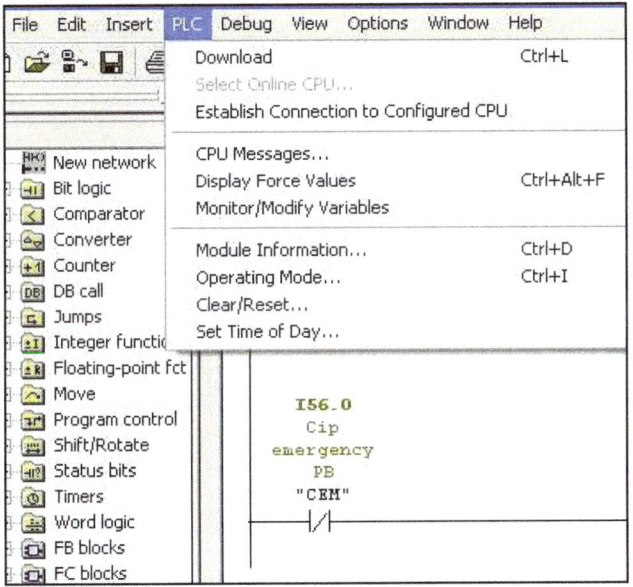

Installing and activating forces is generally a multi-step process. The force is installed in one step, and then activated as a separate action. This is because forces can be very dangerous if implemented incorrectly; you are telling the PLC to do something unnatural, outside of its coding. A force can, however, be helpful in the troubleshooting of a system.

The image to the left shows how forces are accessed in Siemens Step 7 software. From the editor, a Force table is opened from the PLC menu. Force addresses are entered into the table, and then activated.

When active, a red "F" appears by the address indicating that a force is being used.

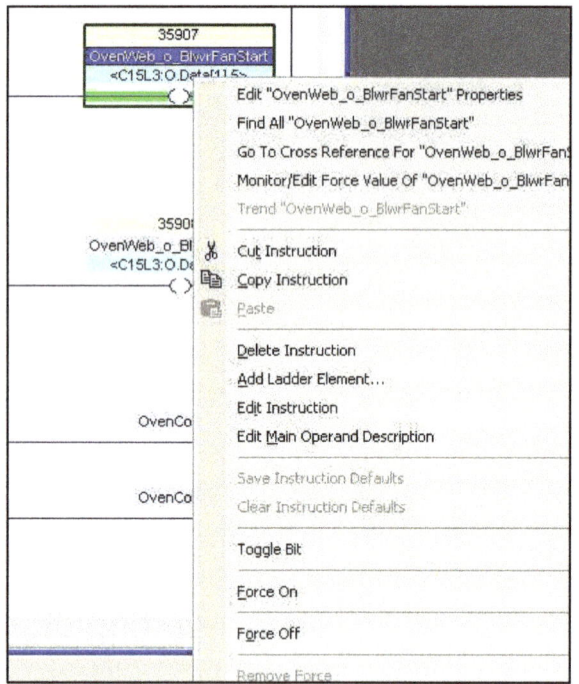

In Allen-Bradley's software, forces are installed by right clicking on an address in the program. After installing the force, forces must be enabled using the dialog shown below:

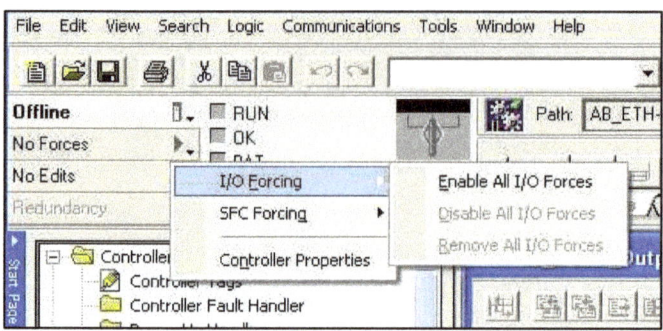

This two-step process ensures that the programmer truly intends to create a force.

Searching and Cross-Referencing

In order to diagnose problems in a machine, it may be necessary to trace logic through a program. There are a number of tools available to determine the location of addresses, check whether addresses have been used, and substitute one address for another. Searches can also locate words in the comments of a program.

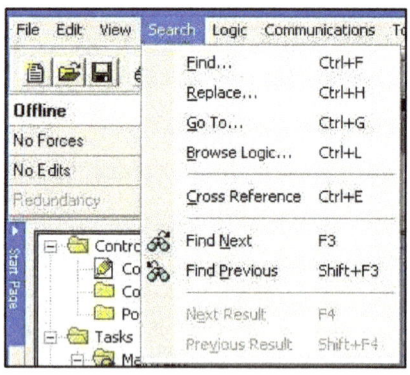

There is usually a tab in the software that will allow various "search" or "Go To" options as shown. Right-clicking an address in the program will also often bring up a selection allowing other instances of the address to be found.

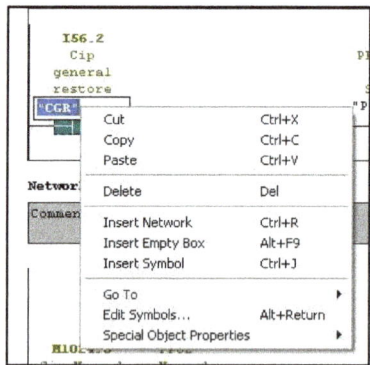

One of the most useful tools when trying to determine why an output is not being activated is the **Cross-Reference**. A cross-reference shows all of the places in a program where an address has been used.

Usually, troubleshooting starts with finding the coil of an address. Right clicking the address brings up a selection for cross-reference or "Find All"; this in turn brings up a list of all of the places where the address is used in the program.

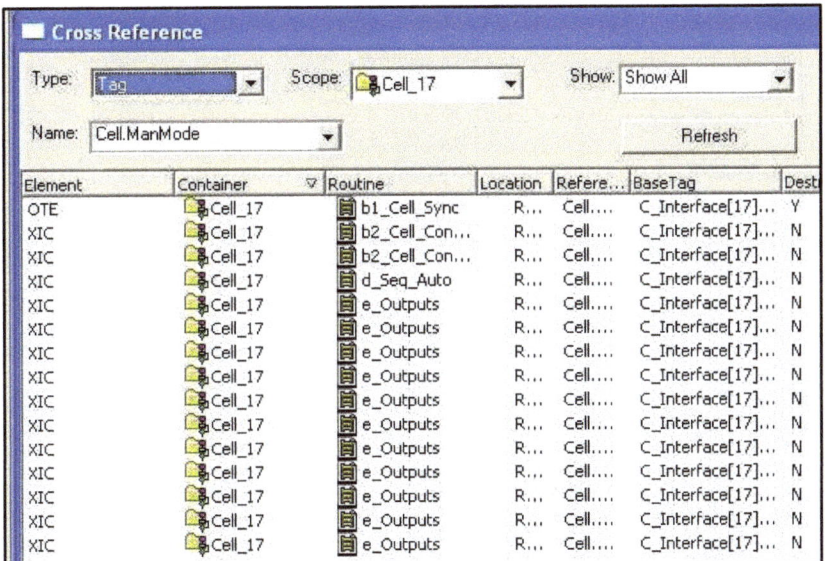

Selecting the location of the coil takes you to the rung or network where the coil (OTE) is activated. With proper programming technique, there should only be one place where a coil will be located for any address!

Addresses can then be traced from rung to rung until finally the cause of the problem can be identified.

The following rungs illustrate tracing the cause of an output failing to come on:

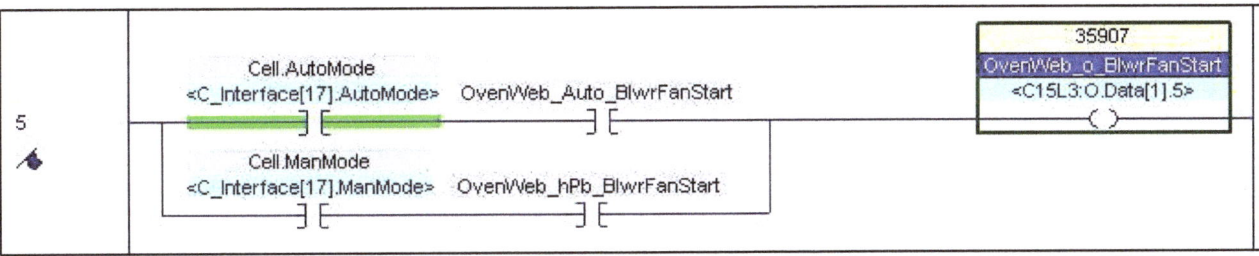

The rung above shows that the Oven Web Blower output (the coil) is not energized. Since the AutoMode contact is energized, a cross-reference of OvenWeb_Auto_BlwrFanStart, is executed.

Searching for the coil brings up this rung:

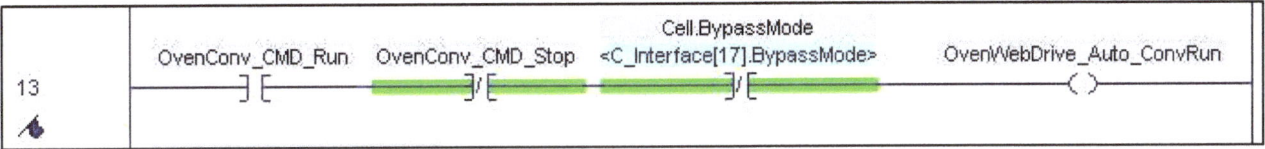

This in turn prompts the searcher to look for: OvenWebDrive_Auto_ConvRun

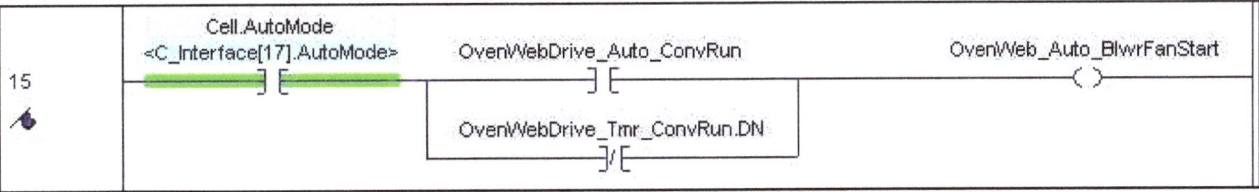

Which in turn brings up: OvenCnv_CMD_Run

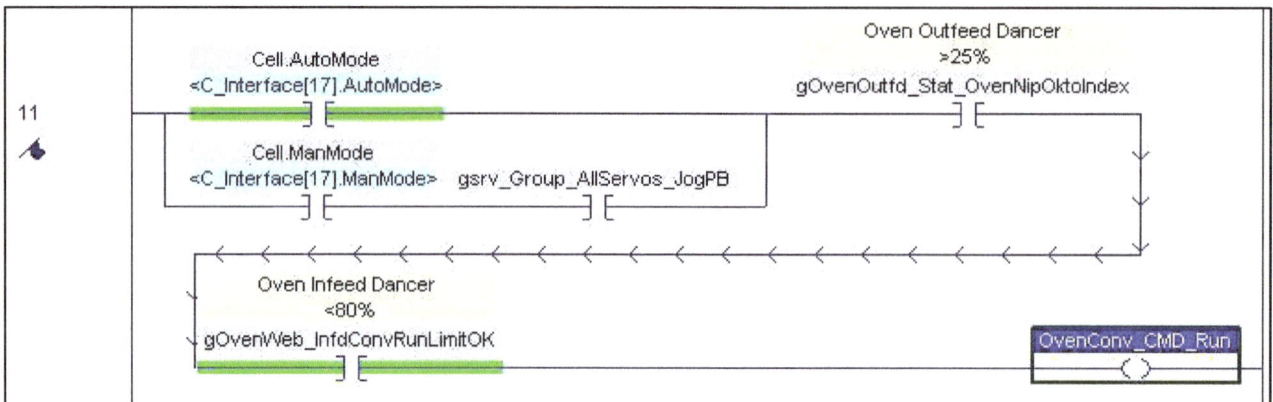

And on to the next address, which, in this case, is not a coil:

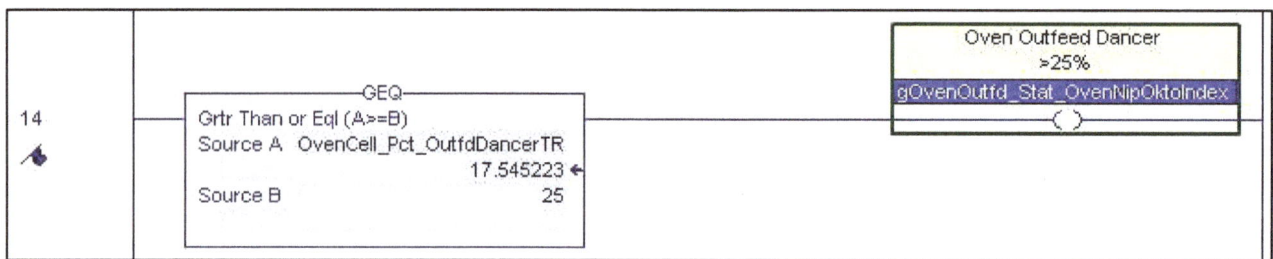

This rung shows that a physical address C15L3:2:I.Ch2InputData (an analog input value) is moved into the variable that we are looking for. Cross-referencing and searching will always end at either a physical input point, or at an address with no coil or address that has been changed by the program. This last would mean that the signal comes from outside the controller, such as an HMI, SCADA or even a signal from another PLC.

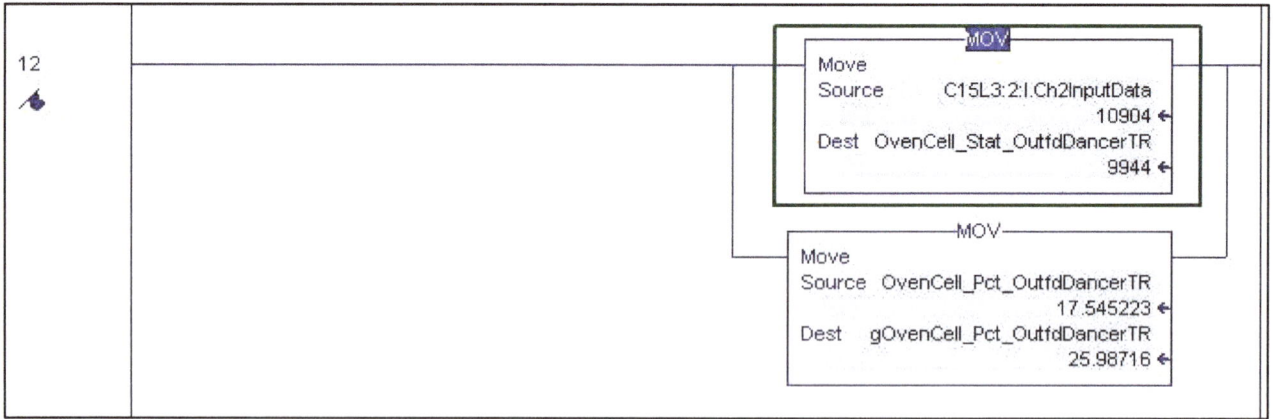

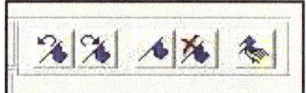 Did you notice the little flags next to the rungs? Allen-Bradley's ControlLogix platform has a toolbar called "Bookmarks" that allow a programmer to mark rungs and then index through them. Very handy!

Siemens also has a useful tool that allows one to look at which registers have been used. This can also be used to find addresses with no symbol or symbols with no address. Notice that several bits have been assigned that are also Word addresses (MB1026 & MB1027). This can be a useful to spot address interferences.

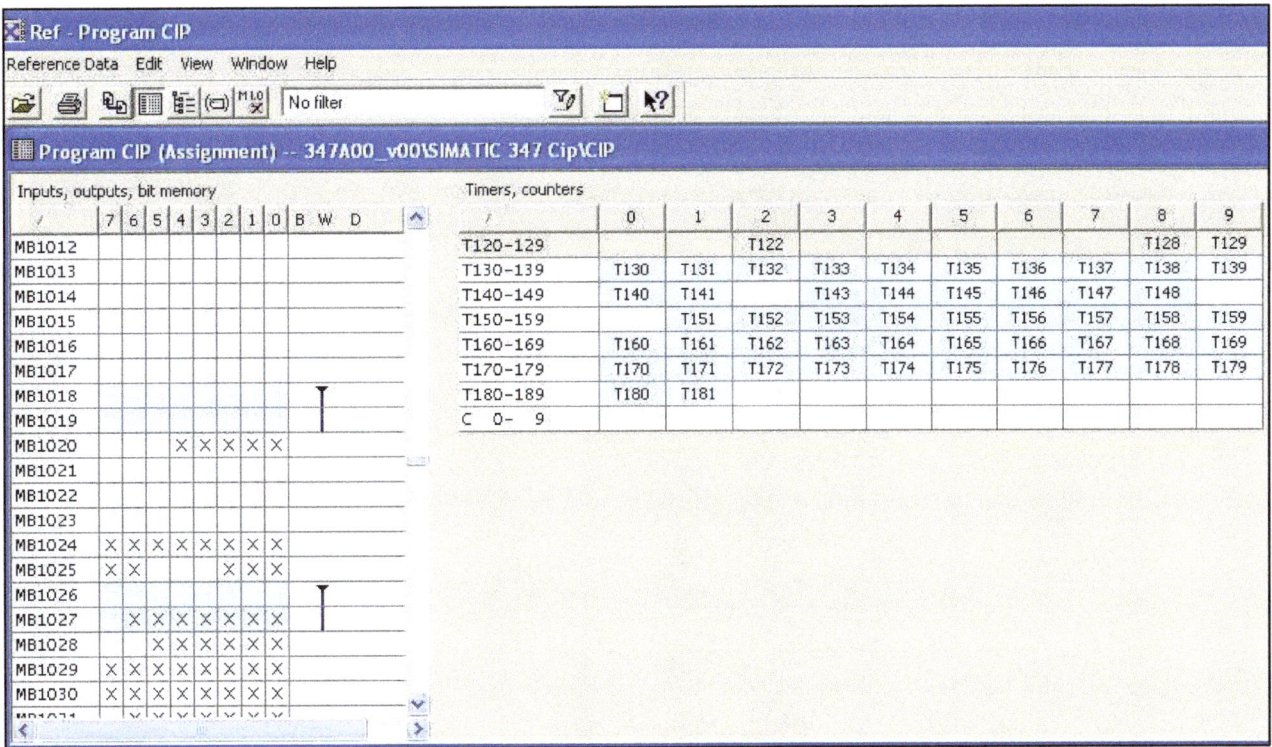

Advanced PLC Hardware & Programming

 Exercise 12

1. Forcing an input applies voltage to the physical input.   True: _____ False: _____

2. Forcing an output applies voltage to the physical output.   True: _____ False: _____

3. In the diagram at right, if Q98.4 "PP03" is forced, what will the state of Q98.5 "PP04" be? (Assuming that input I56.4 is off)
   _____
   _____
   _____

4. If the force on Q98.4 is removed and I56.4 is forced, what will the state of Q98.5 be?
   _____
   _____
   _____

5. To trace a signal through a program, what types of instructions should you look for?
   _____
   _____
   _____

6. What two types of addresses will represent the end of a search or cross reference?
   _____
   _____
   _____

The Art of Programming

The Art of Programming - Overview

The second part of this book covers many of the programming techniques used in industrial machinery, including program organization, types of routines, and common methods for accomplishing control tasks.

Up to this point, the material in this book has primarily discussed the instructions that are used in PLC programming. Program organization has also been mentioned, but only from the perspective of how the different platforms separate sections of code into routines, functions, tasks and the like.

Programming is an art. There are many ways to write code that will accomplish the job, but a well written, well organized PLC program can be beautiful to those who are knowledgeable in the field. While there are many "right" ways to write a program, there are also many wrong ways. This section will discuss some of the different methods and techniques that are used in industry and also explain *why* they are used.

Machine Control vs. Process Control

Before beginning programming, it is important to discuss some of the common types of industries that use PLCs. Machine control is used in the assembly of products and usually includes moving actuators. Pneumatic and hydraulic cylinders, servo motors and conveyor belts are often used to move items or tooling to specific positions; if these actuators aren't properly sensed and sequenced, they can cause damage to the product or to the machine itself. Movements may be to discrete positions, usually at the end of the actuator's stroke or to variable positions in the case of servos and conveyors. Often these movements are very fast, where control is done on a millisecond timescale. An important part of programming machinery is the sequencing of these movements; this is discussed in the section on Auto Sequences. When motions aren't completed properly, it is also very important to detect and report the problem so that it can be remedied, discussed in the Fault and Alarm section.

Process control often involves the handling of fluids and control of variables such as volume/level, temperature, and pressure. Objects may also be heated or cooled for periods of time, and there is often the ability to manually control different parts of the system while other areas or processes proceed automatically. Timescales for process control are often much longer than those of machines; it may take hours to empty a tank of product. As such, process control typically involves more analog I/O than machine control does.

Machine control may also include elements of process control, such as controlling ovens or applying pressure to a product, as with a mechanical press. This involves analog signals, usually sensors indicating positions or other physical attributes and outputs controlling drives or valve positions.

Programs may use techniques from both areas of industry. It is important to understand both of these types of programming.

Advanced Topics

Some of the topics in this section cover material that is not directly applicable to all platforms. Some older PLCs did not have User Defined Data Types (UDTs) or Arrays. Material in the Part Tracking and Recipe sections make extensive use of these concepts and use pointers for indirect addressing.

Program Organization

Before starting a PLC program, it is necessary to understand the operation of the machine or system that you will be controlling. A written overview of what the system does is often a good starting point.

Microsoft Excel is a good tool to create some of the lists that you will need. External devices or systems that need to be interfaced, I/O and operational sequences are all examples of lists that you will want to create. A flow chart is often used to visually describe the operations; this can be drawn by hand or by using a graphical software package such as Microsoft Visio or AutoCAD. Even keeping a handwritten set of notes in a folder or notebook is helpful until information can be typed into a retrievable and sharable file. If this information is not provided, you will need to generate it yourself.

Following is some of the information you will need before beginning a program:

- I/O List (tag names and addresses)
- Alarms or Faults
- Messages or Warnings
- External devices to be communicated with (HMIs, Machine Vision, Robots, Testers)
- Interface Protocols and Addresses (Serial, Ethernet, DeviceNet, etc.)
- Sequences of Operation for Stations or Cells
- Hardware/Parts List

Structure

Despite its importance, there are no rigid, set-in-stone rules for PLC programming organization. The organizational method described below is one that is commonly used by large machine builders; however, there are other techniques you can choose to employ. The key objective in program organization is to make a machine easy to troubleshoot and modify.

After gathering information about the system to be programmed, the organization and layout of the program itself can begin. There are two ways to separate the sections of the program. The first is by the physical areas of the machinery, and the second is by the type of code. Small mechanical or logical sections of a machine, such as a pick and place, conveyor or dial table will be referred to as "stations". Stations can then be grouped into larger groups.

Process control programs are often divided into functional areas, such as tanks or lines. Process control stations can often be operated independently in Manual or "Hand" operation.

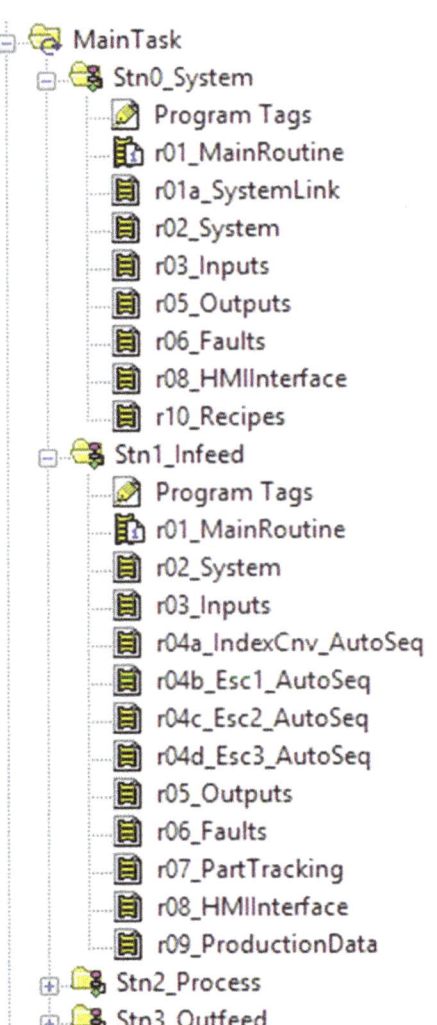

Zones can be used as a logical division of groups of stations. All the mechanism-operating auto sequences can be placed within the station groups. The primary reason for dividing a machine into zones is to allow the zones, or groups of stations, to act autonomously. In this way, zones can be thought of like separate machines. Zones will often each have their own safety circuits so if a guard door opens, for example, it only shuts down the group of stations in that zone. Subsequently, there may be routines within the group called Zone_1, Zone_2, and so forth. Within the groups may be Input, Output and Fault routines, as well as routines that link to all of the stations assigned to the zone. Data acquisition and productivity information may also be assembled at the zone level. Generally, the inputs and outputs would only be those that affect all of the stations, such as safety or indication I/O. A zone may also have an HMI that only controls devices within the zone; HMI bit mapping could also then be contained in the Zone program and routines.

If a platform allows for multiple programs, each with its own group of routines, stations can each have their own program containing standardized routines for system functions, inputs, auto sequences, outputs, and faults or

Figure 2 - Programs and Routines

alarms. An example of this from Allen-Bradley's RSLogix 5000 platform is shown in Figure 2. Each program also contains tags or addresses that are local to the program.

The programs and routines in Figure 2 do not necessarily reflect the order in which the routines are called. For Allen-Bradley, the routines appear in alphabetical order, so the programmer has prepended the prefixes r01, r02 etc. to the routine names. They are organized so that they reflect the order in which they are called.

Other PLC platforms may not allow for routines to be grouped together. Figure 3 shows a list of blocks and routines in a Siemens S7 PLC.

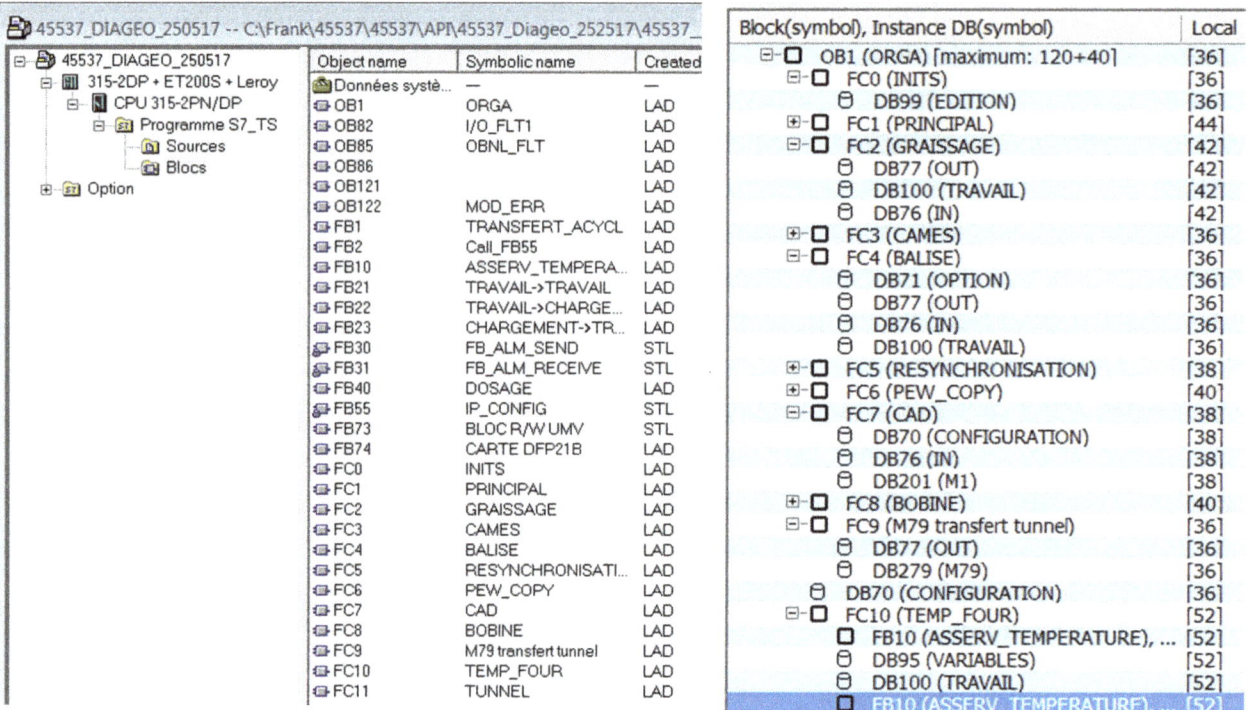

Figure 3 - S7 Organization and Dependencies

Since the blocks and functions can only appear in numerical and alphabetical order, an additional cross-reference utility is available to show the order in which the functions (routines) are called, and the dependencies of the data blocks. This is shown on the right side of Figure 3.

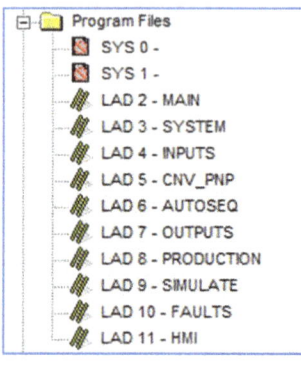

The Allen-Bradley RSLogix500 platform is also organized into numerically numbered routines called ladders (LAD) as shown in Figure 4. As with the Siemens functions, it is up to the programmer to ensure that routines are called in the same order as they are listed. There is no cross-reference utility to show the call structure for the RSLogix500 platform.

Figure 4 - RSLogix500

Common Routine Types

System Routines

System functions include modes of operation and machine states. These are usually modes that describe the condition of the whole machine or zones. Auto Mode, Manual Mode and Auto Cycle are common examples of this. A System Routine will usually contain control logic for these types of functions, along with overall system "housekeeping".

Auto Mode: This mode is an idle state that is a prerequisite for placing a fully-automated machine into Auto Cycle. It usually eliminates the ability to operate individual actuators by pushbuttons or switches, but rather allows a machine to sequence through automated motions.

Manual, Maintenance or "Hand" Mode: This allows actuators and devices to be actuated by means of switches or pushbuttons, outside of the automatic functions. In process control operation, devices are often individually able to be placed in "Hand" mode, even while other devices or systems are being controlled automatically.

Auto Cycle: This mode is a subset of Auto Mode. A machine is usually placed into Auto Cycle by holding a button for a period of time while a warning signal sounds. This allows personnel in the immediate area to stop the operator from starting the machine's operation if a hazard is present. The warning signal often pulses or "beeps".

Other requirements, such as only allowing Auto Cycle start when a machine is at a home or origin position, only allowing the progression of an auto sequence when the machine is in Auto Cycle, and only allowing cycle stop when the machine or sequences are in a specific position or condition, are common.

Other optional modes for machinery:

Dry Cycle: Allows machine to run in Auto Cycle with no parts. Often used to "exercise" a machine during Factory Acceptance Testing (FAT) or Site Acceptance Tests (SAT).

Single Cycle: Allows machines to process parts one at a time in Auto Cycle or run automatic sequences, such as pick and places, in Manual Mode.

Single Step: Allows automatic sequences to run one step at a time. Typically only used for debug. Requires a single-step button in addition to mode control enable/disable.

Purge or Runout: Allows a machine to empty without bringing new parts into the system. Usually used during product changeover.

Homing: Allows actuators to return to their starting or origin location automatically. May be done in Auto or Manual mode depending on preference. Typically, a sequence is used to ensure that actuators are moved in the correct order; see the Auto Sequence section for how to write this.

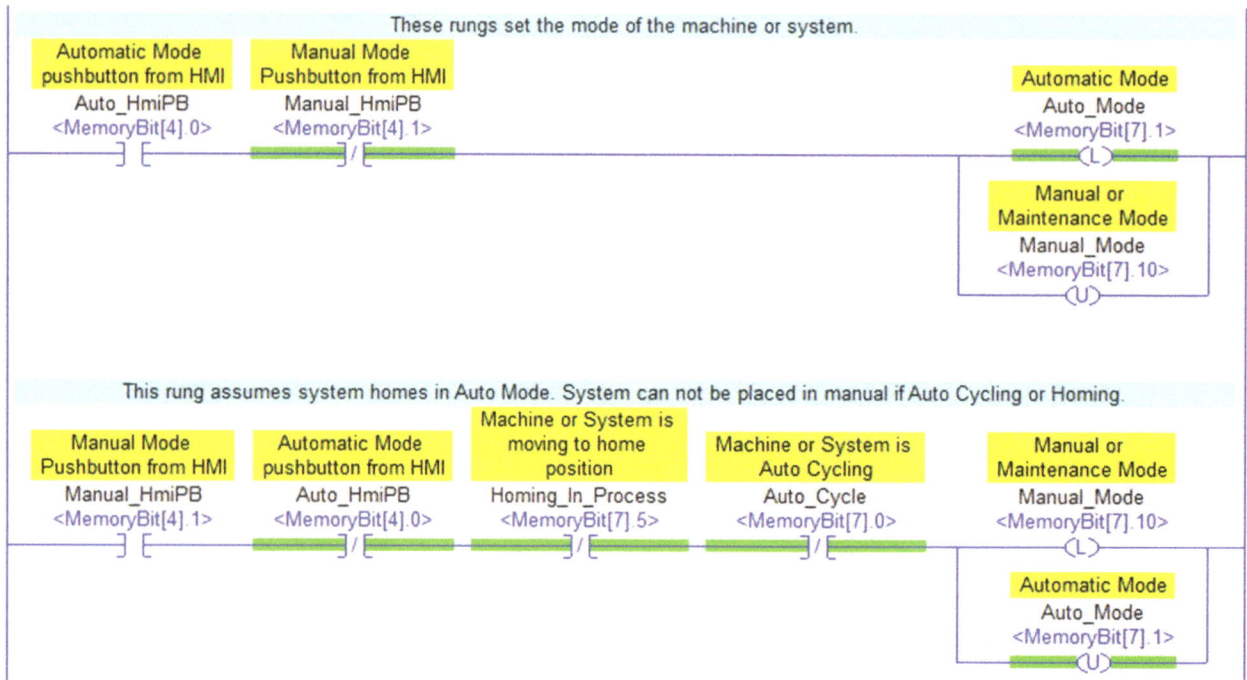

Figure 5 - Auto and Manual Modes

Figure 5 shows some of the conditions for changing modes in a system routine. In this illustration, homing of actuators is done in Auto Mode, so the machine can't be placed in Manual Mode while in Auto Cycle or while homing.

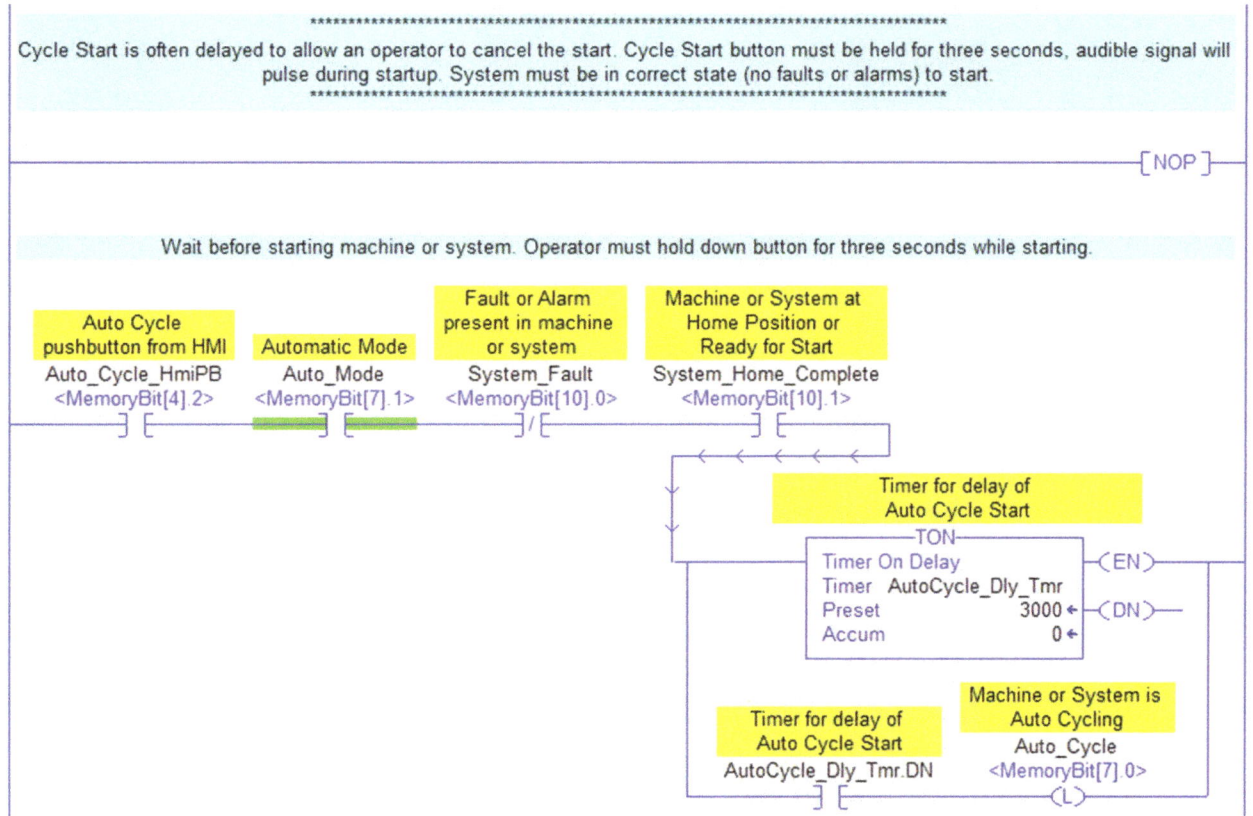

Figure 6 - Auto Cycle Start

In Figure 6, a timer is used to delay the start of a machine for three seconds. Usually, there will be an audible alarm that will sound while the operator presses the button, warning people in the area that the machine is about to start. This gives the operator a chance to remove his finger from the button and abort the startup if necessary.

Auto Cycle is used to allow a machine to automatically process components or material, it will usually do so until either a malfunction occurs or until an operator stops the process.

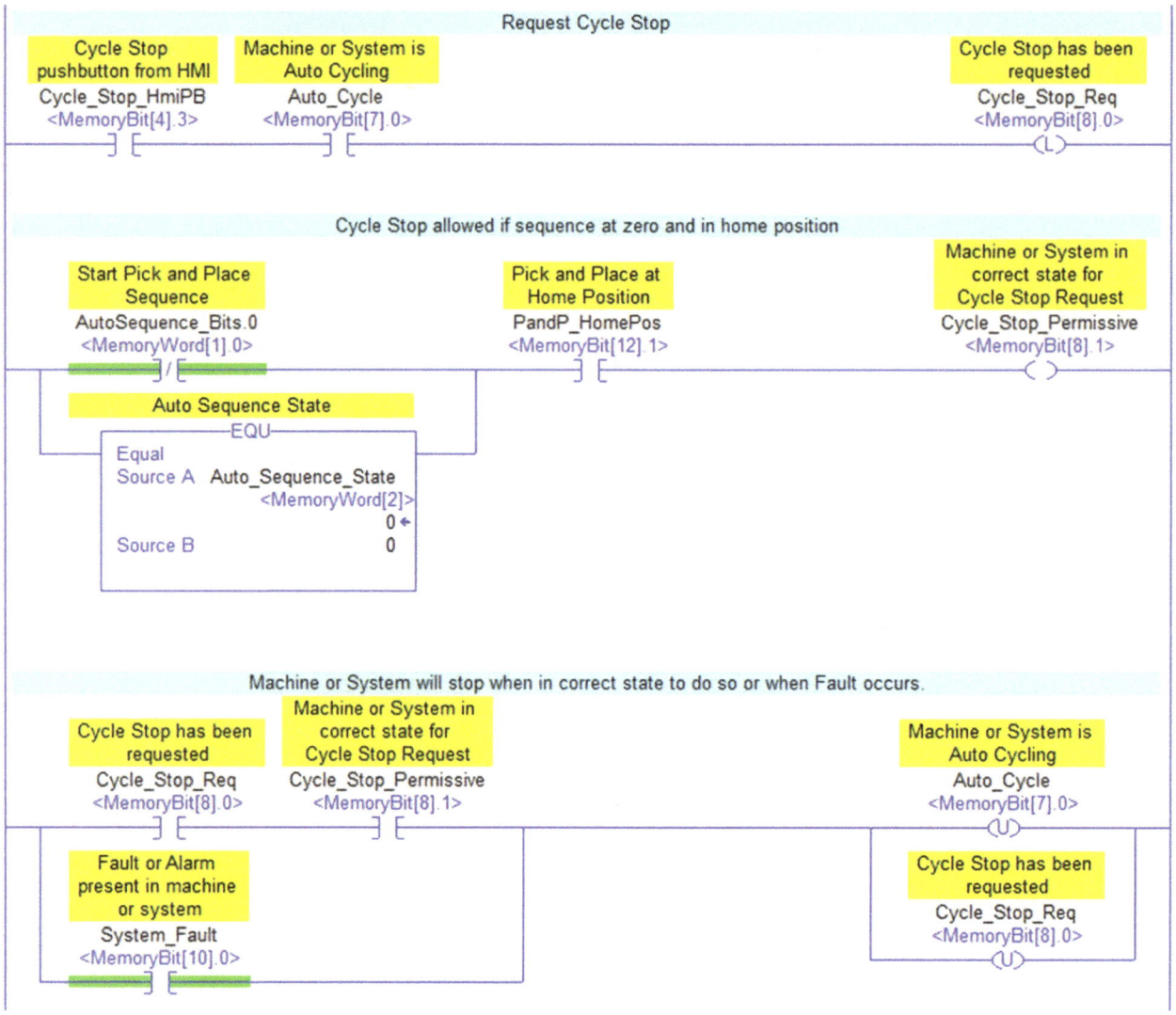

Figure 7 - Auto Cycle Stop

Figure 7 shows logic for an operator to stop the system. If the system or machine is not in the correct state or position to stop, it will continue in Auto Cycle until the desired conditions or positions are satisfied; this requires that a "Request to Stop" bit is latched so that the system can remember the request. In this example, a fault in the system will stop the machine immediately.

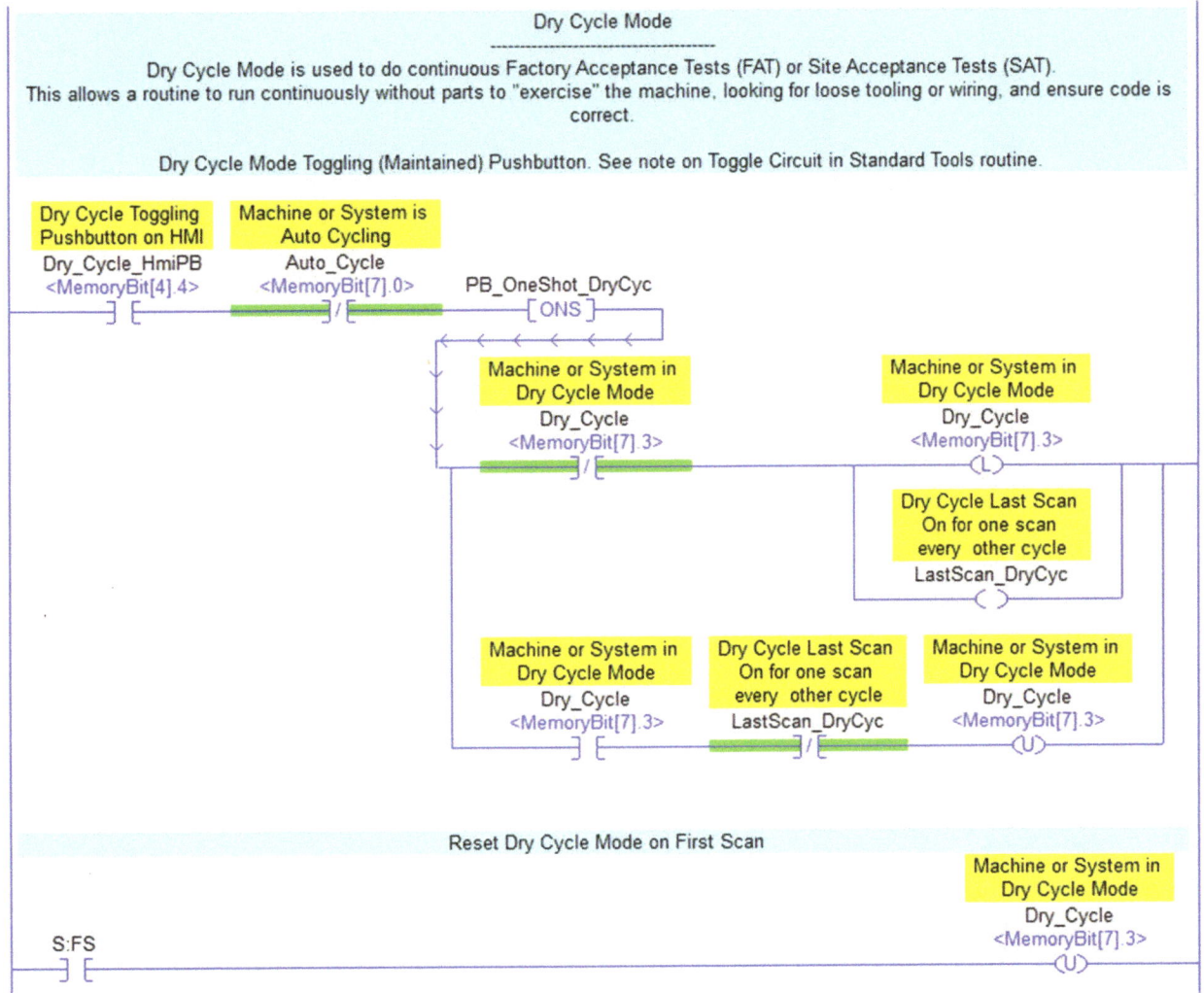

Figure 8 - Dry Cycle

Dry Cycle is typically used in machine runoff conditions by machine builders, but the logic may be left in the program for the customer. The idea behind Dry Cycle is that the machine can run continuously without parts to "exercise" its actuators. The logic in Figure 8 incorporates a toggling pushbutton that will place the machine in Dry Cycle mode and take it back out with a single button.

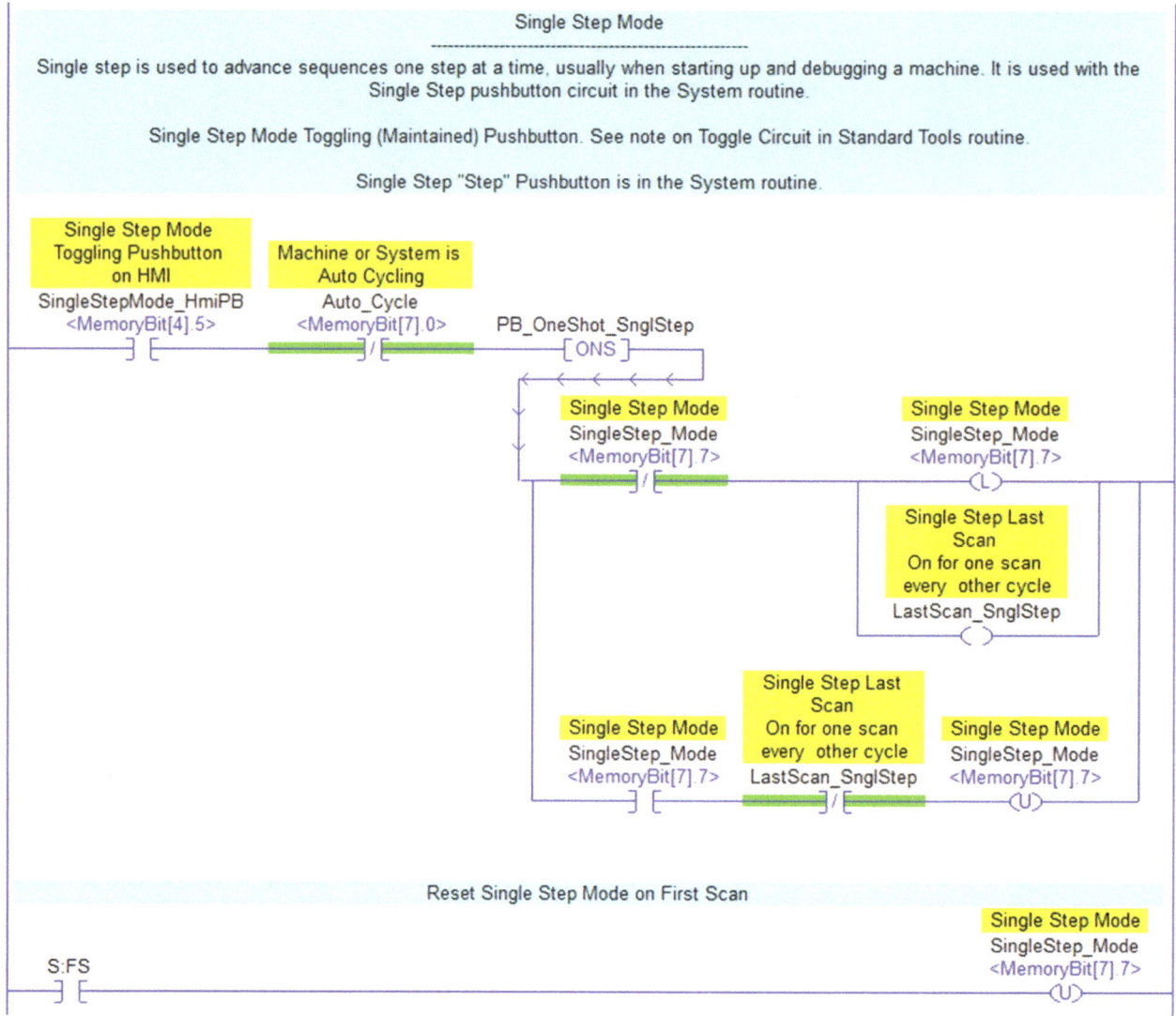

Figure 9 -Mode Control for Single Step

The logic in Figure 9 is similar to that of Dry Cycle mode control, it uses a single button to switch in and out of Single Step mode. Additional logic is used to create the bit that is placed in an auto sequence to step the sequence manually.

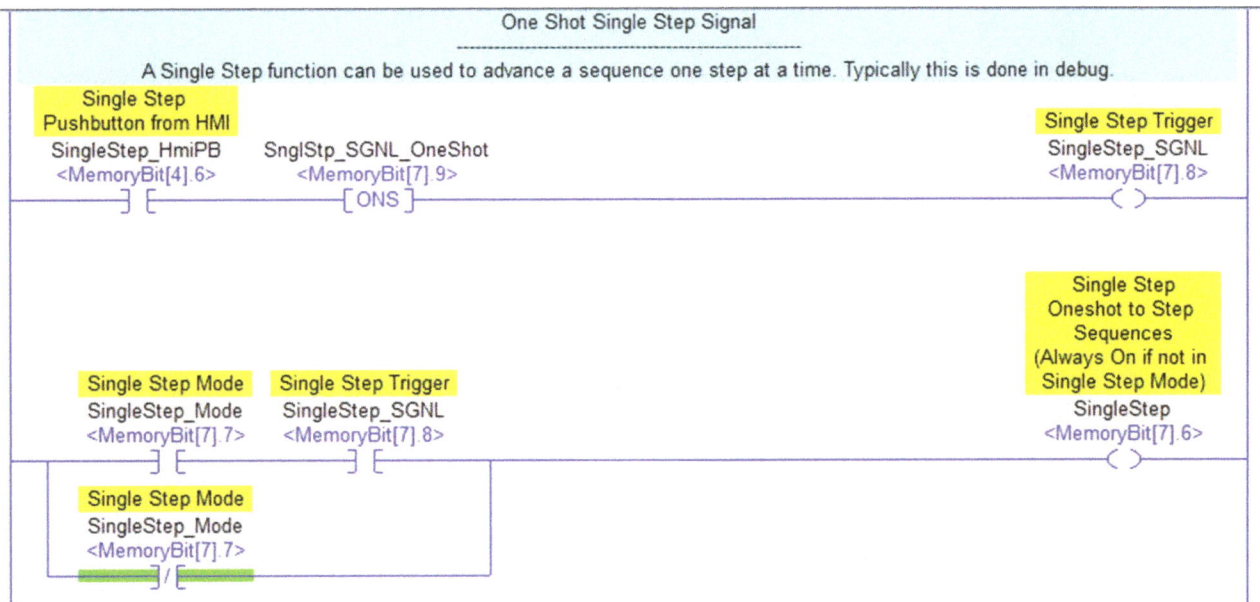

Figure 10 - Single Step Signal Logic

Figure 10 shows how the signal that is used in the actual auto routine operates. When not in Single Step mode, the SingleStep bit is always on, and logic proceeds through the sequence as usual. When the mode is active, a pushbutton is used to move to the next step, provided the other conditions in the sequence step are satisfied.

For both Dry Cycle and Single Step modes, a complete understanding of how a sequence operates is necessary. Read through the Auto Sequence Routine descriptions and then come back to this logic.

Input Routines

An Input Routine is used to modify the status of physical inputs for use in the program. Following are examples:

Back Checking. This is used for complementary pairs of sensors where they physically cannot be on at the same time.

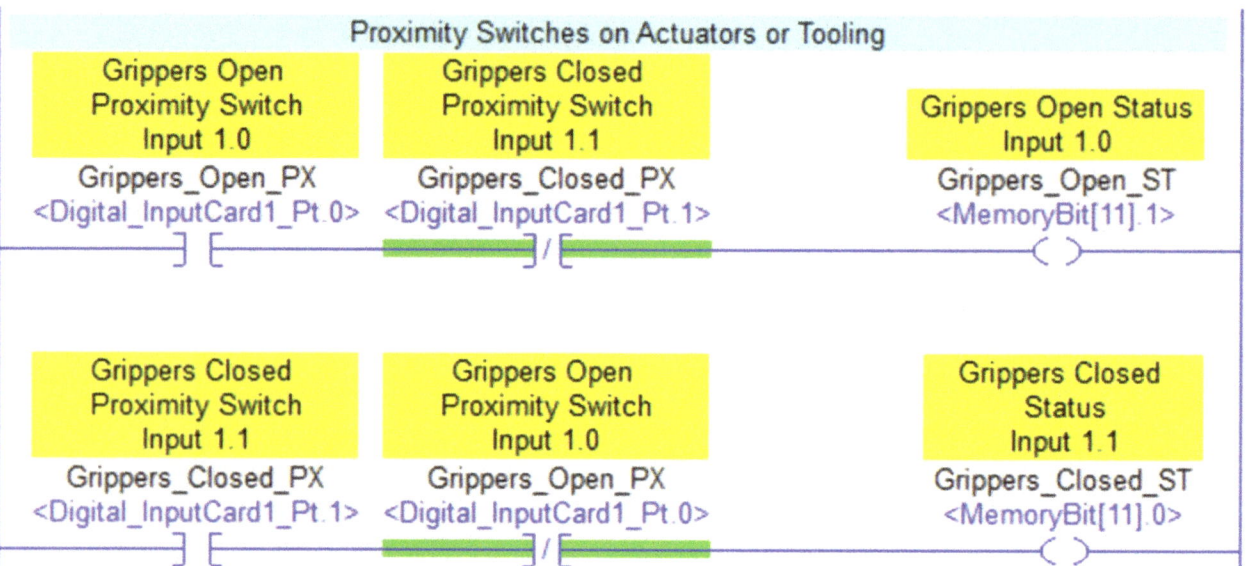

Figure 11 - "Back Checking" Logic

The ST or Status bits are used in the program rather than the physical inputs. This means the logic will not react if the incorrect signals are received due to a wiring or physical error.

Debouncing. This is used both to ignore intermittent, accidental signals and to place a delay on a signal to ensure that a part is settled into place before reacting.

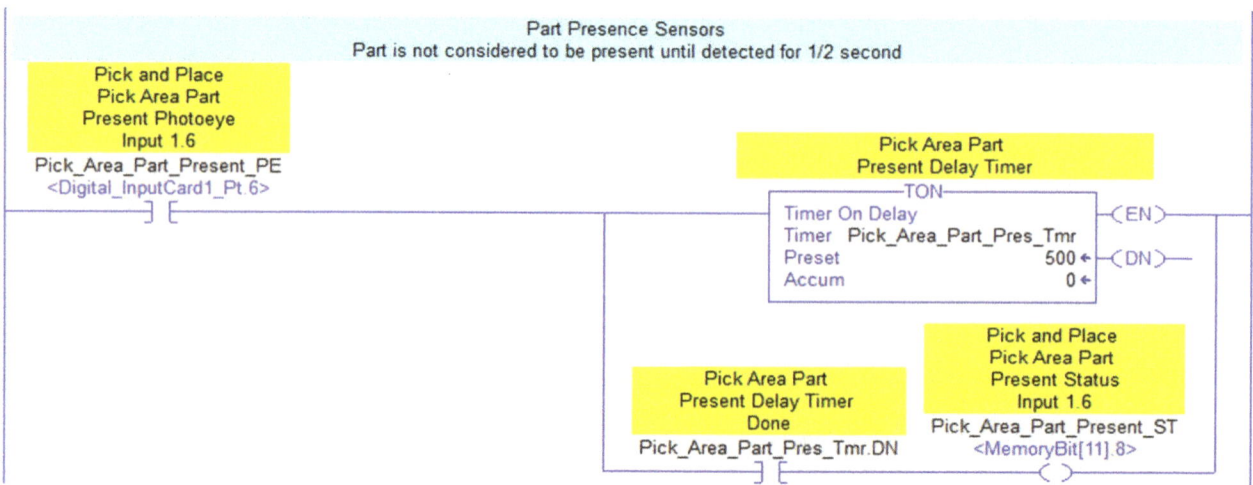

Figure 12 - "Debounce" Logic

This is often used on sensors in conveyor systems where the part is conveyed against a stop before pushing it.

Listing the inputs and statuses in one routine also allows them to easily be located for monitoring or copying into other rungs of logic.

Mapping

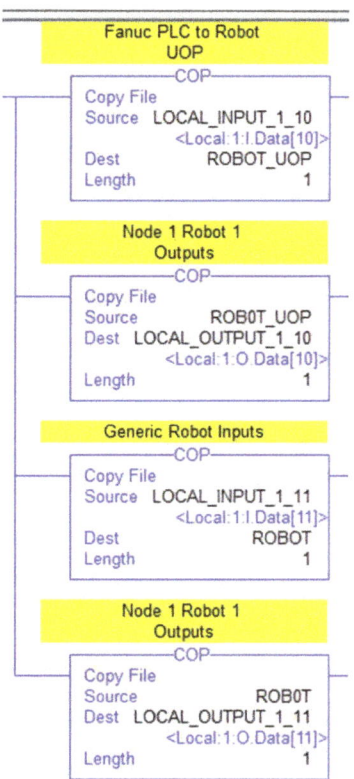

Mapping of I/O points is often done in order to interface a PLC with an external system. Files that represent complex systems, such as robots, variable frequency drives (VFDs) and machine vision are often imported as part of the hardware configuration. In order to add devices to the PLC hardware, a file such as an .eds (Electronic Data Sheet, Allen-Bradley) or .gsd (General Station Description, Siemens) file is imported into the software. This is a text file, which among other things defines the input and output parameters of a device and their data types. While in some PLCs this automatically creates a tag with the I/O element of the device contained in it, sometimes a UDT needs to be created or the individual elements need to be mapped to PLC addresses.

Mapping is also common for interfaces between devices that communicate in different numerical bases, such as Siemens (byte-based) and Allen-Bradley (16 or 32 bit). Modbus communications also often require mapping.

Figure 13 - Robot Word Mapping

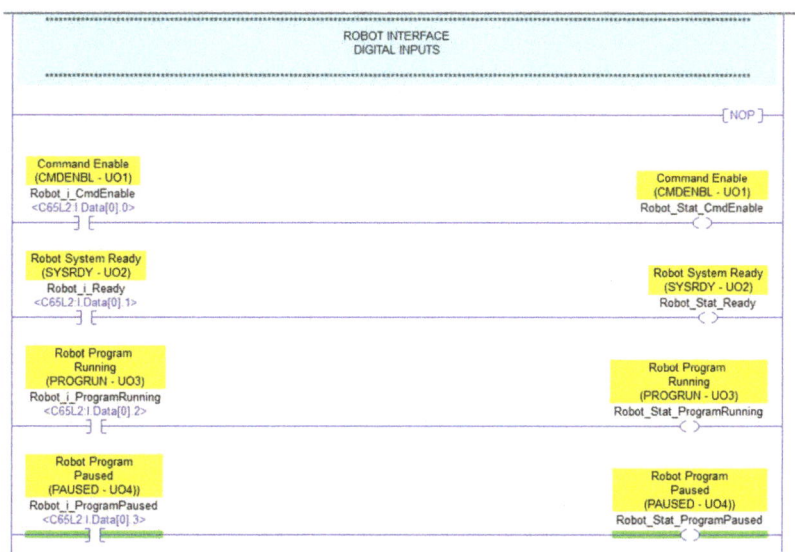

The Output Routine also usually contains mapping logic, as does an HMI Interface routine. HMI Indicator bits are often mapped when an "HMI Word" structure is used for individual indicators. HMIs may also have a list of internal tags that have to be mapped to PLC addresses.

Fault triggers are also often mapped into bits of words for HMI Alarms and Messages.

Figure 14 - Bit Mapping

Output Routines

Because specific coils should only be placed in one location in a program, it is necessary to combine all of the different modes of operating the physical outputs in one location.

It is also important to ensure that outputs can't be energized if they can cause damage to equipment. Permissives are used to prevent outputs from activating if doing so is hazardous.

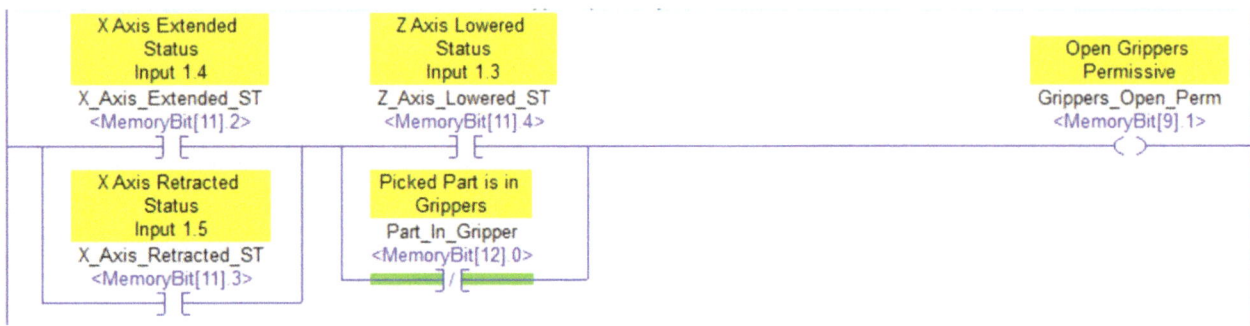

Figure 15 - Permissive Logic

The permissive shown in Figure 15 prevents a gripper from opening unless it is at one end or the other of an X axis actuator. It also ensures that the part can't be dropped by inadvertently opening the gripper when it is raised. Permissives usually only apply in Manual Mode, since automatic functions have code preventing actions from happening in the wrong place.

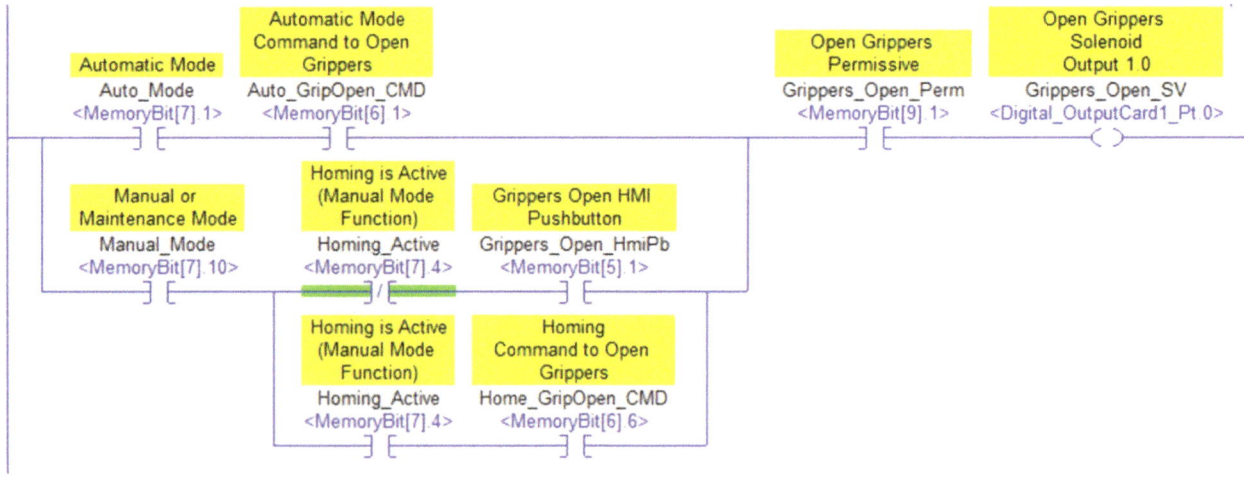

Figure 16 - Output Logic Structure

The output rung itself combines all of the modes' commands into one location along with the permissive. In this case, homing is done in Manual Mode. The Auto Mode Command bit comes from the Auto Sequence, usually in its own routine.

As with the Input Routine, placing all of the output rungs in one routine allows them to be easily located in the program. Both inputs and outputs should be listed in numerical order to make it easier for programmers and maintenance personnel to quickly find them.

Scaling functions for I/O should also be placed in the Input and Output Routines. Output mapping functions and sometimes HMI indicator mapping may also be placed here.

Fault and Alarm Routines

As with the I/O routines, faults and alarms are often placed in their own routine so that they can be easily located. There are many different types of faults and alarms, and also different levels of severity. A fault may shut down the entire machine or system, or disable only part it. Note that in the output structure used in Figure 16, the Auto Mode bit was used rather than the Auto Cycle bit. This means that if AutoCycle is reset, the actuator will still stay energized. The Auto Sequence will not progress however.

Alarms are also displayed on an HMI or SCADA screen. There are utilities in the HMI or SCADA software that allow for archiving of Faults (Alarm History), and acknowledging a fault, which silences alarms without clearing the fault condition. It is important to take this into account when creating the fault logic.

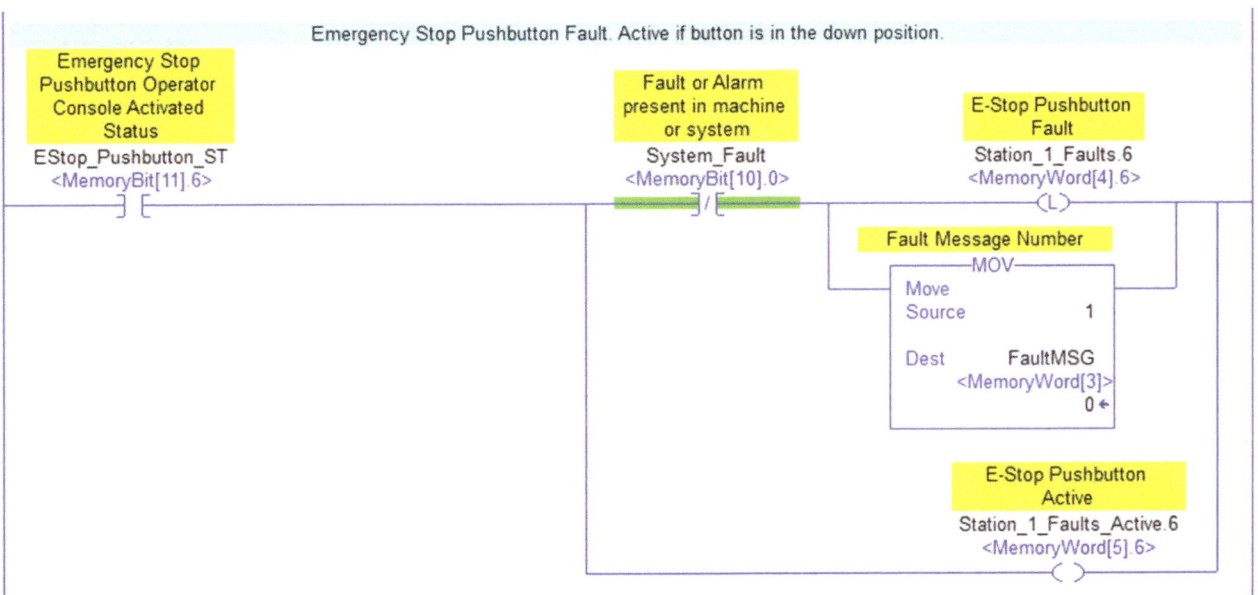

Figure 17 - Typical Fault Logic

The fault logic shown in Figure 17 accomplishes several purposes.

1. It latches or sets a bit in the Station 1 Faults register. Even if the button is pulled back up, the fault stays latched.

2. It moves a number into the Fault message register; this is used to bring up a message in the HMI or SCADA system. The Station 1 Faults Active register is used to indicate whether the cause of the fault has been remedied. If any bit in the Station 1 Faults register is set, then the number will be non-zero; this is used to activate the System Fault bit, which is used in many different places in the program. Notice that in this logic, once the System Fault bit is true, it will prevent any new faults from being latched if it is placed into every fault rung.

It is important to ensure that the original cause of what might be a series of faults is retained. For instance, if an air cylinder jams on an operation, it will latch a fault. If someone opens a door or presses the Emergency Stop to address a safety issue, it will not create a new fault or move a message. If the air cylinder fault is reset after being corrected, a new fault will appear for the E-stop or door.

A common fault is to activate a timer when an output is on and stop it when the corresponding sensor is made.

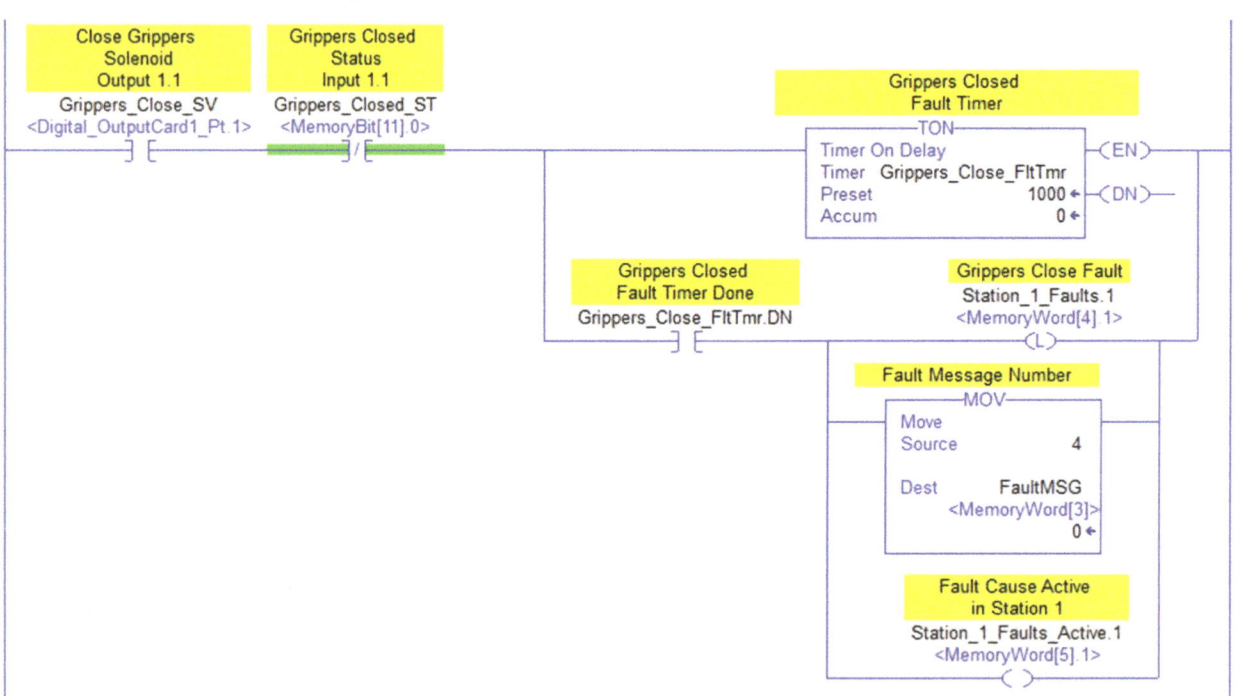

Figure 18 - Gripper Sensor and Actuator Fault

By using the back checked status bit mentioned in the Input Routine section instead of using the actual input address, both the closed and open sensors can be checked for faults.

If a normally closed System Fault bit is placed in series with the timer, as shown in the E-Stop Fault, the Active bit cannot be used to check the current status of the fault, but the timer will start again when the fault is cleared.

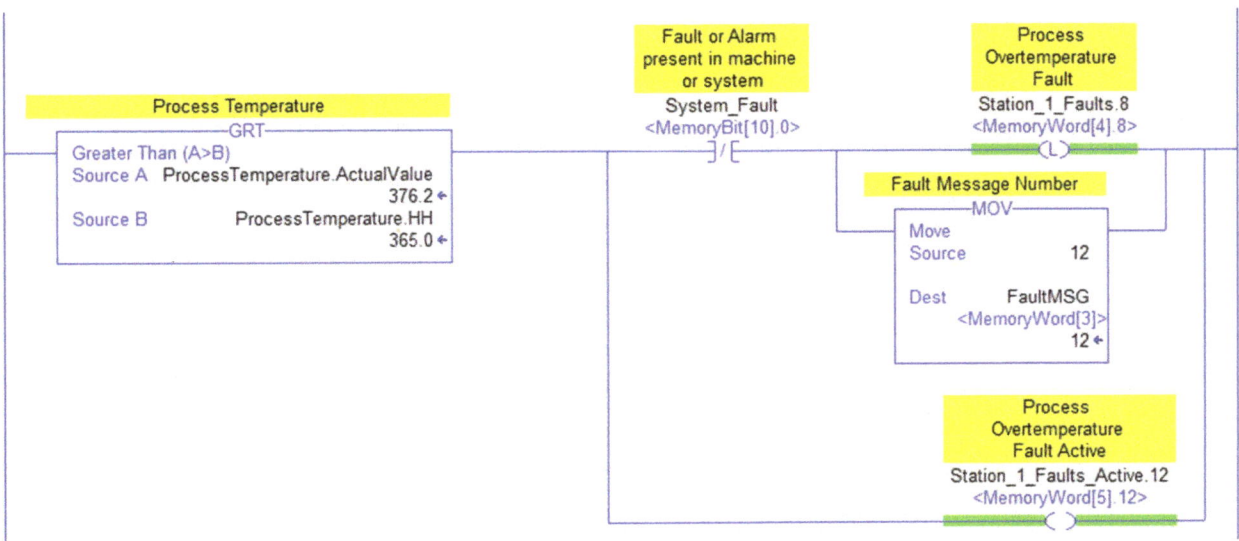

Figure 19 - Process Control Alarm

Process control faults often involve comparison logic that latches a bit if an analog input variable is above a limit. The fault again cannot be reset until the value has dropped below the alarm limit, as shown in Figure 19.

Faults and alarms often have an audible alarm associated with them to alert an operator of the condition, as well as logic that prevents output devices from being energized and possibly damaging equipment. An operator can silence the alarm by acknowledging it on an HMI or SCADA screen.

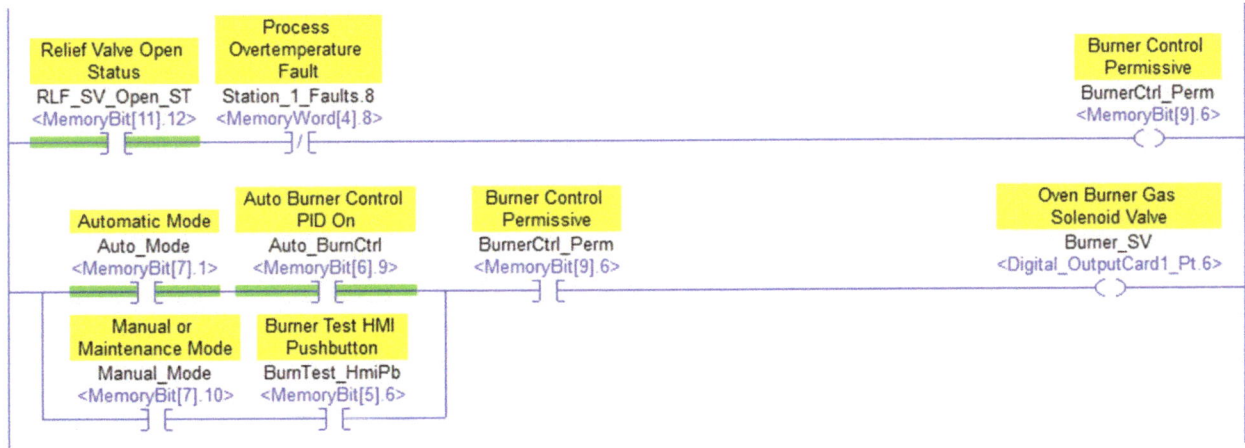

Figure 20 - Burner Control and Permissive

The logic in Figure 20 shows that the valve cannot provide gas to the burner if the fault is present. The Auto control bit is active, but the burner does not operate.

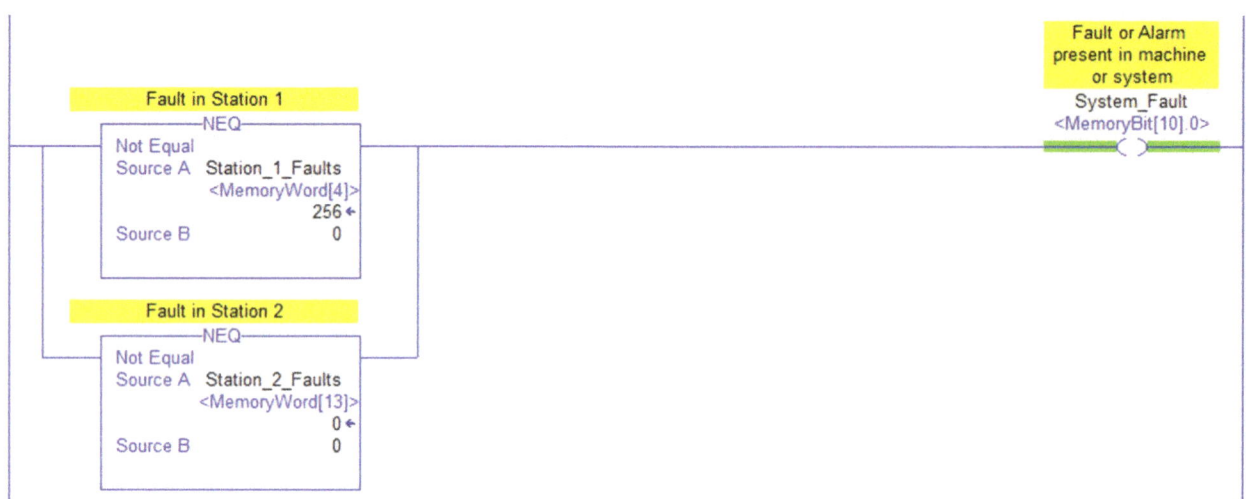

Figure 21- Fault Status Logic

The overall status of faults and alarms can be summarized by comparing the bits as a group to zero as shown in Figure 21. Faults are also often grouped by station so that if it is necessary for one part of a machine system to continue operating, it can be separated from another part.

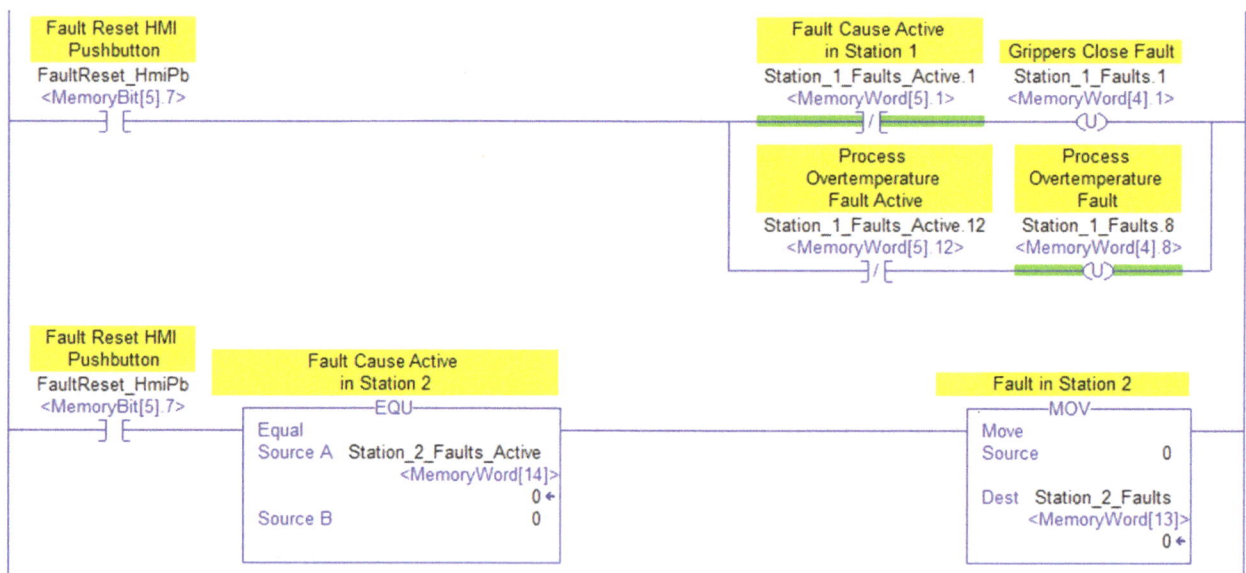

Figure 22 – Fault Reset Methods

Faults and alarms can be reset individually based on their active conditions, or the can be reset as a group based on the station's condition. The programmer must do a detailed analysis of the effects of resetting faults and alarms.

A zero can also be moved into the FaultMSG register when the fault is reset. The message register would then display a message such as "No Faults In System". If there are multiple faults in a system or machine, scrolling message displays or an alarm history screen can be used.

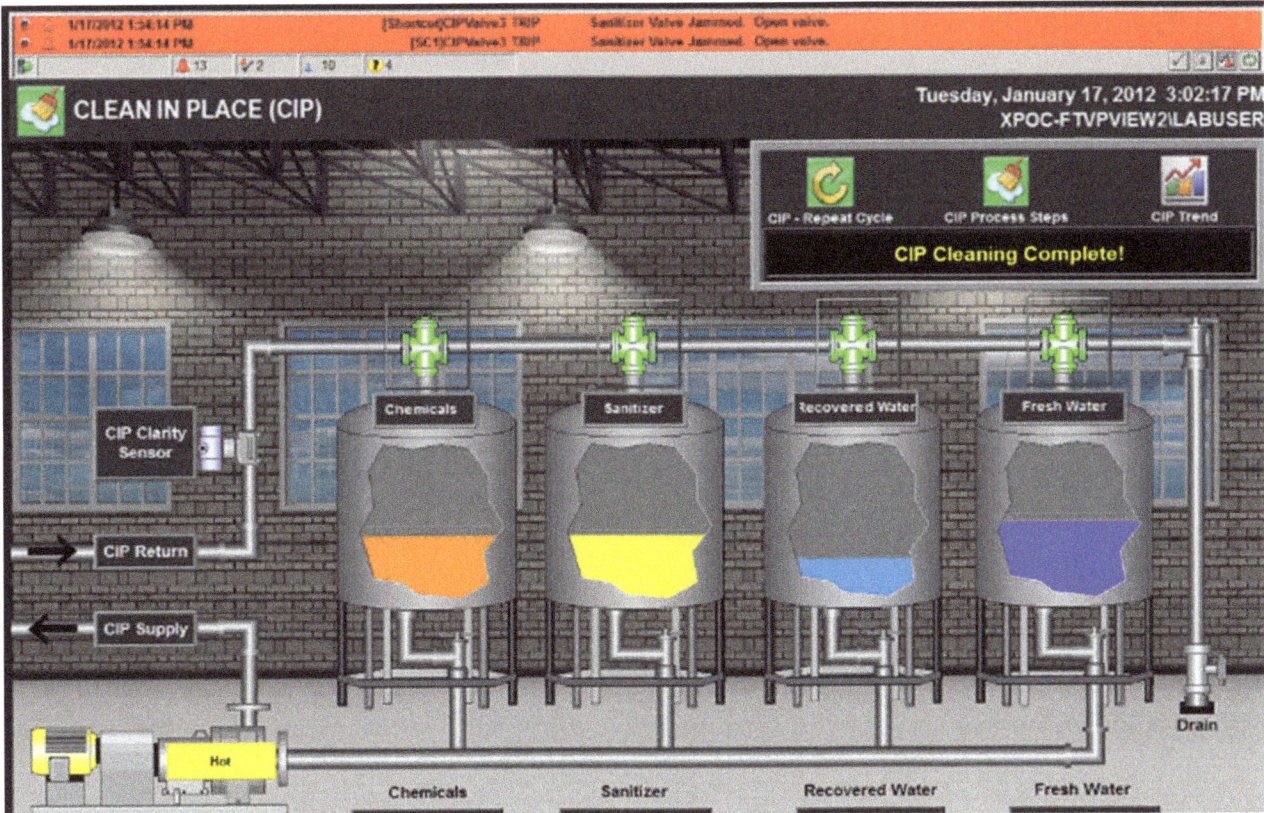

Figure 23 - CIP Screen with Alarm Display

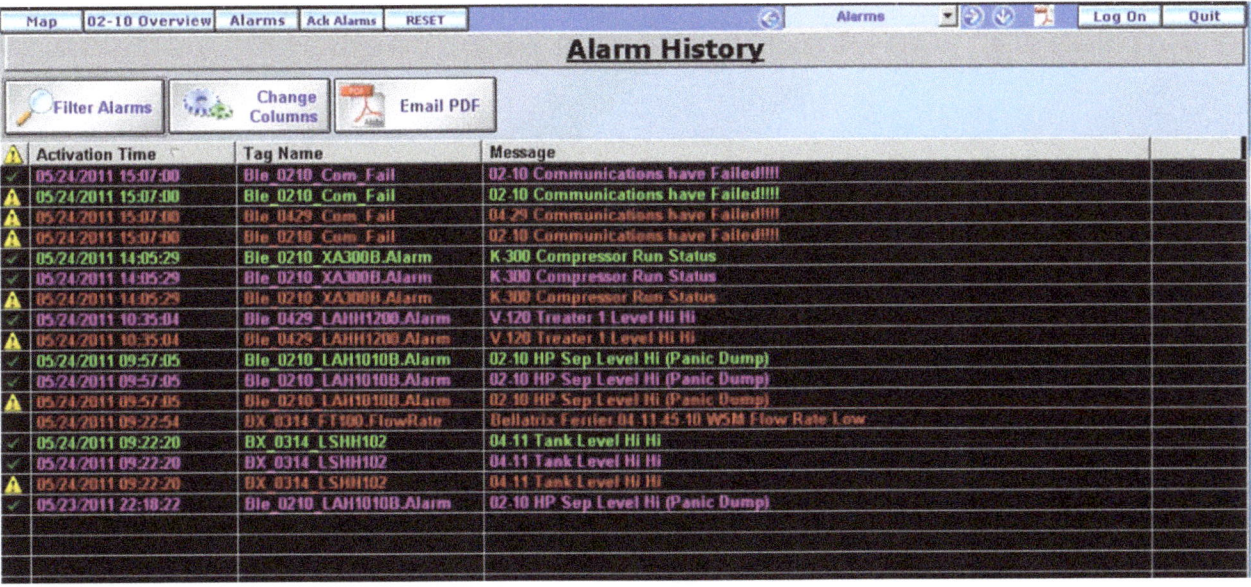

Figure 24 - Alarm History Screen

Many alarm management tools are available in the HMI or SCADA software. The main purpose of the Fault or Alarm Routine is to simply make events available to the operator interface along with their current status.

Auto Sequence Routines

The Auto Sequence is the heart of an automated control system. It requires more planning and analysis than the other types of routines, and there are several methods of programming them.

The first method is often used by beginner programmers, as it is easy to understand and implement.

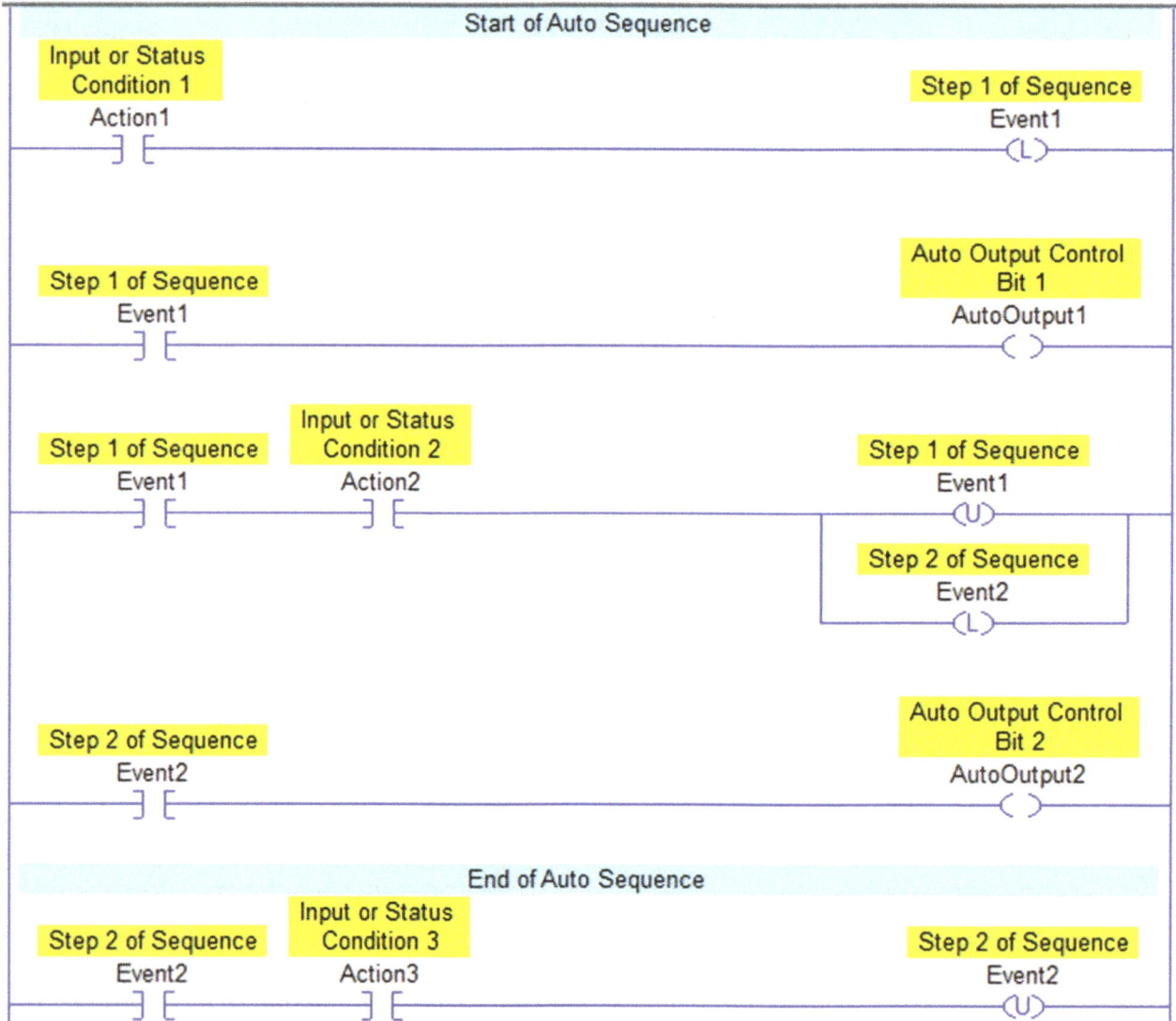

Figure 25 - Auto Bit-Latching Sequence Concept

In Figure 25, a bit is set or latched based on a sensor or other condition. While the bit is on, another bit, the Auto Output bit, is used to energize an output while in Autocycle or Auto Mode. Once another input or status condition is active, often due to the output being energized in the previous step or event, the step is unlatched or reset and the next step is latched on.

The Auto Output bits in this example are used in the position of the CMD or command bits in the section on output routines.

Programmers often latch and unlatch bits to control outputs in Auto Mode without realizing that they are actually writing a sequence. This can become very confusing if a list of events or flow chart is not used to plan the sequence ahead of time. Note that in this example, each output control is energized during only one step; this can easily be modified by adding conditions.

Bit Sequence: The previous example is a simplified version of what is sometimes called a "bit sequence". If the bits of a word or double word are used, the sequence can be easier to control and follow.

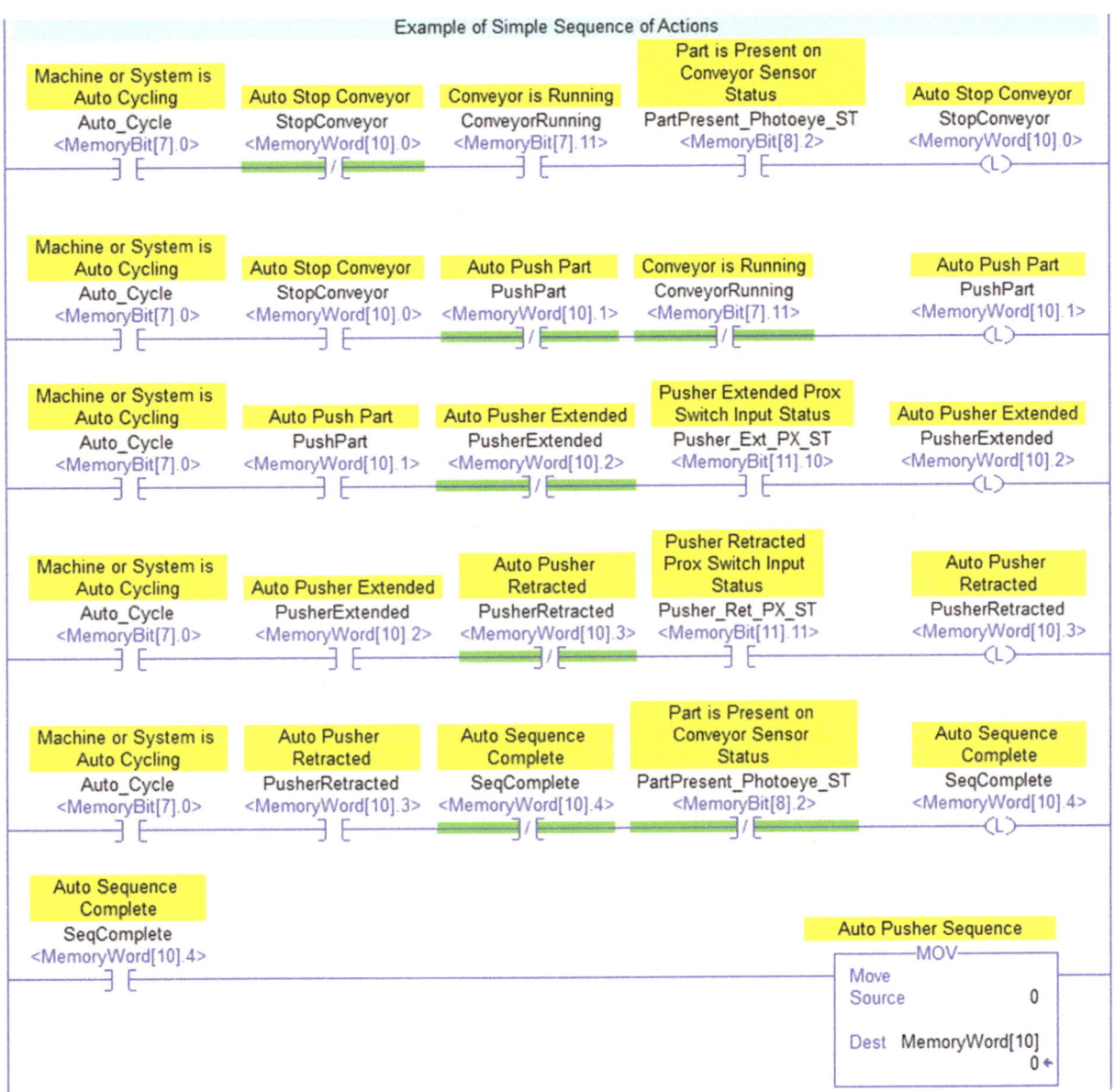

Figure 26 - Example of a Bit Sequence

In the example in Figure 26, a pneumatic actuator is used to push a part off of a conveyor when detected by a photoeye. Notice that the sequence bits are all in the same memory word, making it possible to unlatch all of the bits with the MOVE command at the end. The previous bits are not unlatched at each step as in Figure 25; instead, they are set or latched sequentially and then cleared at the end of the sequence.

Using the *Auto_Cycle* bit in every rung ensures that the sequence will not proceed if a fault unlatches the bit. By placing the normally closed contact of the next step bit the sequence will not evaluate the logic of previous steps. This makes it operate similarly to the Word Sequence described next.

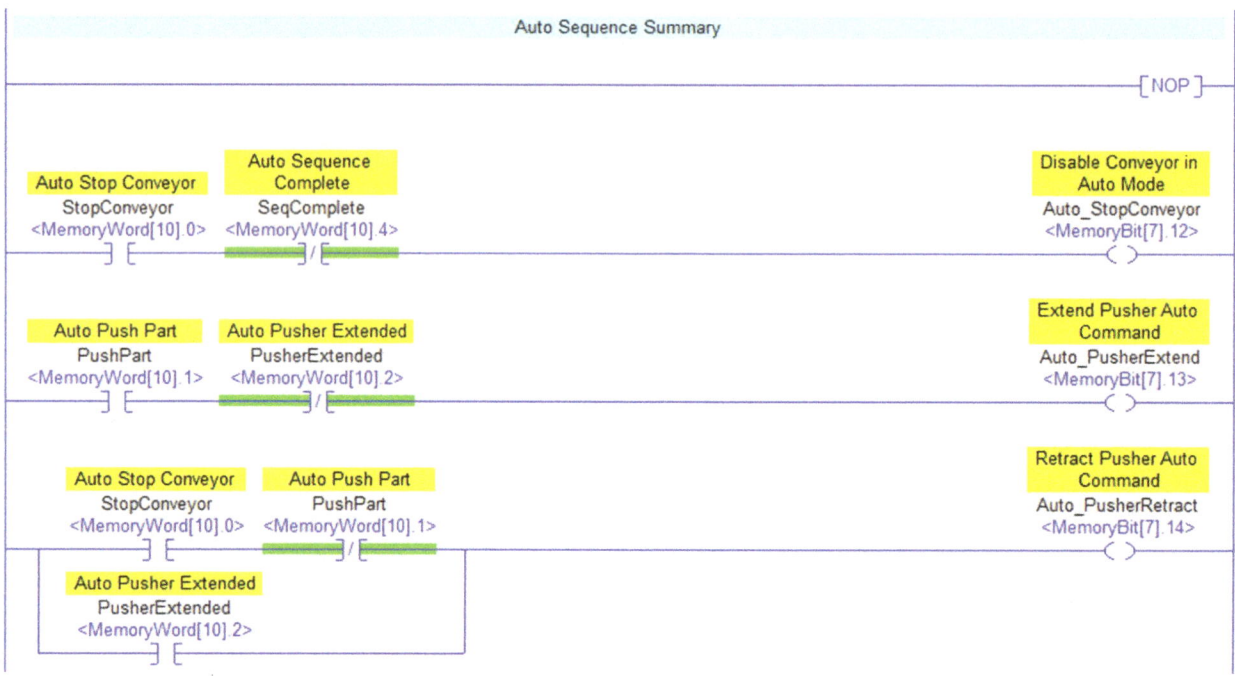

Figure 27 - Bit Sequence Summary

The control bits used in the Output Routine are often summarized after the sequence itself. This can make troubleshooting the sequence easier and improves "readability" by not separating the steps.

Bit sequences can still be somewhat confusing because it is possible to be in more than one step or state at a time. This can be both an advantage and a disadvantage; it is possible to reset bits in part of the sequence while leaving other bits latched. It is important to fully document this type of sequence completely, explaining what the steps are doing.

Word Sequence: Another common method of writing sequences is to move values into an integer register.

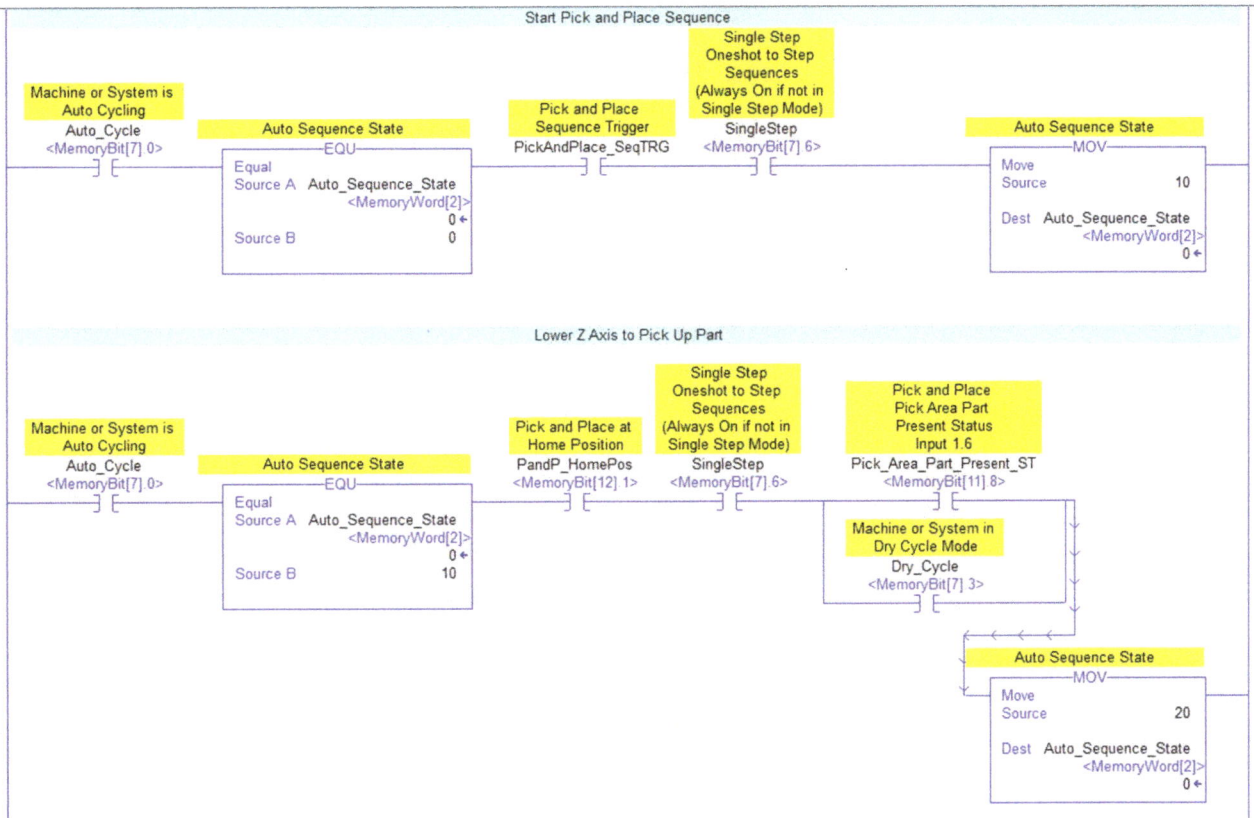

Figure 28 - Steps 10 and 20 of Auto Sequence

The Pick and Place logic in Figures 28-33 shows how the sequence progresses. Rather than latching bits, numbers are moved into a register defining the current state.

This sequence also shows the use of Single Step and Dry Cycle modes. Single Step allows an operator to step the sequence one stage at a time, while Dry Cycle lets the sequence run without parts being present.

Once again, if the machine faults, the system will drop out of AutoCycle and will not advance. It will maintain its current state unless the fault is cleared and placed back into AutoCycle or it is homed.

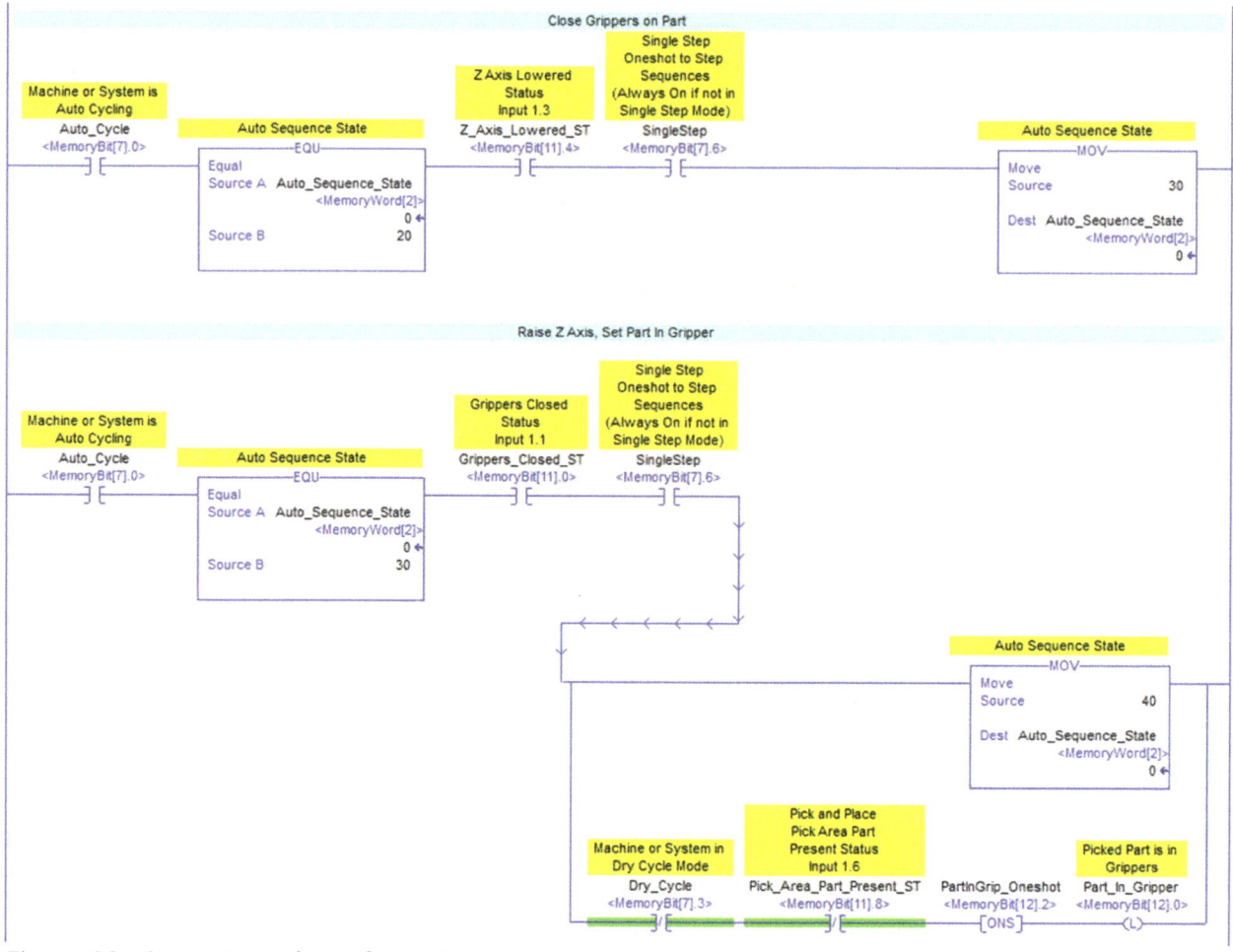

Figure 29 - Steps 30 and 40 of Auto Sequence

Note that during Step 30, a bit is set, verifying that the part was removed and is now held in the grippers. This can be important when determining whether it is possible to open the grippers during a homing routine or in part tracking.

Gripper sensors are also sometimes positioned such that if a part is present, the grippers don't close all the way and the sensor stays on. If the part is not present, the grippers pass the sensor; because of this, a "debounce" timer is often used for gripper sensors.

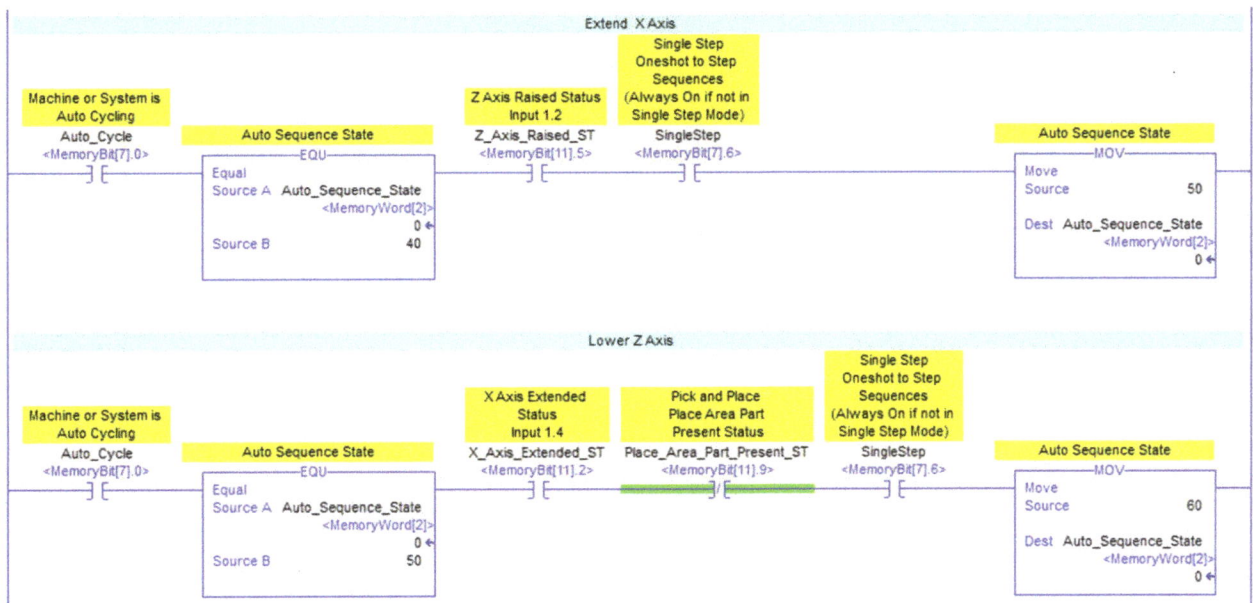

Figure 30 - Steps 50 and 60 of Auto Sequence

Step 50 ensures that a part is not already present in the location where the part is to be placed. This acts as a permissive for setting the part down. If there is already a part there, the sequence will wait to advance to Step 60 until it is removed. This is done by an operation outside of this sequence; multiple sequences are often written to operate machine actuators that may have to interface with each other.

It is unnecessary to use the Dry Cycle mode bit to bypass the Part Present sensor, since the part should not be there.

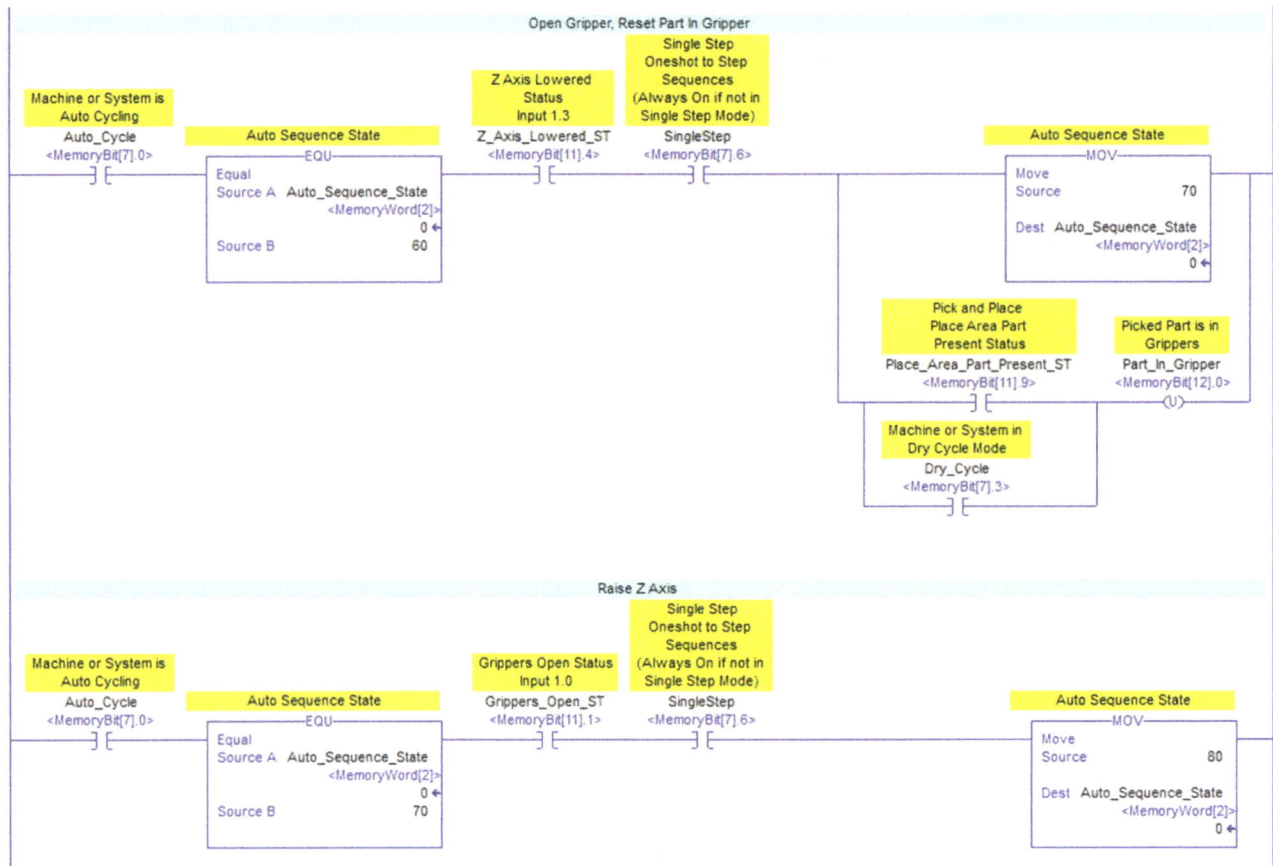

Figure 31 - Steps 70 and 80 of Auto Sequence

As with Step 30, a memory bit is used to determine whether a part is still in the grippers. The bit is unlatched in Step 60 once the part is set down.

In Dry Cycle Mode, the Part Present sensor is bypassed and ignored.

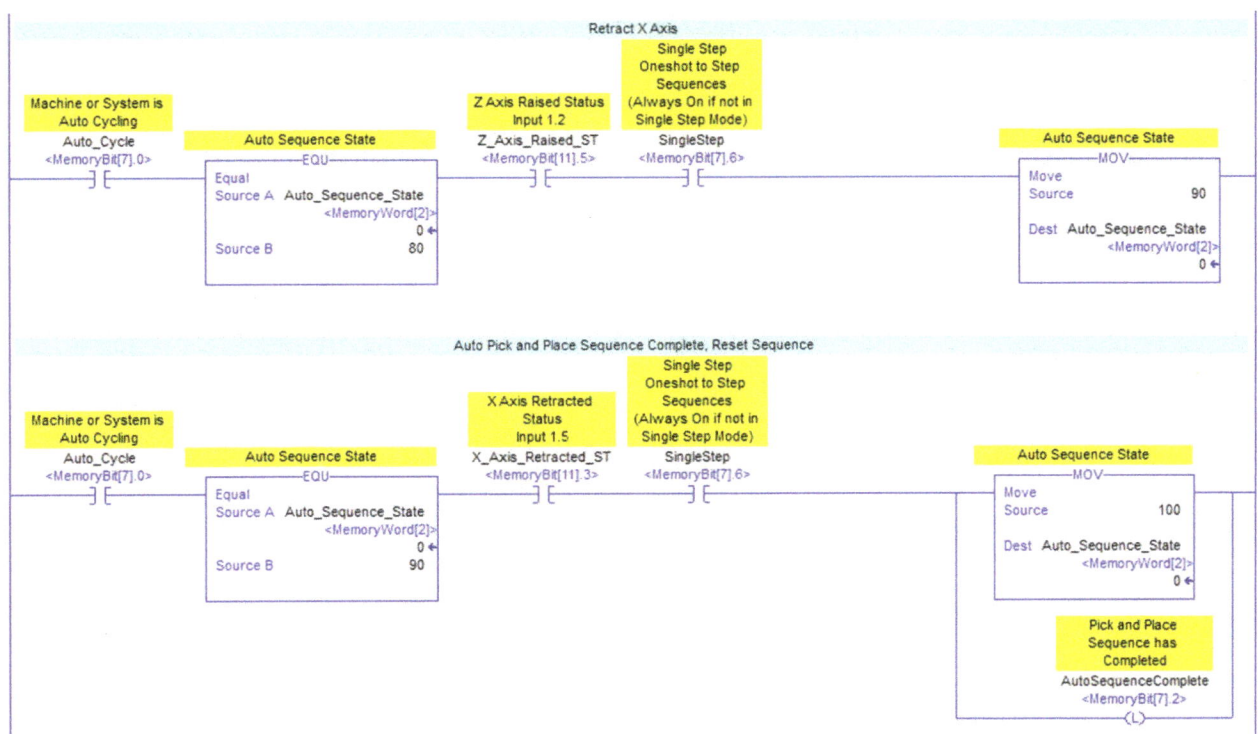

Figure 32 - Steps 90 and 100 of Auto Sequence

The sequence ends with the Pick and Place mechanism back at the home position. A Sequence Complete bit is latched and can serve as a trigger for other operations or sequences.

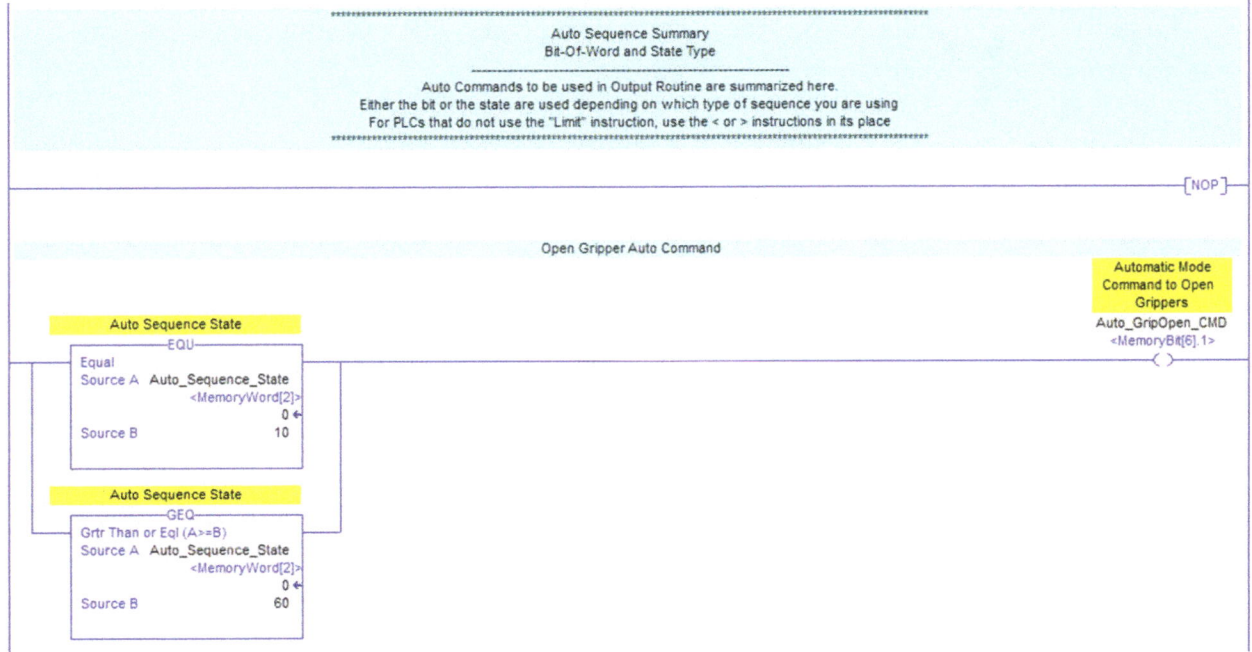

Figure 33 - Example Summary Bit for Word Sequence

As with the bit sequence, summary bits are often listed after the sequence and are then used to drive outputs in the Output Routine.

Word sequences are different than bit sequences in several ways. In a word sequence, there is only one operational state active at any time. By sequencing in steps of ten or even one hundred, spare states can be built into the sequence.

In bit sequences, the states or steps can be individually named, but a machine can have any number of bits latched or unlatched at any given time. This can make them more flexible, but harder to follow the logic and debug.

Similar to a word sequence is a **State Machine**. Modes and sequence steps are sometimes combined to describe not only the step of the operation, but also conditions such as Starting, Started, Aborting, Aborted, Stopping, Stopped and Faulted. Ranges of numbers may be assigned to these states such as 5000-6000 for Auto Cycle states, 9000 for Faults, and <1000 for startup and initialization. There are several templates used by large manufacturers that use this technique. It is also a common method for robot control.

Homing Routines

Homing is used to ensure that actuators and machine conditions are brought to a known and appropriate starting position. Often when machinery is in AutoCycle and a fault condition occurs, operators may need to move actuators manually or remove in-process parts. This means that the Auto Sequence is no longer in the correct condition to continue. In this case there are several things that must be considered:

1. The out of position or missing part condition must be detected. For actuators, this can be done by determining the correct bit pattern for a given sequence step or state and comparing it with the actual input state, as shown in Figure 34. Another method is to check whether the machine was placed into manual mode or if manual operation using pushbuttons occurred while the Auto Sequence was active.

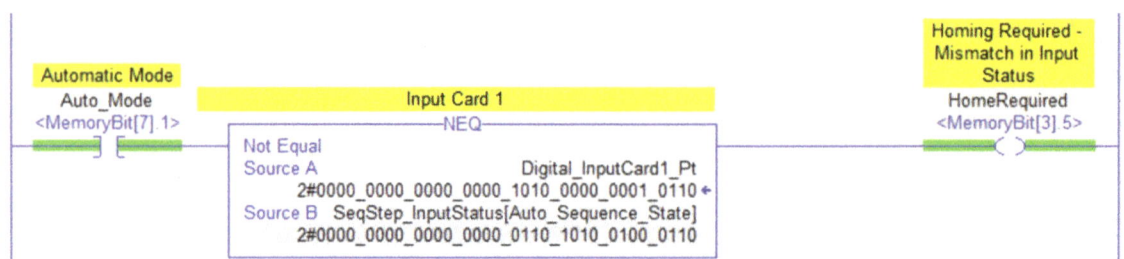

Figure 34 - Mismatch in Required Sensor Position

2. The Auto Sequence affected must be aborted or brought back to the correct state, usually zero.

3. The programmer needs to determine whether the homing operation will occur in Auto Mode, Manual Mode or both, or if an entirely new Homing Mode needs to be defined.

4. A Homing Sequence needs to be written. This uses the same technique previously described in the Auto Sequence section. Actuators need to be moved in order, returning them to their appropriate positions so as not to damage equipment. For example, in the previous Pick and Place sequence, the Z axis must be raised before the X axis is returned. If there is a part in the grippers, a decision must be made as to whether the operator must remove the part manually, or if the part can safely be placed in either the pick location or the place location.

Homing Sequences are every bit as complex and necessary as Auto Sequences are. Down time in machinery is often traced to operators not being able to diagnose why the system can't continue, which is also related to insufficient diagnostic messaging.

Recipes

A recipe is not necessarily a list of ingredients in a process like most people are familiar with in the kitchen. It also contains machine states, timer setpoints, bits that determine whether a part needs to be worked on in a station, and, of course, process values.

Name	Type	Description
⊟ Recipes	udt_Recipe[8]	Recipes
⊞ Recipes[0]	udt_Recipe	Recipes Recipe File for Part
⊟ Recipes[1]	udt_Recipe	Recipes Recipe File for Part
⊞ Recipes[1].Name	STRING	Recipe File for Part Recipe Name
⊞ Recipes[1].C1_PartNbr	DINT	Recipe File for Part Assembly Part Number
⊞ Recipes[1].Component	SINT	Recipe File for Part Component Bits
⊞ Recipes[1].Weight	REAL[8]	Recipe File for Part Component Weights
⊞ Recipes[1].C2_PartNbr	DINT	Recipe File for Part Component 2 Part Number
Recipes[1].C3_Torque	REAL	Recipe File for Part Component 3 Torque
⊞ Recipes[1].C6_Color	INT	Recipe File for Part Component 6 Color
⊞ Recipes[2]	udt_Recipe	Recipes Recipe File for Part
⊞ Recipes[3]	udt_Recipe	Recipes Recipe File for Part
⊞ Recipes[4]	udt_Recipe	Recipes Recipe File for Part
⊞ Recipes[5]	udt_Recipe	Recipes Recipe File for Part

Figure 35 - Part Recipe Array using UDT

Figure 35 shows an array of 8 recipes using a UDT (User Defined Data Type) that describes a part. This part is also used in the part tracking routine.

The recipe information includes the name of the recipe itself, which in this case is also the part number for the assembled part in STRING format. It also includes decimal part numbers for sub-components in DINT form, a target torque value as a REAL number, and an array of REAL numbers corresponding to the approximate target weight at several assembly stations in the production line. A part color is represented by an integer, and a byte or SINT is used to track component presence in bit form.

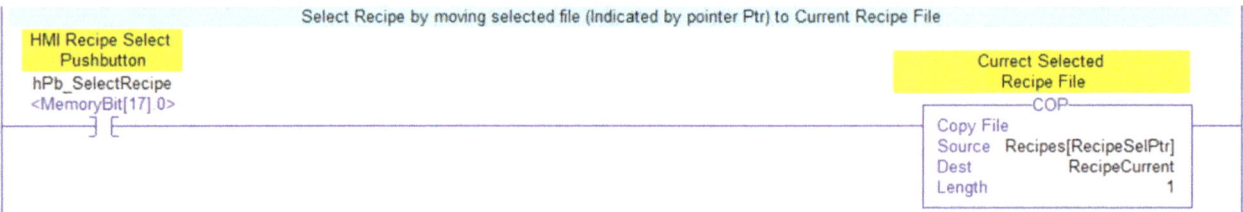

Figure 36 - Recipe Select Logic

The logic in Figure 36 uses a pointer called *RecipeSelPtr* to indicate which Recipe file is moved into *RecipeCurrent*, which is then used in production logic in other routines. This is an example of indirect addressing. When the HMI button is pressed, the selected recipe is moved. The selection is an integer numeric entry field on the HMI screen that identifies the recipe by number. A corresponding ASCII field on the HMI displays the *Recipe[ ].Name* field to verify the recipe selection.

If the recipe needs to be changed or edited from the HMI, the logic in Figure 37 can be used.

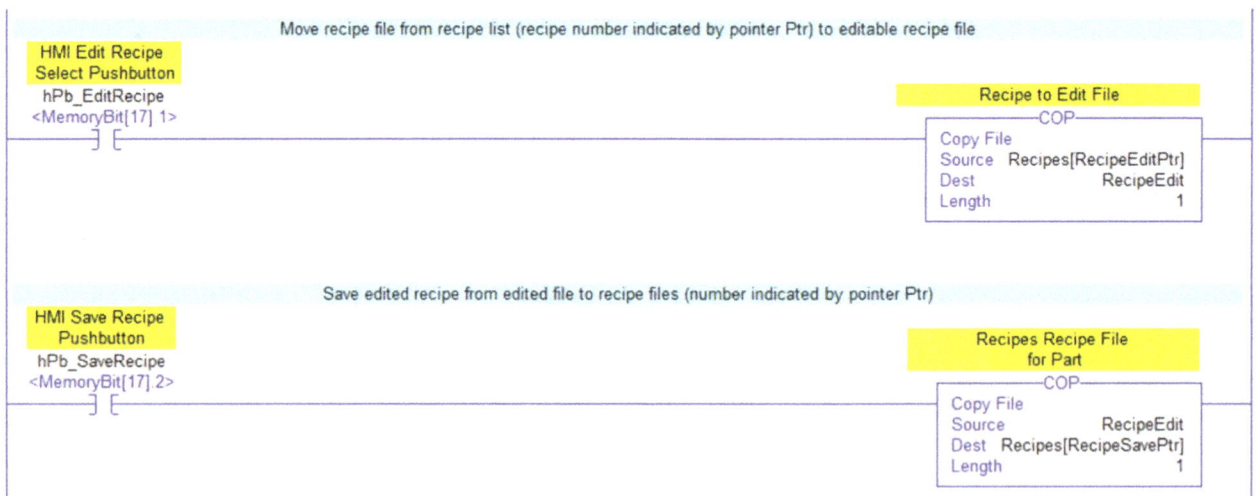

Figure 37 - Recipe Edit and Save Logic

The *RecipeEdit* tag is another instance of the *Recipe* UDT. Entry fields on the HMI correspond to each of the items in the recipe. After changes have been made, the Save Recipe pushbutton moves the edited recipe into the recipe number designated by the Recipe Save pointer (*RecipeSavePtr*)

In most PLC platforms, entering a number outside of the limits of the array dimensions into the pointer will fault the processor. The logic in Figure 38 prevents this from happening.

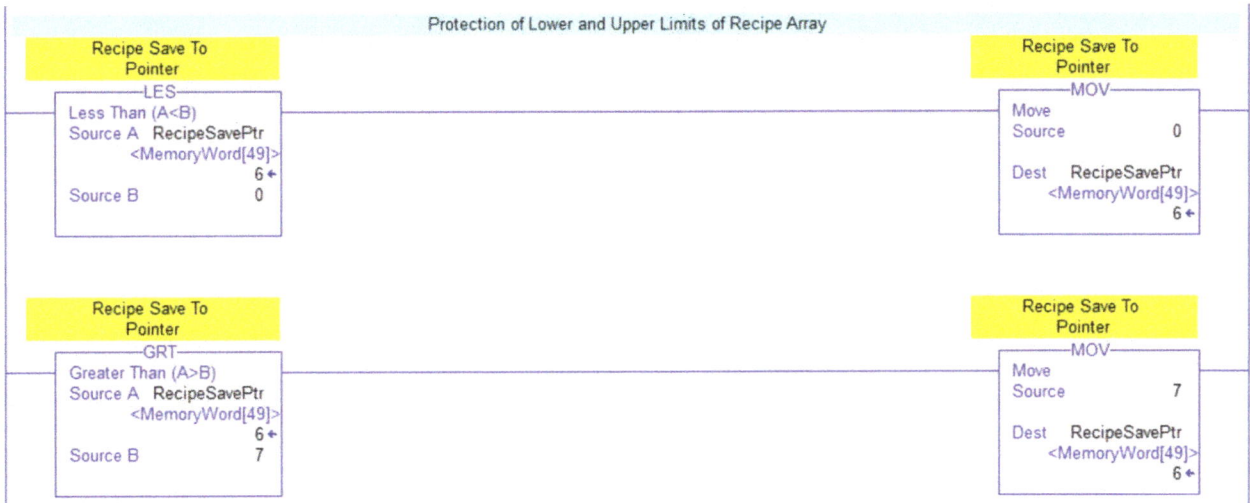

Figure 38 - Ensuring that pointer is within limits

Many HMI and SCADA platforms have recipe management utilities built into their software. Recipe parameters and files also are often comma delimited .csv files that can be edited with external packages such as Microsoft Excel. This simple method of recipe management works well for simpler HMIs. For PLCs that don't have UDT or Array capability, moves need to be made on an individual register basis.

Part Tracking

For machines that process discrete parts in assembly operations, it is important to ensure that data associated with the part is carefully documented and that the data follows the part through the system. It is also not always possible to use bar codes or other identification systems to read the part at every station. Because of this, various techniques can be used to move data as the part proceeds from station to station.

⊟ StationPart	udt_Part[10]	Part by Station Array
⊞ StationPart[0]	udt_Part	Load Station 1
⊞ StationPart[1]	udt_Part	Assembly Station 2
⊟ StationPart[2]	udt_Part	Torque Station 3
StationPart[2].Weight	REAL	Torque Station 3 Weight of Component
⊞ StationPart[2].Component	SINT	Torque Station 3 Component Installed Bits
⊞ StationPart[2].PartNbr	DINT	Torque Station 3 Part Number
⊞ StationPart[2].PartCode	STRING	Torque Station 3 Serial Number from Bar Code
⊞ StationPart[2].FailCode	SINT	Torque Station 3 Failure Code Byte
⊞ StationPart[2].Station	DINT	Torque Station 3 Station Complete
StationPart[2].OK	BOOL	Torque Station 3 Part OK Inspection Pass
StationPart[2].NG	BOOL	Torque Station 3 Part NG Inspection Fail
StationPart[2].InProc	BOOL	Torque Station 3 Part In Process
StationPart[2].Complete	BOOL	Torque Station 3 Part Complete
⊞ StationPart[3]	udt_Part	Torque Test Station 4
⊞ StationPart[4]	udt_Part	Reject Station 5
⊞ StationPart[5]	udt_Part	Assembly Station 6
⊞ StationPart[6]	udt_Part	Gasket Station 7
⊞ StationPart[7]	udt_Part	Inspect Station 8
⊞ StationPart[8]	udt_Part	Reject Station 9
⊞ StationPart[9]	udt_Part	Unload Station 10

Figure 39 - Part Information Array using UDT

The information in the array not only needs to be moved from station to station as the part moves through the assembly process, it also needs to be modified at each station. Figure 39 shows the structure of the part data. The Weight variable is updated at each station as the pallet is weighed; this information is used to help ensure that the proper components have been added. As components are added, the Component bits are set at the corresponding station. The *PartCode* variable is obtained from a bar code reader at the first station, which corresponds with the part number in the recipe. The *OK* and *NG* bits are updated at each inspection station; if the NG bit is set, then the part will be rejected at Stations 5 and 9. The cause of the failure can be encoded into the *FailCode* byte (SINT).

The InProc and Complete elements of the UDT are used to determine when the movement of parts will occur. For this example, a simple 10-station assembly line is used to illustrate some of the concepts in tracking parts. Station names and functions are listed in Figure 39.

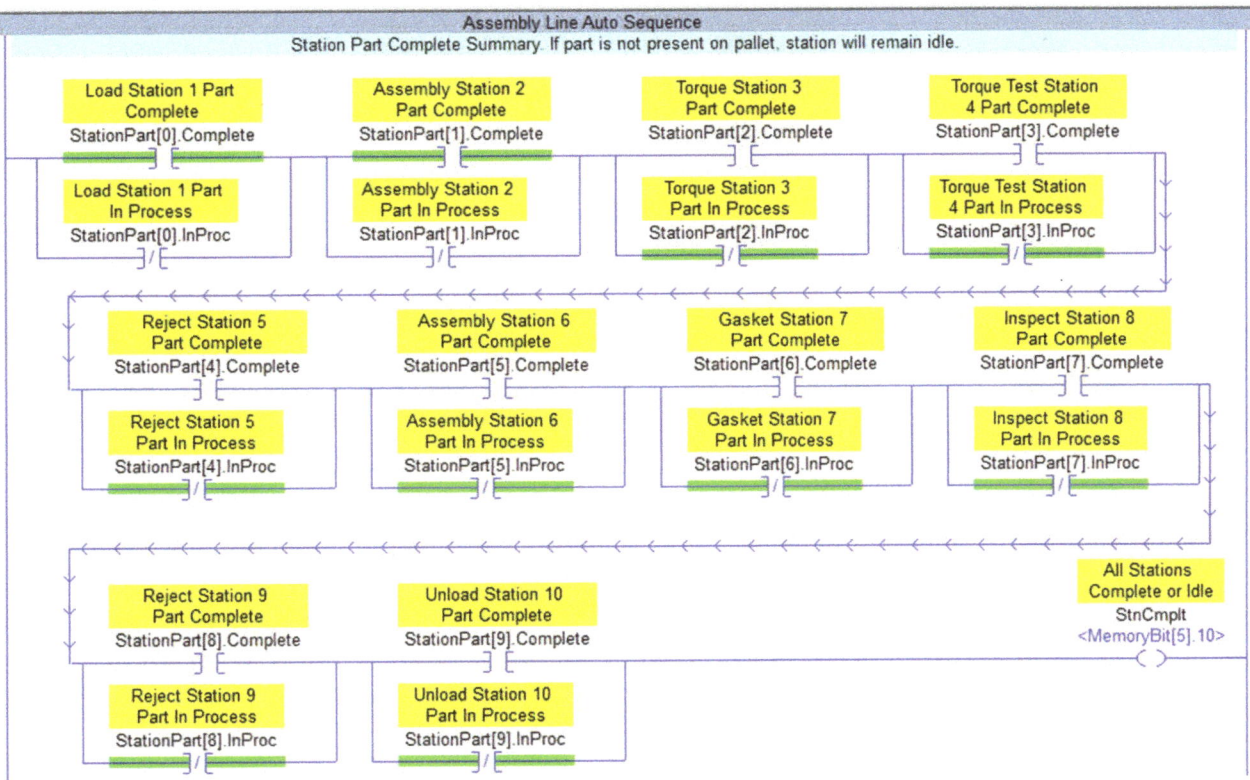

Figure 40 - Station Status Summary using Part Status elements

As each part enters a station, the *InProc* bit is set in the corresponding file, indicating whether a part is present on the pallet. A station that is not In Process is considered idle. When the operation at the station is finished, the *Complete* bit is set indicating that it is okay to index the line.

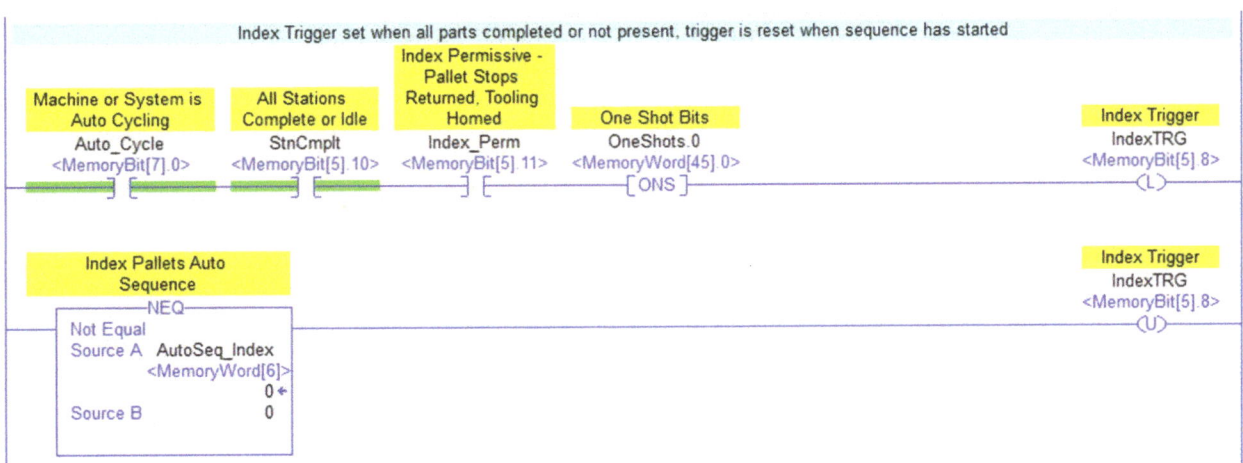

Figure 41 - Index Trigger logic

The Index Trigger is used to start the pallet movement sequence when all products' status are correct and when mechanical tooling is in the correct position (*Index_Perm*).

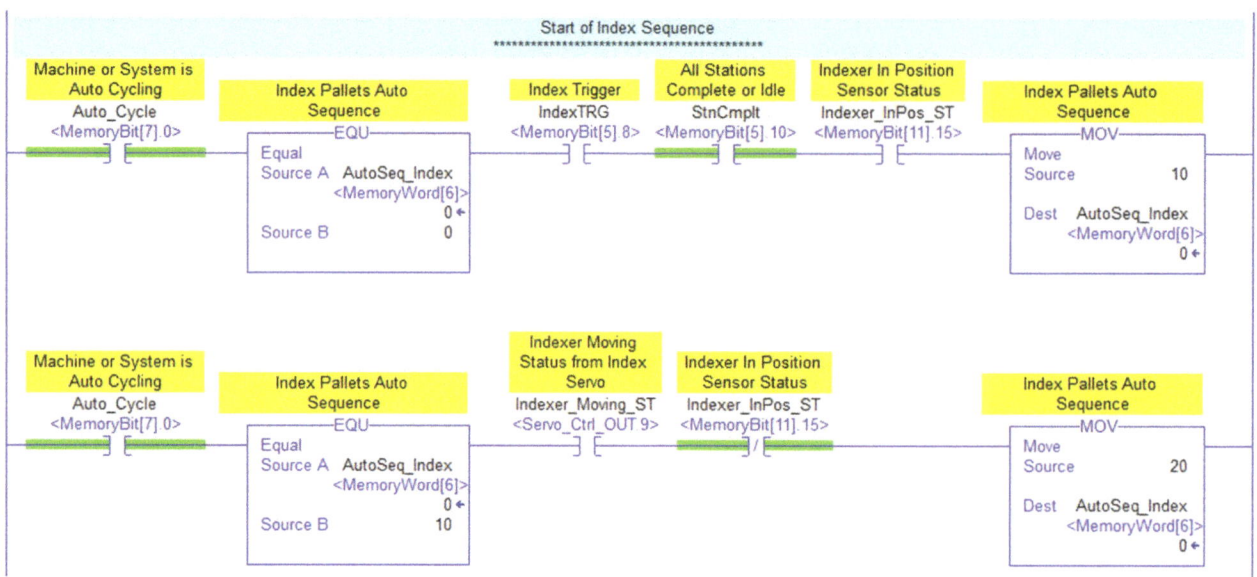

Figure 42 – Pallet Index Sequence Steps 10 and 20

The Index Sequence itself is similar to other Word Sequences discussed in this manual. Step 10 is used to tell a servo type pallet indexer to begin movement. When motion has started, the sequence moves to Step 20, removing the signal to the servo, and when movement is completed, the sequence moves to Step 30. At this time, an *IndexComplete* signal is also set. This bit is used to move the part records.

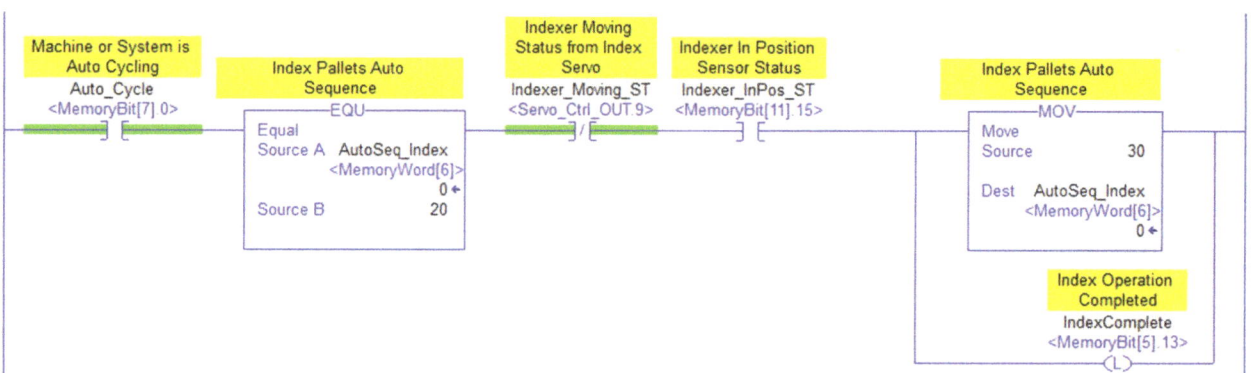

Figure 43 - Index Sequence Step 30 and Index Complete signal

The end of movement also resets all of the Part Complete bits in the part records, as shown in Figure 44. When all of the part records have been moved in the part tracking routine, the *IndexComplete* bit is reset. This bit serves as a handshake between the servo indexer's movement and the part tracking.

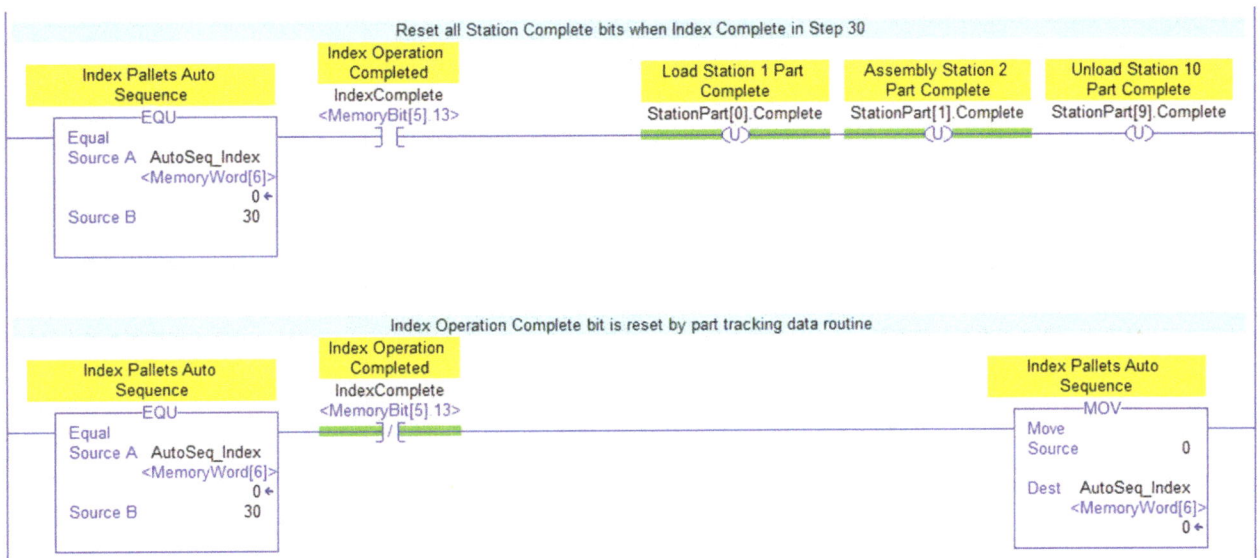

Figure 44 - Index Sequence complete and interface with Part Tracking routine

As mentioned in the section on Recipes and data movement, not every PLC platform supports arrays or UDTs. This can make data movement much more labor intensive, since numbers and other data needs to be addressed one element at a time.

Some PLC brands also only allow data files to be moved one file at a time, while others allow "Block Moves".

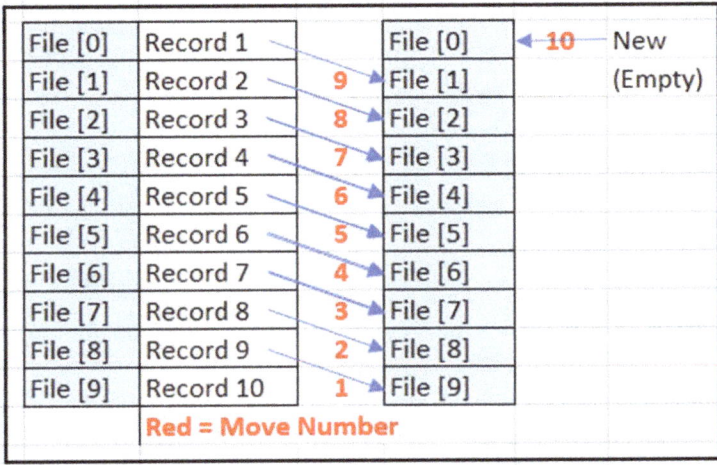

Cascading File Moves: If files must be moved one at a time, it is necessary to move the last part records first so that the newest data is not overwritten. Figure 43 shows a 10-element array, where the 9th record is moved into the 10th spot, the 8th into the 9th spot, and so on. In the final or 10th sequential move, a new empty record is moved into location 1 and is ready to be updated.

Figure 45 - Individual File Moves

The logic for this is sometimes called "cascading" logic, because records cascade from one location to the next.

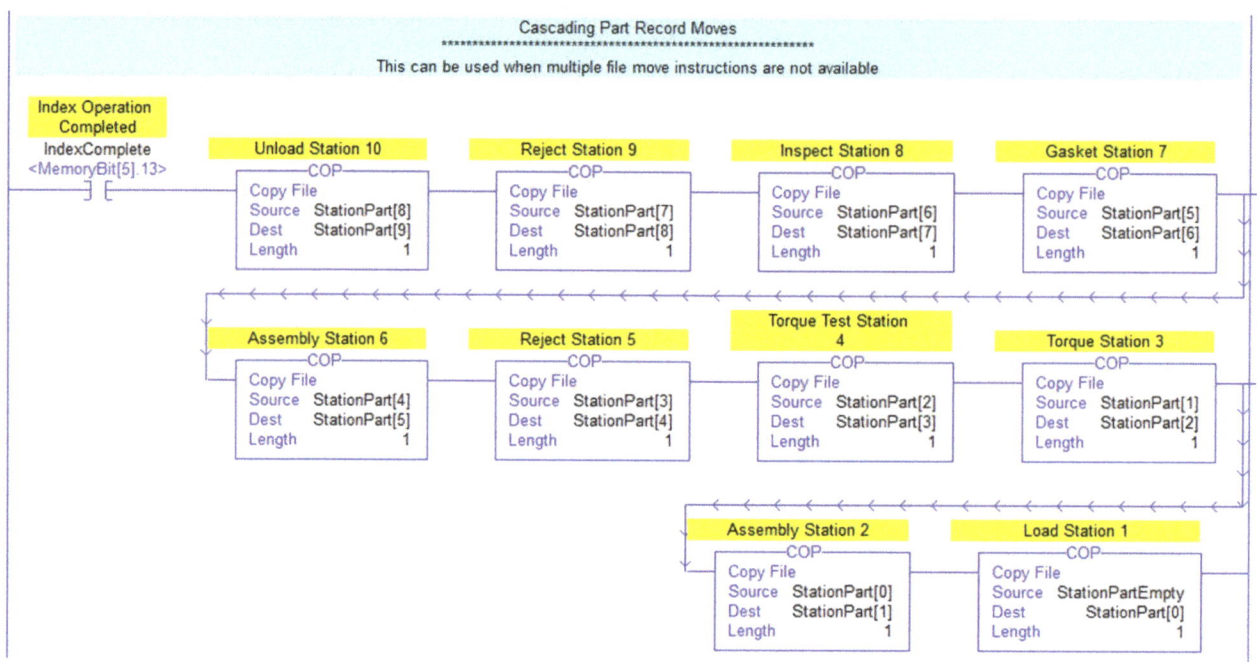

Figure 46 - Cascading Sequential Record Moves

Figure 46 shows individual sequential Copy instructions being used to move data. In the ControlLogix PLC platform, these output type COP instructions can be placed in series. This logic is triggered by the *IndexComplete* signal discussed previously.

Block File Moves: Multiple files can also be moved with the COP instruction. This requires that the data is in array form and also requires an additional "dummy" array the same size as the *StationPart* array.

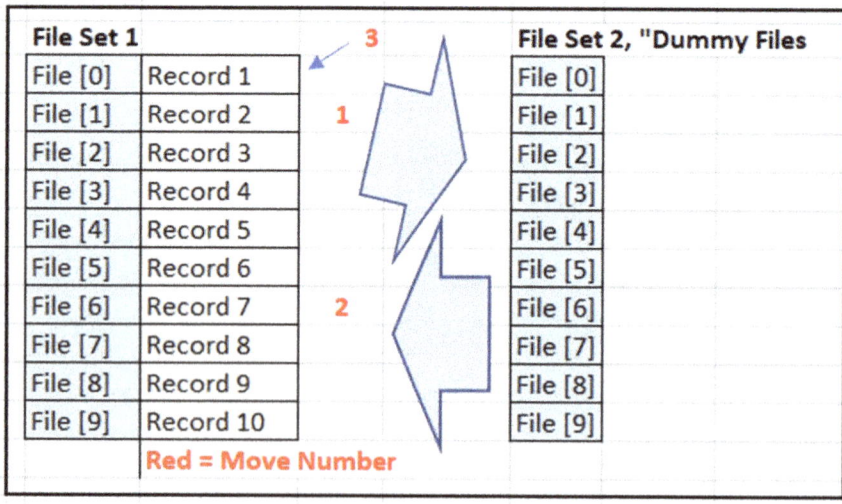

Figure 47 - Block Movement of Data

The first operation moves records 1-9 into positions 2-10 in the dummy array. This is because files cannot overwrite themselves on block type moves.

The second operation moves the records in positions 2-10 of the dummy array back into the original array, also into positions 2-10.

The third operation moves the new empty file into position 1 of the record; it is ready to be updated.

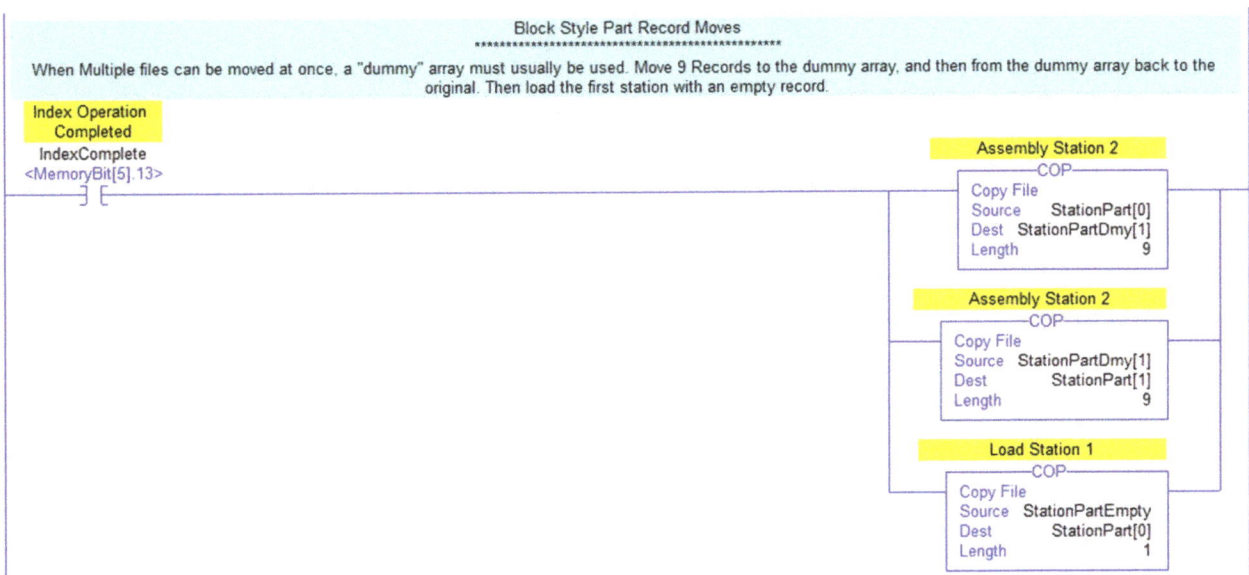

Figure 48 - Block Data Move Logic

Figure 48 shows the COP instruction being used to move multiple files at once. A Synchronous Copy (CPS) instruction also exists for the Allen-Bradley ControlLogix PLC that ensures that data can't change while the data is being moved. This Block Move method is preferred if a large number of files or array elements need to be handled, though it does require more memory for the additional "dummy" array.

Tips and Tricks

Following are some useful methods used in PLC Programming.

Hold-In or Latching Circuit

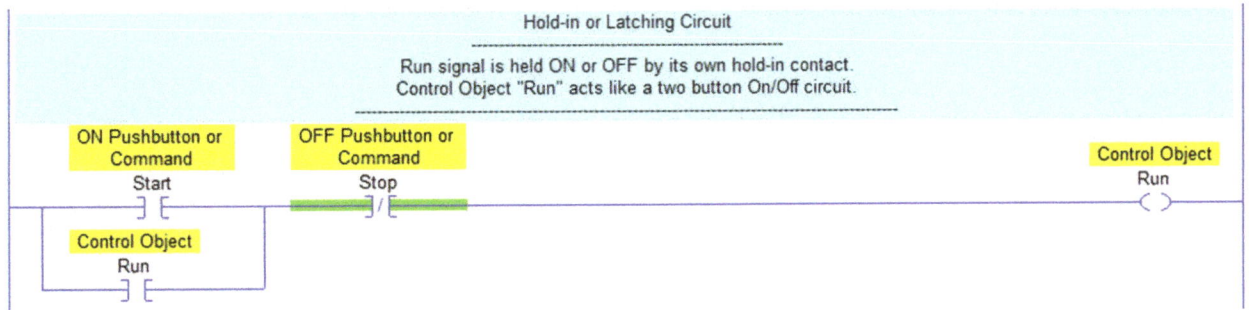

Figure 49 - Hold-In or Latching Circuit

This logic acts as a set and reset pair and is commonly used with various modifications. If the coil address Run's contact is parallel with the logic, it seals in the circuit even if the Start button is not on.

Toggling Pushbutton

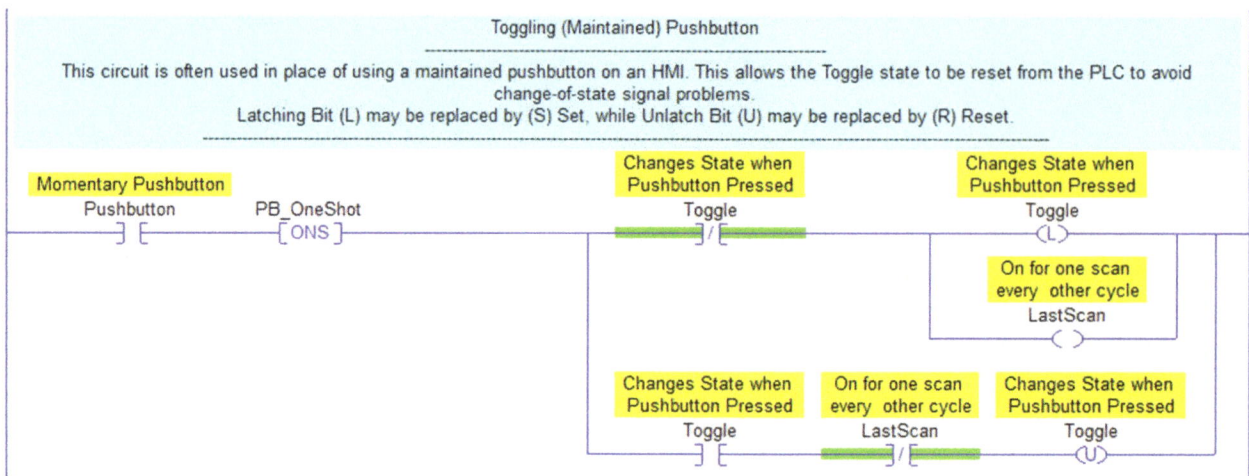

Figure 50 - Toggling or Maintained Pushbutton

This circuit is used to imitate a maintained or "cammed" pushbutton. When the momentary pushbutton is pressed, it sets the Toggle bit. If it is pressed again, the Toggle bit resets. This can be done in other PLC languages with fewer instructions, as it is essentially an Exclusive OR (XOR) instruction.

"Free Running" Timer

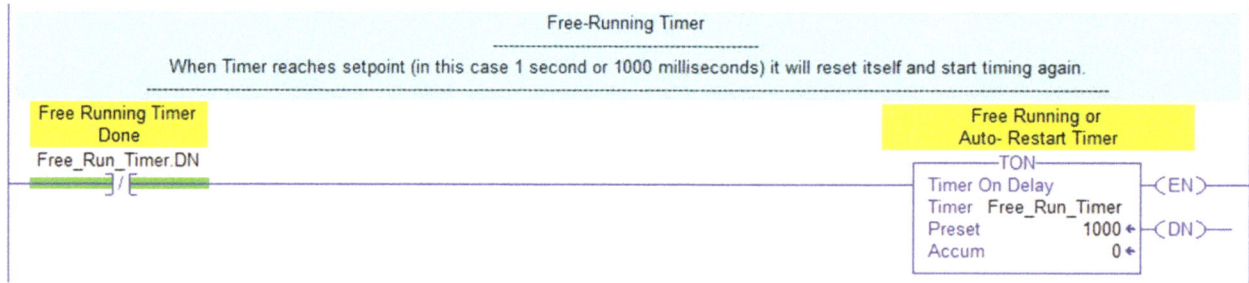

Figure 51 - Free Running (Repeating) Timer

This timer counts up to the setpoint and then resets itself. The accumulator value can be used in many applications, such as creating ramps, and various math functions can be applied. The bits of the accumulator can be used to flash lights at various frequencies.

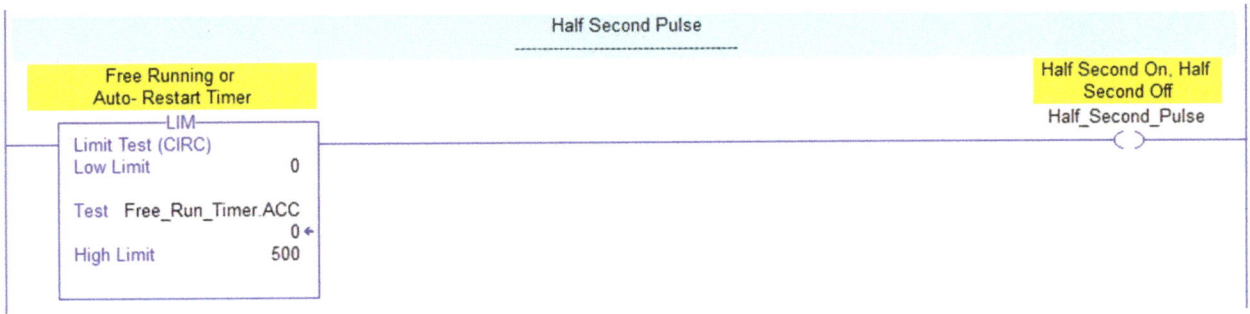

Figure 52 - Half Second Pulse from Free Running Timer

Blink Timers

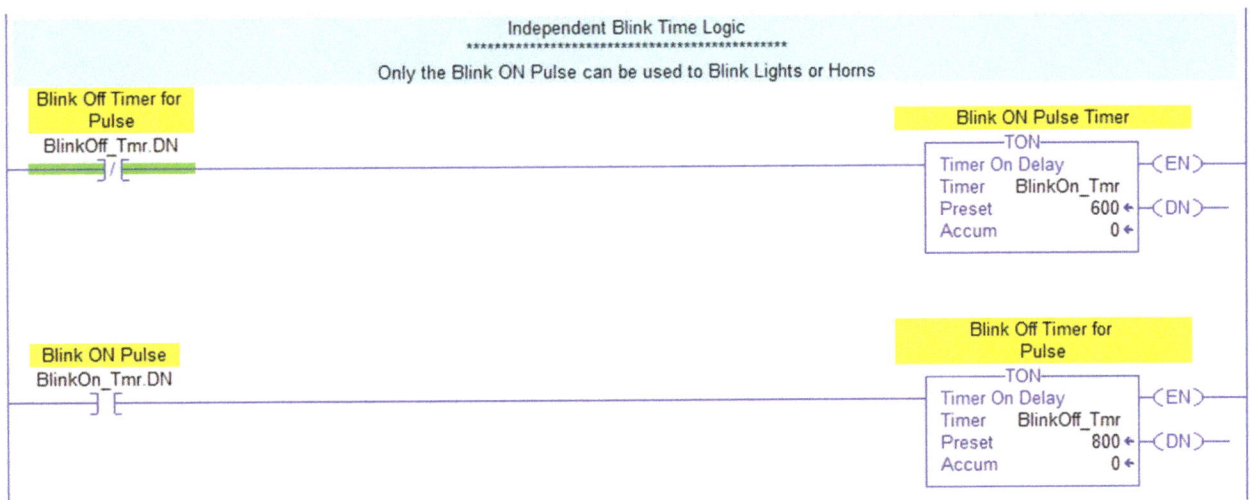

Figure 53 - Blinking Timer Logic

A Blinking Timer pair can be used when specific on and off pulse times are desired. Only the first timer's Done bit can be used; the Blink Off timer's Done bit is a one-shot.

Accumulators and Decumulators

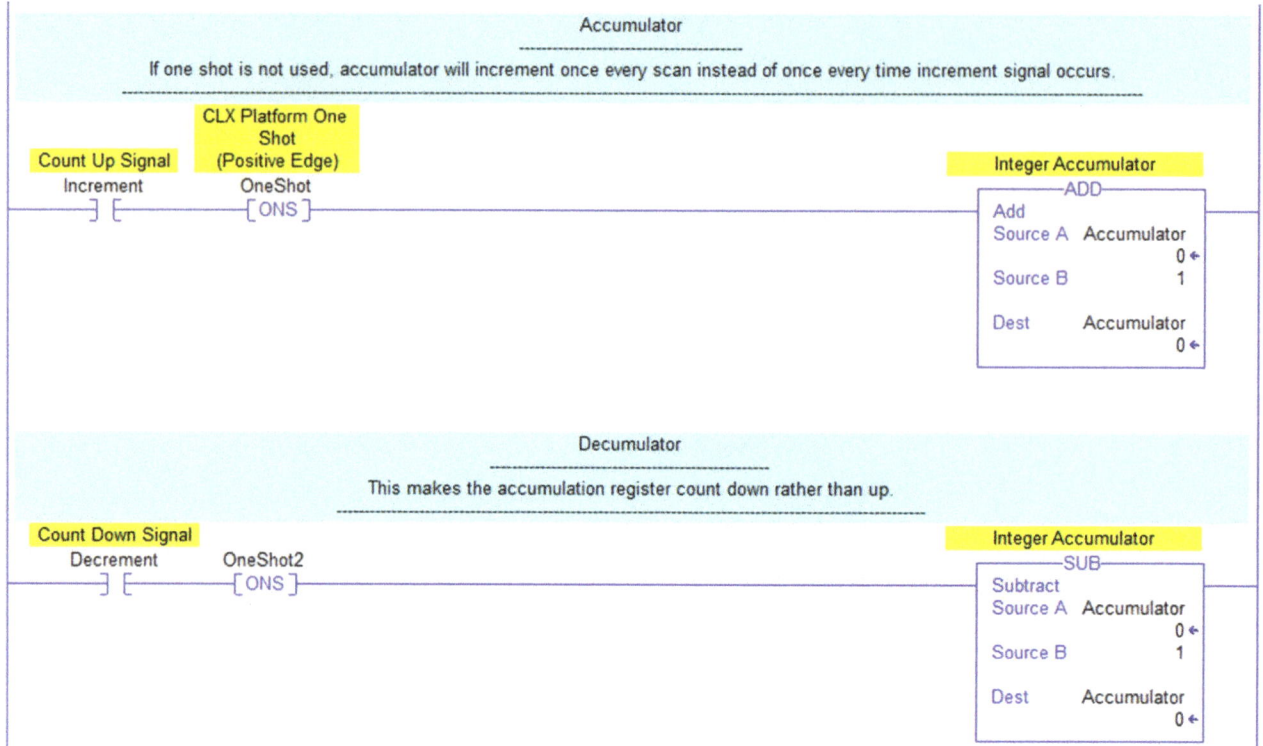

Figure 54 - Accumulation and Decumulation Logic

A number can be added to a register and then placed back into itself, as shown in Figure 54. These rungs act like an Up-Down Counter. Some PLC platforms have limitations, such as a low maximum setpoint value, the done bit changing state at zero rather than at the setpoint, or they use BCD instead of decimal numbers. This logic allows the programmer to make his own counter.

To create a done bit, use a Greater Than or Equal Instruction on the value; to reset the counter, move a zero into the value.

Trainers and Simulators

As mentioned in the introduction of this book, there are different ways that industrial PLC training is approached. Most courses are instruction set based; that is, the instructions for a particular platform are presented and then students do exercises using these instructions. The material in the first part of this book is presented in this way; Bit logic, Timers, Counters, Math, Comparisons, and possibly some advanced instructions are covered. This is a modular way of approaching training, and the best approach for week-long classes.

Standard Trainers

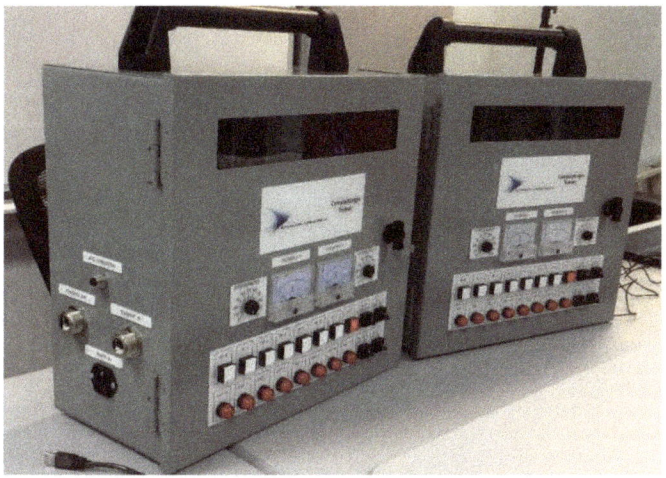

Figure 55 - Standard PLC Trainers (Courtesy of Automation NTH)

Instruction-based training is a good way to learn the basics, and most training courses approach PLC training this way. The trainers that are used for this usually have pushbuttons and lights for digital inputs and outputs and potentiometers and gauges for analog I/O. Figure 55 shows the trainers used for standard Allen-Bradley RSLogix 5000 training in the author's training facility. Note that there are buttons (including a red normally closed), switches, indicator lights, potentiometers and gauges. This trainer also has ports for remote devices.

"Suitcase" Trainers

These kinds of trainers are also often contained in cases so that they can easily be shipped for training classes, as shown in Figure 56. Automation Training provides Allen-Bradley and Siemens classes across North America, shipping this type of trainer along with laptops to regional and onsite (factory) locations. Since they are packed and unpacked often, they need to be rugged and shock resistant.

Figure 56 - Omron Suitcase Trainer (Courtesy of Automation Training)

Simulation Trainers

Another type of trainer that is used is a simulation trainer. This type of trainer can interface with external simulations such as Fischertechnik's factory models or the conveyors shown in the introduction of this book. If buttons and pilot lights are used, they are often accessories that can be plugged into the trainer. To properly simulate industrial equipment, it is useful to have an Emergency Stop with an MCR (Master Control Relay) circuit and a stacklight simulator to indicate the mode or status of the simulated machine.

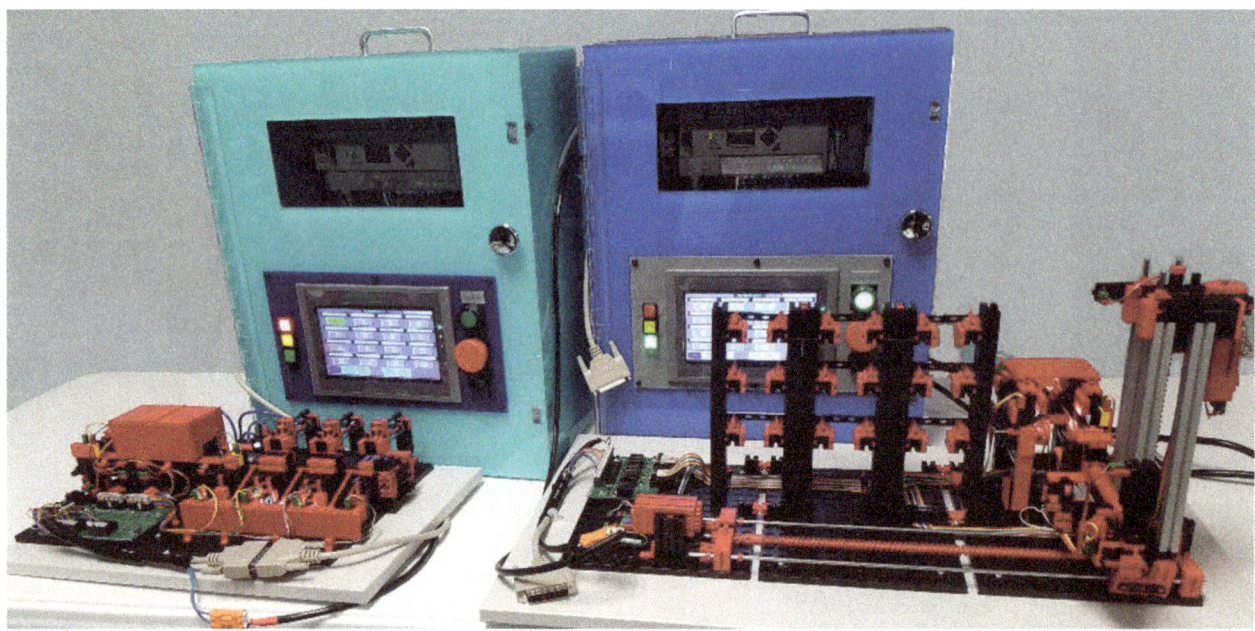

Figure 57 - Simulation Trainer with Fischertechnik Accessories (Courtesy of Automation Consulting, LLC)

Simulation trainers are used for longer duration "project-based" courses like those found in technical schools and universities. It takes a long time to write a complete program for a factory simulation like the ones shown above, and students in these courses are usually more experienced than those in a standard instruction set based class.

Simulation trainers do not have to be used with physical equipment; they can also have simulation software built into the program.

Software Simulation

For software simulation, routines are created that interact with the code that the programmer writes. If an actuator is simulated, when the output controlling the actuator is activated, a timer starts with a setpoint approximating the time it takes the actuator to reach a sensor. The simulated input is activated by the Done bit of the timer. If the actuator has analog positional feedback, a ramping accumulator can be used to add to a value over a time period (combining the free-running timer with the accumulator/decumulator in the previous section). This kind of logic can be used to step an auto sequence through its stages.

Most newer HMI software also has the ability to "animate" objects and show them in different positions. Rather than using physical models, a picture of the actuators can be used to show the status of the machinery. For this reason, simulation trainers usually have an HMI.

Simulation is often used by programmers to test the links between HMI objects, such as pushbuttons and indicators and the PLC program. This is easier than testing the auto sequence logic as described here. Bits and numbers can simply be modified in the program to change the HMI display.

Writing code to simulate the entire program is much more complicated than simply testing links between objects. It requires mapping physical I/O to simulated I/O, since the physical inputs can't be modified. Input feedback needs timers and ramps for every point used in the program. This code should be kept in separate routines to isolate it from the operational part of the program; it can always be deleted or deactivated later. Outputs can be used directly as long as they aren't connected to operating equipment!

Programming Lab: Fischertechnik Color Identification and Conveyor with Bins

This section makes use of a MicroLogix 1400 Simulation Trainer with touchscreen and a Fischertechnik conveyor accessory, as described below. It is included in this book as a tutorial on writing a full program using the techniques described in this section. It is recommended that the programmer read the following section on Allen-Bradley PLCs, particularly the section on SLC and MicroLogix platforms before doing this lab.

The Trainer

The PLC Trainer is comprised of a 20"x 16"x 8.5" plastic enclosure containing a PLC and supporting power components. It has an HMI mounted on the front, programmed with pushbuttons, indicators and numerical entry registers already mapped to addresses inside the PLC.

There is a simulated red/yellow/green stack light, an E-Stop button with MCR (Master Control Relay), and a power button to enable the outputs. This simulates actual industrial equipment used in many plants. The power button also doubles as a fault reset signal.

This trainer is wired with an external multiconductor cable and 25-pin d-sub connector. The cable interfaces with various simulation components. Small factory accessories from Fischertechnik and others allow programming of real-world simulations; I/O addresses are common for all of these simulators.

PLC and I/O

The PLC can be any brand or type, subject to fitting into the plastic enclosure. An Allen-Bradley MicroLogix 1400 (RSLogix500 software) is currently available and used in this lab description. I/O required to interface with accessories consists of 18 DC discrete inputs and 14 discrete outputs, all 24vdc. One or two analog 0-10v inputs are also required for some accessories, as well as two high-speed counters (HSC) for encoders.

The PLC program has registers pre-assigned to the buttons, indicators and numeric registers of the HMI. This frees up the I/O for interfacing with external accessories.

Emergency Stop and MCR

The E-Stop and Power buttons are wired into a relay simulating a safety circuit. When the E-Stop is pressed, power is removed from the power feeds to the outputs, similar to systems in industrial plants. They are also wired to inputs for monitoring. The Power button is also wired to an input, to be used as a Fault Reset pushbutton.

Emergency Stops and MCRs (Master Control Relays) are a standard feature on industrial automated machinery and systems. Interfacing with the I/O from these circuits is an important part of PLC and automated system training. Full instructions on how to program the faults, alarms and mode effects resulting from the state of the circuit are included.

Stack Light and Horn/Buzzer

Outputs are wired to a red/yellow/green stack light simulator and a buzzer for audible alarming and startup warning signals. A stack light is a standard part of most industrial machines, usually indicating mode status (Auto, Auto Cycle, Manual or Maintenance mode, and Alarms or Faults).

An audible alarm is also a part of most control systems, alerting the operator to fault/alarm or startup conditions. Interfacing with these indicators and creating the mode control necessary for this is included with the instructions.

The HMI

The operator interface is a Samkoon 7" color touchscreen. It is pre-programmed with four pushbutton/Indicator screens (64 of each), four Integer input/output screens (64 of each), and four REAL input/output screens (32 of each). Each of the components on the screen allows the user to type in a label for each device by means of a pop-up keyboard. The label registers are not retentive! They only keep the label content until power is removed!

The HMI has communication drivers for all major PLC brands.

Software

All trainers contain routines and registers or tags linked to addresses of the devices on the HMI. A simulation routine in the program with instructions on how to create interactive programs is built into the software. In addition, simulations can be created for lab exercises and made available to the instructor.

The following worksheets are used to record the programmer's assignments to the HMI addresses, which are already present in the trainer.

Allen-Bradley MicroLogix

Bulletin 1766-L32BWA
Ethernet I/P Port address:192.168.0.11

I/O Worksheet

Pushbutton/Indicator Screens 1 and 2

Pushbuttons		Indicators		Description
hPb1	B11:0/0	hInd1	B11:4/0	
hPb2	B11:0/1	hInd2	B11:4/1	
hPb3	B11:0/2	hInd3	B11:4/2	
hPb4	B11:0/3	hInd4	B11:4/3	
hPb5	B11:0/4	hInd5	B11:4/4	
hPb6	B11:0/5	hInd6	B11:4/5	
hPb7	B11:0/6	hInd7	B11:4/6	
hPb8	B11:0/7	hInd8	B11:4/7	
hPb9	B11:0/8	hInd9	B11:4/8	
hPb10	B11:0/9	hInd10	B11:4/9	
hPb11	B11:0/10	hInd11	B11:4/10	
hPb12	B11:0/11	hInd12	B11:4/11	
hPb13	B11:0/12	hInd13	B11:4/12	
hPb14	B11:0/13	hInd14	B11:4/13	
hPb15	B11:0/14	hInd15	B11:4/14	
hPb16	B11:0/15	hInd16	B11:4/15	
hPb17	B11:1/0	hInd17	B11:5/0	
hPb18	B11:1/1	hInd18	B11:5/1	
hPb19	B11:1/2	hInd19	B11:5/2	
hPb20	B11:1/3	hInd20	B11:5/3	
hPb21	B11:1/4	hInd21	B11:5/4	
hPb22	B11:1/5	hInd22	B11:5/5	
hPb23	B11:1/6	hInd23	B11:5/6	
hPb24	B11:1/7	hInd24	B11:5/7	
hPb25	B11:1/8	hInd25	B11:5/8	
hPb26	B11:1/9	hInd26	B11:5/9	
hPb27	B11:1/10	hInd27	B11:5/10	
hPb28	B11:1/11	hInd28	B11:5/11	
hPb29	B11:1/12	hInd29	B11:5/12	
hPb30	B11:1/13	hInd30	B11:5/13	
hPb31	B11:1/14	hInd31	B11:5/14	
hPb32	B11:1/15	hInd32	B11:5/15	

I/O Worksheet

Pushbutton/Indicator Screens 3 and 4

Pushbuttons		Indicators		Description
hPb33	B11:2/0	hInd33	B11:6/0	
hPb34	B11:2/1	hInd34	B11:6/1	
hPb35	B11:2/2	hInd35	B11:6/2	
hPb36	B11:2/3	hInd36	B11:6/3	
hPb37	B11:2/4	hInd37	B11:6/4	
hPb38	B11:2/5	hInd38	B11:6/5	
hPb39	B11:2/6	hInd39	B11:6/6	
hPb40	B11:2/7	hInd40	B11:6/7	
hPb41	B11:2/8	hInd41	B11:6/8	
hPb42	B11:2/9	hInd42	B11:6/9	
hPb43	B11:2/10	hInd43	B11:6/10	
hPb44	B11:2/11	hInd44	B11:6/11	
hPb45	B11:2/12	hInd45	B11:6/12	
hPb46	B11:2/13	hInd46	B11:6/13	
hPb47	B11:2/14	hInd47	B11:6/14	
hPb48	B11:2/15	hInd48	B11:6/15	
hPb49	B11:3/0	hInd49	B11:7/0	
hPb50	B11:3/1	hInd50	B11:7/1	
hPb51	B11:3/2	hInd51	B11:7/2	
hPb52	B11:3/3	hInd52	B11:7/3	
hPb53	B11:3/4	hInd53	B11:7/4	
hPb54	B11:3/5	hInd54	B11:7/5	
hPb55	B11:3/6	hInd55	B11:7/6	
hPb56	B11:3/7	hInd56	B11:7/7	
hPb57	B11:3/8	hInd57	B11:7/8	
hPb58	B11:3/9	hInd58	B11:7/9	
hPb59	B11:3/10	hInd59	B11:7/10	
hPb60	B11:3/11	hInd60	B11:7/11	
hPb61	B11:3/12	hInd61	B11:7/12	
hPb62	B11:3/13	hInd62	B11:7/13	
hPb63	B11:3/14	hInd63	B11:7/14	
hPb64	B11:3/15	hInd64	B11:7/15	

Advanced PLC Hardware & Programming

I/O Worksheet

Integer Screens 1 and 2

Symbol	Address	Description
hInt1	N12:0	
hInt2	N12:1	
hInt3	N12:2	
hInt4	N12:3	
hInt5	N12:4	
hInt6	N12:5	
hInt7	N12:6	
hInt8	N12:7	
hInt9	N12:8	
hInt10	N12:9	
hInt11	N12:10	
hInt12	N12:11	
hInt13	N12:12	
hInt14	N12:13	
hInt15	N12:14	
hInt16	N12:15	
hInt17	N12:16	
hInt18	N12:17	
hInt19	N12:18	
hInt20	N12:19	
hInt21	N12:20	
hInt22	N12:21	
hInt23	N12:22	
hInt24	N12:23	
hInt25	N12:24	
hInt26	N12:25	
hInt27	N12:26	
hInt28	N12:27	
hInt29	N12:28	
hInt30	N12:29	
hInt31	N12:30	
hInt32	N12:31	

*16-bit Signed Integers

I/O Worksheet

Integer Screens 3 and 4

Symbol	Address	Description
hInt33	N12:32	
hInt34	N12:33	
hInt35	N12:34	
hInt36	N12:35	
hInt37	N12:36	
hInt38	N12:37	
hInt39	N12:38	
hInt40	N12:39	
hInt41	N12:40	
hInt42	N12:41	
hInt43	N12:42	
hInt44	N12:43	
hInt45	N12:44	
hInt46	N12:45	
hInt47	N12:46	
hInt48	N12:47	
hInt49	N12:48	
hInt50	N12:49	
hInt51	N12:50	
hInt52	N12:51	
hInt53	N12:52	
hInt54	N12:53	
hInt55	N12:54	
hInt56	N12:55	
hInt57	N12:56	
hInt58	N12:57	
hInt59	N12:58	
hInt60	N12:59	
hInt61	N12:60	
hInt62	N12:61	
hInt63	N12:62	
hInt64	N12:63	

*16-bit Signed Integers

I/O Worksheet

REAL Screens 1 through 4

Symbol	Address	Description
hReal1	F13:0	
hReal2	F13:1	
hReal3	F13:2	
hReal4	F13:3	
hReal5	F13:4	
hReal6	F13:5	
hReal7	F13:6	
hReal8	F13:7	
hReal9	F13:8	
hReal10	F13:9	
hReal11	F13:10	
hReal12	F13:11	
hReal13	F13:12	
hReal14	F13:13	
hReal15	F13:14	
hReal16	F13:15	
hReal17	F13:16	
hReal18	F13:17	
hReal19	F13:18	
hReal20	F13:19	
hReal21	F13:20	
hReal22	F13:21	
hReal23	F13:22	
hReal24	F13:23	
hReal25	F13:24	
hReal26	F13:25	
hReal27	F13:26	
hReal28	F13:27	
hReal29	F13:28	
hReal30	F13:29	
hReal31	F13:30	
hReal32	F13:31	

Advanced PLC Hardware & Programming

Pin	Addr	Color (v1)	PLC SYM	Color Sorter	Warehouse High Bay	Machine Center
1	I:0/0	Brown	IN0	Conveyor Pulse	Ref. Sw. Horizontal	Pusher 1 Extend
2	I:0/1	Red	IN1	Infeed Eye	PE Inside	Pusher 1 Retract
3	I:0/2	Orange	IN2	Outfeed Eye	PE Outside	Pusher 2 Extend
4	I:0/3	Pink	IN3	Bin 1 Eye	Ref. Sw. Vertical (Top)	Pusher 2 Retract
5	I:0/4	Yellow	IN4	Bin 2 Eye	Spare	PE Pusher 1
6	I:0/5	Green	IN5	Bin 3 Eye	Spare	PE Mill
7	I:0/6	Aqua	IN6		Encoder Horiz. P1	PE Load Station
8	I:0/7	Blue	IN7		Encoder Horiz. P2	PE Drill
9	I:0/8	Light Blue	IN8		Cantilever Front	PE Exit Conveyor
10	I:0/9	Violet	IN9		Cantilever Back	
11	I:0/10	Gray	IN10		Encoder Vert. P1	
12	I:0/11	White	IN11		Encoder Vert. P2	
	I:0/12		IN12			
	I:0/13	Blue	MCR			
	I:0/14	Green	ESTOP			
	I:0/15	Gray	RESETPB			
	I:0/16		IN16			
	I:0/17		IN17			
	I:0/18		IN18			
	I:0/19		IN19			
25	I:2.0	White-Black	ANINCH0	Color Sensor		
14	I:2.1	Brown-Black	ANINCH1			
13	O:2.0	Black	ANOUTCH0			
	O:0/0	White	Buzzer			
	O:0/1		Green Lt			
	O:0/2		Yellow Lt			
	O:0/3		Red Lt			
15	O:0/4	Red-Black	OUT4	Conveyor Fwd.	Conveyor Forward	Pusher 1 Fwd. Motor
16	O:0/5	Orange-Black	OUT5	Compressor	Conveyor Reverse	Pusher 1 Rev. Motor
17	O:0/6	Pink-Black	OUT6	Ejector 1	Horiz. Forward	Pusher 2 Fwd. Motor
18	O:0/7	Yellow-Black	OUT7	Ejector 2	Horiz. Reverse	Pusher 2 Rev. Motor
19	O:0/8	Green-Black	OUT8	Ejector 3	Vert. Down	Feed Conveyor Motor
20	O:0/9	Aqua-Black	OUT9		Vert. Up	Mill Conveyor Motor
21	O:0/10	Blue-Black	OUT10		Cantilever Forward	Mill Motor
22	O:0/11	Light Blue-Black	OUT11		Cantilever Reverse	Mill Conveyor Motor
23	O:1/0	Violet-Black	OUT1_0			Drill Motor
24	O:1/1	Gray-Black	OUT1_1			Exit Conveyor Motor

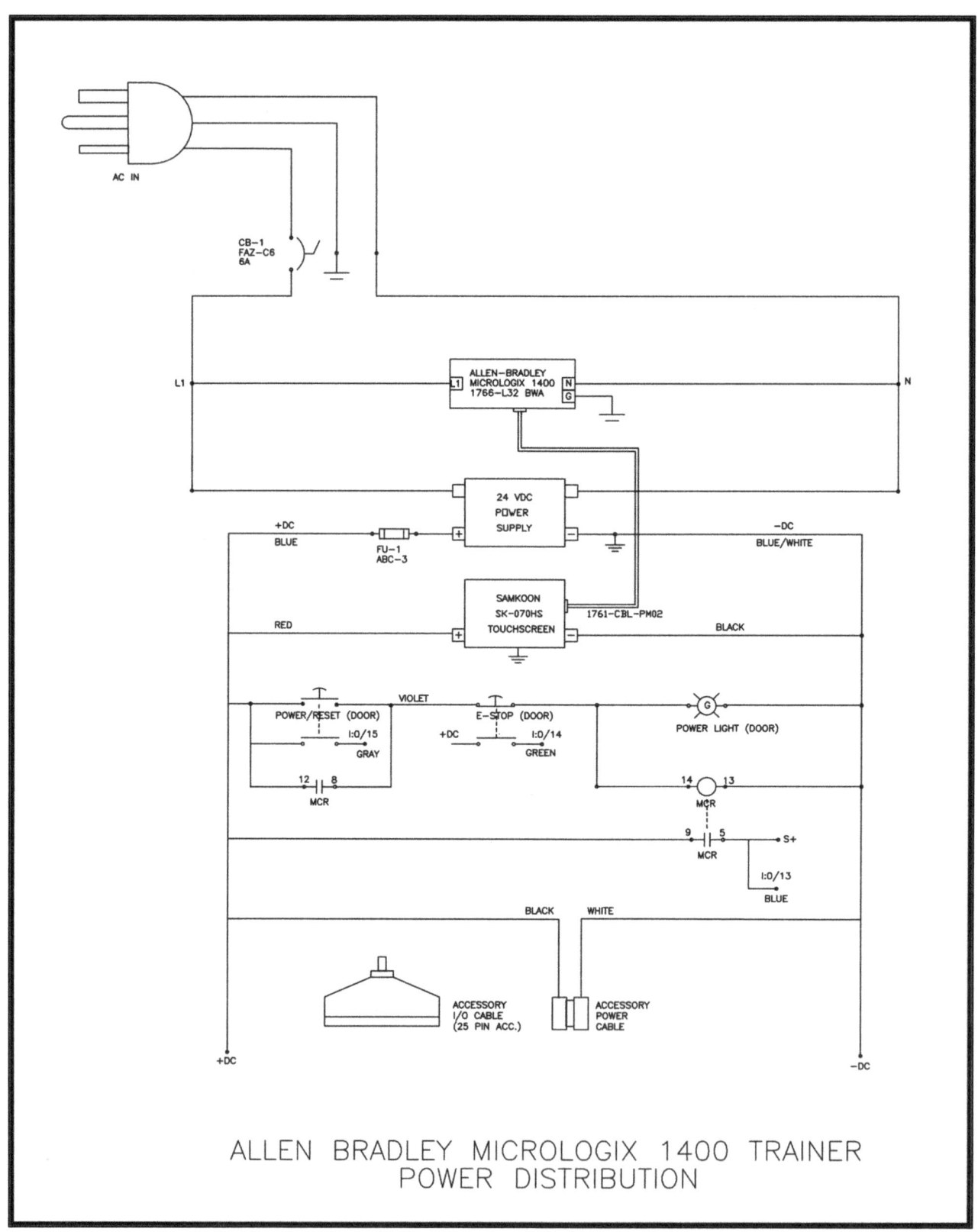

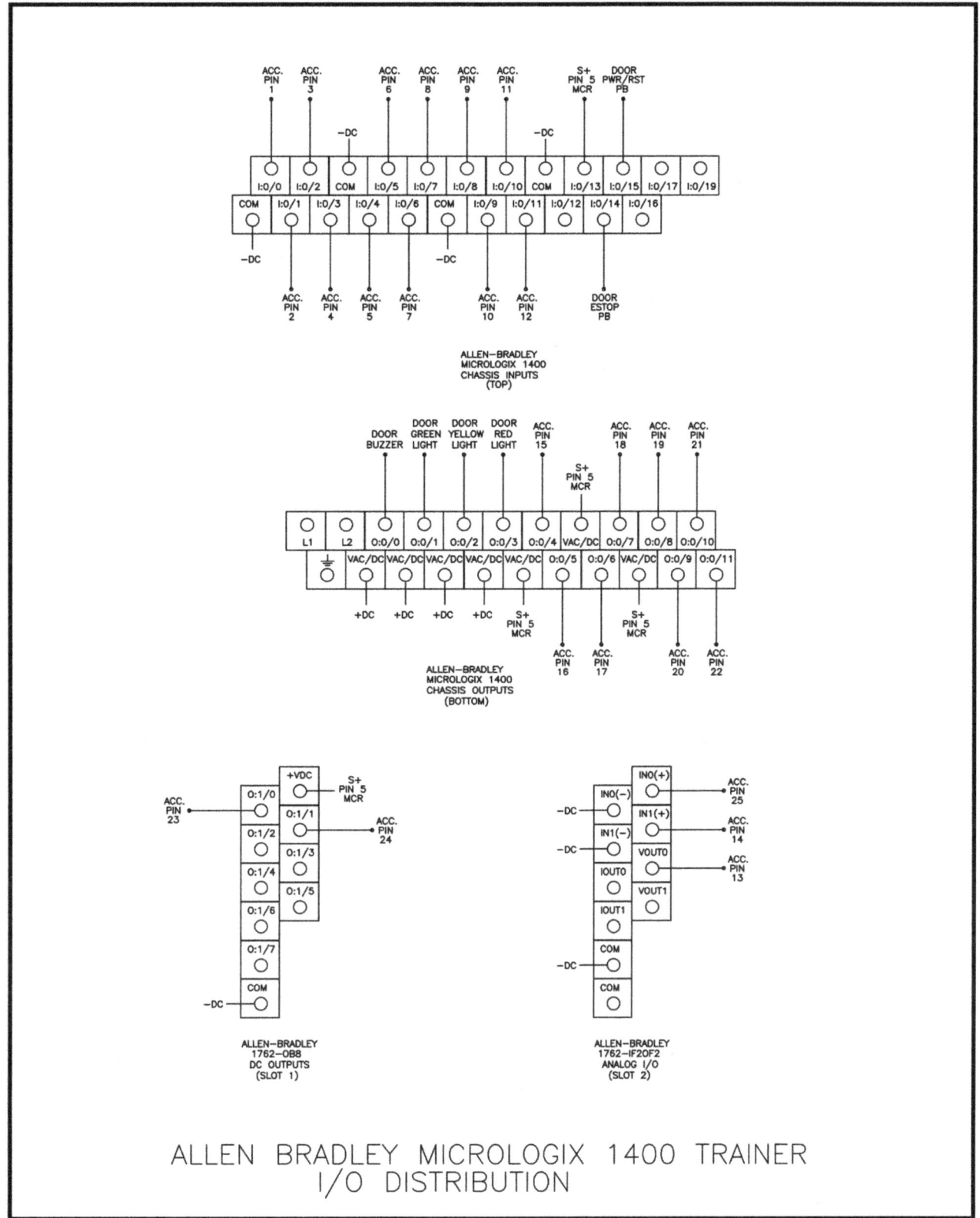

ALLEN BRADLEY MICROLOGIX 1400 TRAINER
I/O DISTRIBUTION

Color Identification and Conveyor with Bins

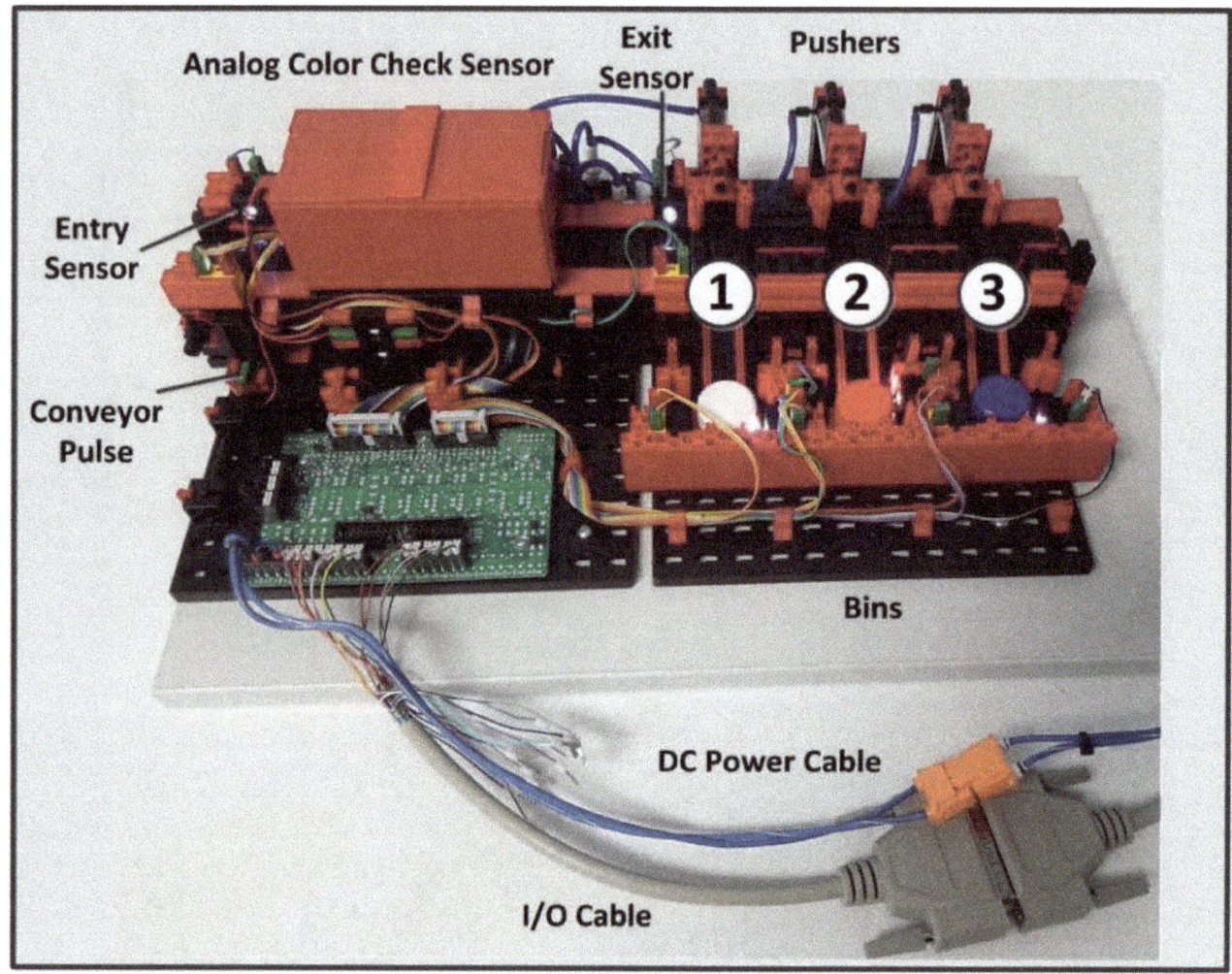

This training accessory has a conveyor, three pneumatic pushers with a small compressor, and an analog color sensor for detecting different colored parts. 6 Inputs, 5 Outputs. 1.1A, 24VDC.

The Color Sorter accessory is a conveyor with an enclosed color detection sensor. There are also three pushers and bins located after the color detection area. A small compressor is used to provide air to the pushers. There is also a microswitch that detects rotations of the conveyor motor. The I/O cable connects to the PLC Trainer, a DC power cable brings 24vdc power from the PLC Trainer to the pins of the connection board.

Inputs and Outputs are listed below:

Input	Description	Output	Description
IN0	Conveyor Pulse	OUT4	Conveyor Run
IN1	Entry Sensor	OUT5	Compressor On
IN2	Exit Sensor	OUT6	Bin 1 Pusher
IN3	Bin 1 Sensor	OUT7	Bin 2 Pusher
IN4	Bin 2 Sensor	OUT8	Bin 3 Pusher
IN5	Bin 3 Sensor		
ANINCH0	Analog Color Sensor		

The symbol names or tags in the PLC Trainer match the Input and Output names listed above.

The Entry and Exit Sensors are on when no part is present to block the light. The Analog Color Sensor provides a voltage level to the PLC based on the amount of light reflected from the parts as they pass through the color detection tunnel.

In order to energize the pushers, the compressor must be running.

Objective

Write a PLC program to detect the color of parts as they pass through the tunnel. Choose a bin for each color (red, white and blue) and push the detected part into the correct bin. If no color is detected, the part should pass off the end of the conveyor.

System Functions: Create an Auto and Manual Mode, selectable by pushbuttons on the HMI. Create an Auto Cycle state that is enabled by pressing an Auto Cycle button on the HMI. The button should be held for three seconds to activate Auto Cycle. While the button is pressed, a buzzer should pulse, warning that the machine is about to start. Create a Cycle Stop button on the HMI to take the machine out of Auto Cycle.

The machine can only be placed into Auto Cycle while it is in Auto Mode. If it is in Auto Cycle, the cycle must be stopped before allowing the machine into manual mode. A fault condition should also take the machine out of Auto Cycle, create a fault bit that you will use later. If there is a part on the conveyor, the machine should remain in Auto Cycle until the part has been processed completely by either being pushed into the proper bin or by exiting the conveyor.

Stack Light: Illuminate the stack light indicators on the PLC Trainer. When the machine is faulted, activate the red light. When it is in manual mode, activate the yellow light. When in Auto Mode but not in Auto Cycle, blink the green light, and when the machine is in Auto Cycle, the green light should stay on solidly.

Inputs: Map physical inputs to input status bits, as required. For instance, if you wish to use a bit named "Bin 1 Part Present", you will need to use the normally closed contact for IN3.

Outputs: Create an output structure for all of the moving parts. In Manual Mode, a pushbutton is used to activate the Bin Pushers. In Auto Mode, your sequence or control logic will turn on a bit that will activate the output. Don't forget to use permissive bits where necessary; if there is already a part in a bin, the pusher should not be able to energize. Test the pushers.

You may want to create two different control scenarios for the conveyor: one to latch the conveyor on and off, and one to jog it for as long as you hold your finger on the button.

Analysis: Create an Auto Routine and place your testing and analysis logic here.

1. Color Values: Capture values for the following:
    a. No Part Present _____
    b. White Part _____
    c. Red Part _____
    d. Blue Part _____

2. Speed and Distance:
    a. Determine how fast the conveyor runs in Pulses per Second. _____
    b. Determine the time it takes for a part to move 300mm _____
    c. Determine the number of pulses in 300mm _____
    d. What is the speed of the conveyor in mm/second? _____
    e. Pulses from the Exit Sensor to Bin 1 Entry _____
    f. Pulses from the Exit Sensor to Bin 2 Entry _____
    g. Pulses from the Exit Sensor to Bin 3 Entry _____
    h. Time from the Exit Sensor to Bin 1 Entry _____
    i. Time from the Exit Sensor to Bin 2 Entry _____
    j. Time from the Exit Sensor to Bin 3 Entry _____

In order to accomplish the above, it can be useful to write code to capture the number of pulses and the amount of time between when the Exit Sensor is blocked and when the Conveyor is stopped.

Auto Sequence: Create logic that starts the conveyor when a part is present at the entry to the color detection tunnel and a button is pressed. As the part passes through the tunnel, capture the color of the part. When the part arrives at the proper bin location, stop the conveyor and push the part into the correct bin.

Faults: Create logic to detect the following conditions:

1. E-Stop button pressed
2. MCR not activated
3. Bin 1 part already in bin when trying to push
4. Bin 1 part did not arrive in bin when pushed
5. Bin 2 part already in bin when trying to push
6. Bin 2 part did not arrive in bin when pushed
7. Bin 3 part already in bin when trying to push
8. Bin 3 part did not arrive in bin when pushed

When the fault occurs, the machine should drop out of Auto Cycle. Clear the fault with the Fault Reset pushbutton. If the cause of the fault has not been corrected, the fault should not be able to be cleared. After the fault has been cleared and reset, the machine should be able to be returned to Auto Cycle without losing track of the part.

Record the number of the fault. Only the fault that originally caused the machine to stop should be recorded, subsequent faults should have no effect.

Production: Create logic to count the number of each part that has been processed, along with the total number of parts. Create a Count Reset button on the HMI. Display the count values on the Integer screen.

Optional: Create counters for each of the different faults you created in the Fault Routine.

Use the worksheets on the included pages to lay out the buttons and numbers you will use for your project. The names you use for these devices and registers can be entered into the field above the device on the touchscreen using a pop-up keyboard. Do not remove power from the touchscreen; these registers are not retentive!

A solution for this exercise is shown in the back of this book.

PLC Platforms

PLC Platforms - Overview

The third part of this book covers specific PLC platforms, primarily Siemens and Allen-Bradley. Detailed information for these two platforms, including hardware, the instruction set and a tutorial on starting a project on the different platforms is included. There are sections on how IEC 61131 languages are used in Allen-Bradley and Siemens, along with a section on how to connect to the controllers.

After learning the generic concepts of PLC Hardware and Programming presented in the first section of this book, then exploring some of the more advanced techniques in The Art of Programming, this section applies these concepts to the specific platforms. Allen-Bradley and Siemens have very different approaches to both hardware and software; this section should help the programmer understand some of these differences and hopefully allow them to apply this knowledge.

A listing of some of the other larger PLC platforms is also included after the Allen-Bradley and Siemens sections.

Allen-Bradley PLCs

Allen-Bradley traces its origins to 1903 and the formation of the Compression Rheostat Company, founded by Lynde Bradley and Dr. Stanton Allen with an initial investment of $1000. In 1904, 19-year-old Harry Bradley joined his brother in the business, and in 1909, the company was renamed the Allen-Bradley Company.

The company expanded rapidly during World War I in response to government contract work. Its product line at that time included automatic starters and switches, circuit breakers, relays and other electrical equipment. Located in Milwaukee, the first sales office was established in New York.

During the 1970s, the company expanded its production facilities and became a global company. In 1985, the company was purchased by Rockwell International.

In 1994, Rockwell Software was launched. Rockwell also acquired Reliance Electric and Dodge; the collection of companies was marketed as Rockwell Automation.

In 2002, Rockwell International split into two companies. The industrial automation division remained as Rockwell Automation, while the avionics division became Rockwell Collins.

The original large rack PLC system was the PLC, followed by the PLC2, PLC3 and PLC5. The PLC5 is still supported.

In 1991, the SLC (Small Logic Controller) made its debut as a smaller version of the PLC5, using an abbreviated version of the PLC instruction set. The MicroLogix family appeared in 1995, followed by the first ControlLogix tag-based platform in 1997.

Rockwell Software still makes all of the programming and communications software for the Allen-Bradley PLC and PAC (Programmable Automation Controller) families.

Allen-Bradley SLC and MicroLogix Platforms

Rockwell Software – RSLogix 500

The main software package for programming the SLC and MicroLogix families of PLCs is RSLogix 500. The current release as of June 2017 is version 11.0.

Name	Catalog #	Controllers	Description
Professional Edition	9324-RL0700NXENE	All SLC 500 and MicroLogix controllers except 800 series	Online/Offline Programming, Includes RSNetworx for ControlNet/ DeviceNet/ Ethernet/IP and RSLogix Emulate
Standard Edition	9324-RL0300ENE	All SLC 500 and MicroLogix controllers except 800 series	Online/Offline Programming
Starter Edition	9324-RL0100ENE	All SLC 500 and MicroLogix controllers except 800 series	Offline Programming only, no cross referencing, Data usage, Program Compare.
Micro Developer	9324-RLM0800ENE	All MicroLogix controllers except 800 series	Online/Offline Programming
Micro Starter	9324-RL0300ENE	All MicroLogix controllers except 800 series	Online/Offline Programming, no Data usage, Program Compare, Trending, Advanced Diagnostics, Program Report.
Micro Starter Lite	Free	MicroLogix 1000 and 1100 only	Online editing for MicroLogix 1100 only, no Data usage, Program Compare, Trending, Advanced Diagnostics, Program Report.

The Micro800 family of controllers is programmed using the Connected Components Workbench, which is also used for other devices such as VFDs, Servo Drives, Light Curtains and Safety Relays.

Name	Catalog #	Controllers	Description
Developer Edition	9328-CCWDEVENE	All Micro800 products	Offline programming, Run Mode Change, Spy List, UDT, IP Protection
Standard Edition	Free	All Micro800 products	Offline Programming

MicroLogix 800

Software:

Connected Components Workbench, (Standard Edition is free). Ladder, FBD, Structured Text support.

Models:

Micro810: 12 I/O points with 4 high-current relay outputs, DC models allow 4 inputs to function as 0-10v analog inputs, Real Time Clock, Optional 1.5" local LCD for monitoring/modifying application data.

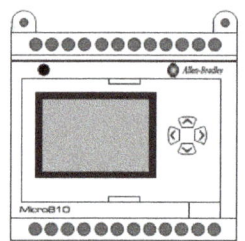

Micro820: Up to 36 I/O points, up to 2 plug-in modules, Ethernet, MicroSD card for recipe or data logging, 5khz PWM output, Real Time Clock, Optional 3.5" LCD Display, analog I/O.

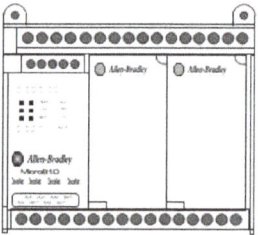

Micro830: Up to 88 I/O points with up to 20 analog inputs, Hi performance I/O, Interrupts, PTO Motion, 2080 expansion I/O.

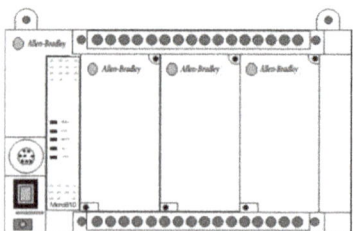

Micro850: Up to 132 I/O points, Hi-performance I/O, Interrupts, PTO Motion, embedded Ethernet, 2085 expansion I/O. Can function as RTU unit for SCADA (Serial Modbus or Ethernet).

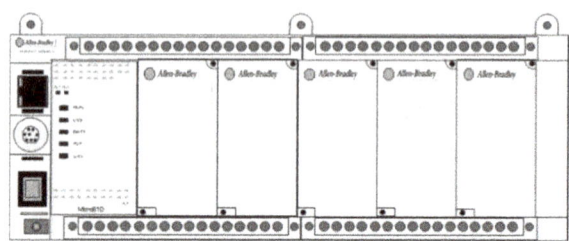

MicroLogix 1000

Software:

RSLogix 500, RSLogix 500 Micro. Ladder Logic only.

Models:

1763-LXX, where XX denotes the number of I/O points. AC and DC power supplies, I/O ranges from 10 to 32 points. AC or DC Inputs; AC, DC and Relay Outputs. Analog available on some models.

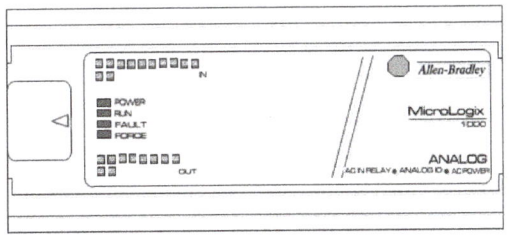

The MicroLogix 1000 is discontinued as of June 30, 2017. Recommended replacement is the MicroLogix 820.

MicroLogix 1100

Software:

RSLogix 500, RSLogix 500 Micro Starter Lite. Online Editing. Ladder Logic only.

Models:

1763-LXX, where XX denotes the number of I/O points. AC and DC power supplies, I/O ranges from 10 to 16 points. Expandable to 144 I/O. AC or DC Inputs; AC, DC and Relay Outputs. Analog available on some models. Ethernet/IP, DH-485 and Modbus RTU capable.

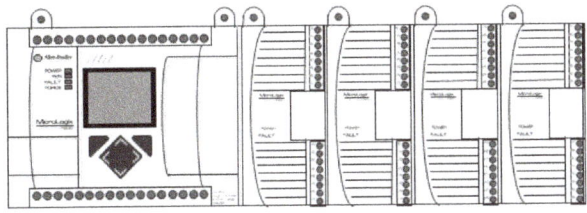

MicroLogix 1200

Software:

RSLogix 500. Ladder Logic only.

Models:

1762-LXX, where XX denotes the number of I/O points. AC and DC power supplies, I/O ranges from 24 to 40 points. Expandable to 136 I/O. AC or DC Inputs; AC, DC and Relay Outputs. Analog available on some models. RS232, RS485 Combo Port. 2 built-in potentiometers.

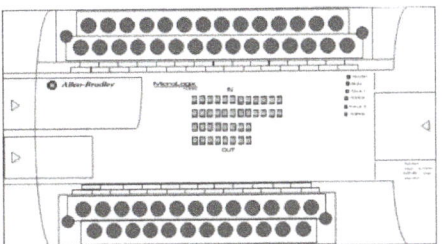

MicroLogix 1400

Software:

RSLogix 500, RSLogix Micro. Online Editing. Ladder Logic only.

Models:

1766-LXX, where XX denotes the number of I/O points. 10K words user program memory, 10K words user data memory

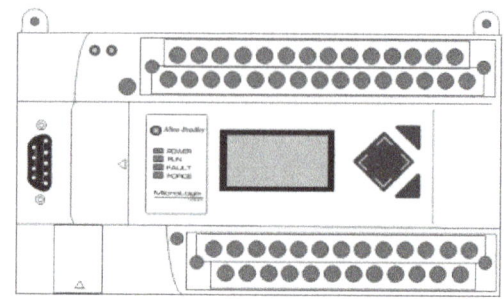

Built-in LCD, AC and DC power supplies, 32 built-in I/O points. Expandable to 256 I/O (7 modules). AC or DC Inputs; AC, DC and Relay Outputs. Analog available on some models. Two built-in serial ports, (DF1/DH485/Modbus RTU/DNP3/ASCII); Ethernet port (EtherNet/IP, Modbus, DNP3). Memory cards for data logging (128K) or recipes (64K).

MicroLogix 1500

Software:

RSLogix 500. Ladder Logic only.

Models:

1769-LXX, where XX denotes the number of I/O points. Built-in LCD, AC and DC power supplies, 32 built-in I/O points. Expandable to 512 I/O. AC or DC Inputs; AC, DC and Relay Outputs. Analog available on some models. Two built-in serial ports, Ethernet port.

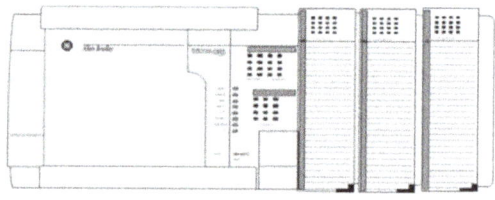

The MicroLogix 1500 is discontinued as of June 30, 2017. Recommended replacement is the MicroLogix 1400 system or the CompactLogix 5370 L1 and L2.

SLC500 Series

Software:

RSLogix 500. Ladder Logic and Structured Text. Advanced instruction sets includes file handling, sequencer, diagnostic, shift register, immediate I/O and program control instructions.

Models:

1747-LXXX, where XXX denotes the family of controllers. The SLC (Small Logic Controller) family is a rack-based system, though the older SLC150 and SLC500 controllers were fixed and had on-board I/O. These have been discontinued.

Processors reside in the leftmost slot (slot 0) of the rack, which comes in 4, 7, 10 and 13 slot sizes. Processors include the 5/01, 5/02, 5/03, 5/04 and 5/05. The 5/01 and 5/02 processors have been discontinued, and the 5/03 8K series will be discontinued as of August 31, 2018. Recommended replacements are the CompactLogix 5370 and 5380 series.

Processor	Communications	Memory	Max I/O	# Instructions
SLC 5/01	DH-485 Slave	1K, 4K	3940	51
SLC 5/02	DH-485	4K	4094	~70
SLC 5/03	DH-485, DF1	8K, 16K	4094	~100
SLC 5/04	DH+, DF1	16K, 32K, 64K	4094	~100
SLC 5/05	Ethernet, DF1	16K, 32K, 64K	4094	~100

Online editing is possible with the 5/03, 5/04 and 5/05 processors.

SLC and MicroLogix Memory Registers

Structure: <File> : <Element Number> <Delimiter> <Bit or Word Number>

File	Number	Name	Example	Description	Comment
O	0	Outputs	O:5/3	Fifth Slot, Fourth Bit	Physical Digital Outputs
			O:6.1	Sixth Slot, Second Channel	Physical Analog Outputs
I	1	Inputs	I:4/6	Fourth Slot, Seventh Bit	
			I:7.3	Seventh Slot, Fourth Channel	
S	2	Status	S:1/15	First Scan Bit	Can't be modified
			S:5/0	Math Overflow Bit	(Read Only)
B	3	Bits	B3:5/0 (B3/80)	Word 5, Bit Zero (Bit 80)	
T	4	Timers	T4:2	Timer 2	
			T4:2.PRE	Timer 2 Preset	Signed Integer
			T4:2.ACC	Timer 2 Accumulator	Signed Integer
			T4:2/DN	Timer 2 Done	
			T4:2/TT	Timer 2 Timing	
C	5	Counters	C5:8	Counter 8	
			C5:8.PRE	Counter 8 Preset	Signed Integer
			C5:8.ACC	Counter 8 Accumulator	Signed Integer
			C5:8/DN	Counter 8 Done Bit	
R	6	Control	R6:1	Control File 1	
			R6:1.LEN	Control File 1 Length	
			R6:1.POS	Control File 1 Position	
			R6:1/ER	Control File 1 Error	
N	7	Integers	N7:20	Integer Word 20	Signed Integer
			N7:20/6	Integer Word 20, Bit 6	
F	8	Float	F8:5	Floating Point Value Number 5	REAL Data type, 32 bits
ST		User String	ST9:1	String 1 (File 9)	
A			ASCII		

For detailed information, see Allen-Bradley Instruction Set Reference Manual 1747-rm001_-en-p

Files above F8 are defined by the Programmer, to a maximum of 255 files.

SLC and MicroLogix Instructions

Basic Instructions

Mnemonic	Name	Purpose
XIC	Examine If Closed, Examine On	Examines a bit for an ON condition
XIO	Examine If Open, Examine Off	Examines a bit for an OFF condition
OTE	Output Energize	Turns a bit ON or OFF
OTL	Output Latch	Sets a bit ON when executed, the bit retains its state until unlatched or the register is cleared
OTU	Output Unlatch	Resets a bit OFF when executed
OSR	One-Shot Rising	Triggers a one time event on leading edge of signal, ON for one scan
OSF	One-Shot Falling	Triggers a one time event on falling edge of signal, ON for one scan
ONS	One-Shot (Rising)	Same as OSR
TON	Timer On-Delay	Counts timebase intervals when the energizing instruction is TRUE
TOF	Timer Off-Delay	Counts timebase intervals when the energizing instruction is FALSE
RTO	Retentive Timer	Counts timebase intervals when the energizing instruction is TRUE, retains accumulated value when energizing instruction is FALSE
CTU	Count Up	Increments the accumulated value at each false to true transition and retains the accumulated value when the instruction goes false or when power cycle occurs
CTD	Count Down	Decrements the accumulated value at each false to true transition and retains the accumulated value when the instruction goes false or when power cycle occurs
HSC	High-speed Counter	Counts high-speed pulses from a fixed controller high-speed input
RES	Reset	Resets the accumulator value and status bits of a timer or counter. *Do not use with TOF Timers!

Comparison Instructions

Mnemonic	Name	Purpose
EQU	Equal	Test whether two values are equal
NEQ	Not Equal	Test whether two values are not equal
LES	Less Than	Test whether one value is less than another value
LEQ	Less Than or Equal	Test whether one value is less than or equal to another value
GRT	Greater Than	Test whether one value is greater than another value
GEQ	Greater Than or Equal	Test whether one value is greater than or equal to another value
MEQ	Masked Equal	Test portions of two values to determine whether they are equal through a mask
LIM	Limit Test	Test whether one value is within the range of two other values

Math Instructions

Mnemonic	Name	Purpose
ADD	Add	Adds source A to source B and stores the result in the destination
SUB	Subtract	Subtracts source B from source A and places the result in the destination
MUL	Multiply	Multiplies source A by source B and places the result in the destination
DIV	Divide	Divides source A by source B and places the result in the destination
DDV	Double Divide	Divides the contents of the math register by the source and stores the result in the destination and the math register
CLR	Clear	Sets all bits of a word to zero
SQR	Square Root	Calculates the square root of the source and places the result in the destination
SCP	Scale with Parameters	Produces a scaled output value that has a linear relationship between the input and scaled values
SCL	Scale Data	Multiplies the source by a specific rate, adds to an oddest value, and stores the result in the destination
RMP	Ramp	Provides the ability to create linear acceleration, deceleration and "S" curve ramp output data waveforms
ABS	Absolute	Calculates the absolute (positive) value of the source and places the result in the destination
CPT	Compute	Evaluates an expression and places the result in the destination
SWP	Swap	Swaps the low and high bytes of a specified number of words in a bit, integer, ASCII or string file
COS	Cosine	Takes the cosine of a number and stores the result in the destination
SIN	Sine	Takes the sine of a number and stores the result in the destination
TAN	Tangent	Takes the tangent of a number and stores the result in the destination

ASN	Arc Sine	Takes the arc sine of a number and stores the result (in radians) in the destination
ACS	Arc Cosine	Takes the arc cosine of a number and stores the result (in radians) in the destination
ATN	Arc Tangent	Takes the arc tangent of a number and stores the result (in radians) in the destination
LN	Natural Log	Takes the natural log of the value in the source and stores the result in the destination
LOG	Log to the base 10	Takes the log base 10 of the value in the source and stores the result in the destination

Data Handling Instructions

Mnemonic	Name	Purpose
TOD	Convert to BCD	Converts the integer source value to BCD format and stores it in the destination
FRD	Convert from BCD	Converts the BCD source value to an integer and stores it in the destination
DEG	Convert from Radians to Degrees	Converts radians (source) to degrees and stores the result in the destination
RAD	Convert from Degrees to Radians	Converts degrees (source) to radians and stores the result in the destination
DCD	Decode 4 to 1 of 16	Decodes a 4-bit value (0 to 15), turning on the corresponding bit in the 16 bit destination
ENC	Encode 1 of 16 to 4	Encodes a 16 bit source to a 4 bit value. Searches the source from the lowest to the highest bit and looks at the first set bit. The corresponding bit position is written to the destination as an integer.
COP	Copy File	Copies data from the source file to the destination file
FLL	Fill File	Loads a source value into each position in the destination file
MOV	Move	Copies the source value to the destination
MVM	Masked Move	Copies the source value to part of the destination
AND	And	Performs a bitwise AND operation
OR	Or	Performs a bitwise OR operation
Mnemonic	Name	Purpose
XOR	Exclusive Or	Performs a bitwise inclusive OR operation

NOT	Not	Performs a NOT (invert) operation
NEG	Negate	Changes the sign of the source and stores it in the destination
FFL	FIFO Load	Loads a word into a FIFO (First In - First Out) stack on each false to true transition. The first word loaded is the first to be unloaded.
FFU	FIFO Unload	Unloads a word from a FIFO (First In - First Out) stack on each false to true transition. The first word loaded is the first to be unloaded.
LFL	LIFO Load	Loads a word into a LIFO (Last In - First Out) stack on each false to true transition. The last word loaded is the first to be unloaded.
LFU	LIFO Unload	Unloads a word from a LIFO (Last In - First Out) stack on each false to true transition. The last word loaded is the first to be unloaded.

Program Flow Instructions

Mnemonic	Name	Purpose
JMP, LBL	Jump to Label and Label	Jump forward or backward to a specified "Label" instruction
JSR	Jump to Subroutine	Jump to a designated subroutine or ladder
SBR	Subroutine label	Designates the start of a subroutine or ladder
RET	Return from subroutine	Returns from a subroutine to the point from which it was called
MCR	Master Control Reset	Turn off all non-retentive outputs in a section of ladder
TND	Temporary End	Mark a temporary end that halts program execution
SUS	Suspend	Identifies specific conditions for program debugging and system troubleshooting
IIM	Immediate Input with Mask	Program an immediate input update using a mask
IOM	Immediate Output with Mask	Program an immediate output update using a mask
REF	Refresh	Interrupt the program scan to update the I/O and service communications

Application Specific Instructions

Mnemonic	Name	Purpose
BSL, BSR	Bit Shift Left or Right	Loads a bit of data into a bit array, shifts the pattern through the array and unloads the last bit of data in the array. BSL shifts data to the left, while BSR shifts it to the right
SQO, SQC	Sequencer Output and Sequencer Compare	Controls sequential machine operations by transferring 16 bit data through a mask to image addresses
SQL	Sequencer Load	Captures referenced conditions by manually stepping the machine through its operating sequences
TDF	Compute Time Difference	Calculates the number of 10 microsecond ticks between any two captured time stamps
FBC	File Bit Compare	Compares bits between two different files
DDT	Diagnostic Detect	Used to monitor machine or process operations to detect malfunctions
RPC	Read Program Checksum	Copies the program checksum from processor memory or from the memory module into the data table

ASCII Instructions

Mnemonic	Name	Purpose
ABL	Test ASCII Buffer for Line	Determine the number of characters in the buffer up to and including the user configured end of line characters
ACB	Number of ASCII Characters in Buffer	Determine the total number of characters in the buffer
ACI	String to Integer	Convert a string to an integer value
ACL	ASCII Clear Buffer	Clear the receive and /or transmit buffers
ACN	String Concatenate	Combine two strings into one
AEX	String Extract	Extract a portion of a sring to create a new string
AHL	ASCII Handshake Lines	Set or reset modem handshake lines
AIC	Integer to String	Convert an integer value into a string
ARD	ASCII Read Characters	Read characters from the input buffer and place them into a string
ARL	ASCII Read Line	Read one line of characters from the input buffer and place them into a string

ASC	String Search	Search a string
ASR	ASCII String Compare	Compare two strings
AWA	ASCII Write with Append	Write a string with user configured characters appended
AWT	ASCII Write	Write a string

Block Transfer and PID Instructions

Mnemonic	Name	Purpose
BTR	Block Transfer Read	Receive data from a remote device
BTW	Block Transfer Write	Send data to a remote device
PID	Proportional Integral Derivative	Controls physical properties such as temperature, pressure, liquid level or flow rate using closed process loops

Interrupt Routine Instructions

Mnemonic	Name	Purpose
	User Fault Routine	Provides the option of preventing a processor shutdown
STI	Selectable Timed Interrupt	Allows you to interrupt the scan of the main program file automatically, on a periodic basis, to scan a specified subroutine file
STD	Selectable Timed Disable	Disables STIs from occurring
STE	Selectable Timed Enable	Enables STIs to occur
STS	Selectable Timed Start	Sets or changes the file number or setpoint frequency of the STI routine
DII	Discrete Input Interrupt	Allows the processor to execute a subroutine when the input pattern of a discrete input card matches a compare value that you programmed.
ISR	I/O Interrupt	Allows a specialty I/O module to interrupt the normal processor operating cycle in order to scan a specific subroutine file
IID	I/O Interrupt Disable	Disables I/O Interrupts from occurring
IIE	I/O Interrupt Enable	Enables I/O Interrupts to occur
RPI	Reset Pending Interrupt	Aborts a pending I/O Interrupt
INT	Interrupt Subroutine	Optional instruction to identify interrupt subroutines

Communication Instructions

Mnemonic	Name	Purpose
SVC	Service Communications	Interrupts the program scan to execute the service communication portion of the operating cycle
MSG	Message Read/Write	Transmits data from one node to another on the network
CEM	ControlNet Explicit Message	Transmits CIP generic commands to other ControlNet nodes via the 1747-SCNR
DEM	DeviceNet Explicit Message	Transmits CIP generic commands to other DeviceNet nodes via the 1747-SDN
EEM	EtherNet/IP Explicit Message	Transmits CIP generic commands to other Ethernet/IP nodes via channel 1

For detailed information, see Allen-Bradley Instruction Set Reference Manual 1747-rm001_-en-p

Starting and Editing a Project with RSLogix 500

Allen Bradley's RSLogix 500 software is installed in the Rockwell Software folder under RSLogix 500 English. It may also be on your desktop as a shortcut, as shown in the icon on the left. The latest version as of June 2017 is version 11.0.

The extension for RSLogix 500 programs is .RSS, double-clicking on a file will also open the programming software. When the software is first opened, it will look as shown in the image below:

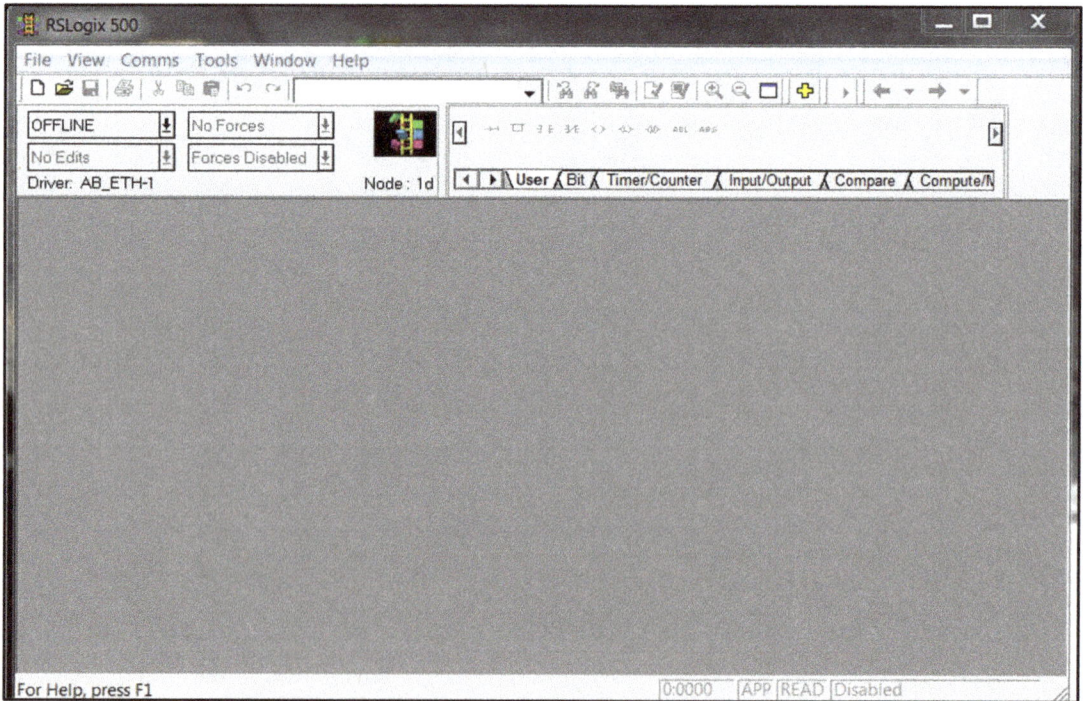

Creating a Project

To create a new project, select "New" under the File tab. The first screen that will appear will ask you to choose and name your processor. The processor name can only be a maximum of 8 characters; a typical name might contain an abbreviation or number for a machine or system, such as "PNTLINE" (Paint Line), "RLMILL42" (Roll Mill 42) or the like.

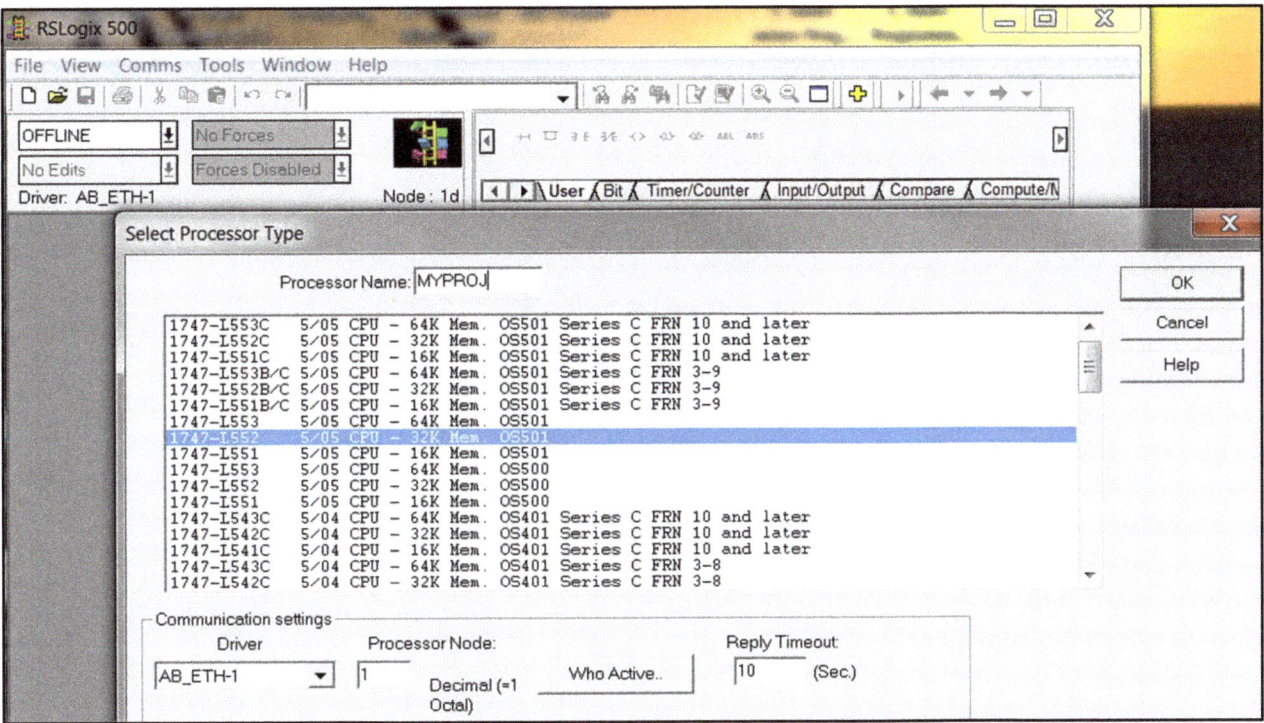

A communication settings dialog is also shown on this screen; don't worry about setting it up yet. This will be done later using the RSLinx communications software.

Until the hardware platform is selected, the software doesn't know how to set up the data tables and properties. It is important to note that the Firmware Revision Number (FRN) selection is part of the hardware selection process.

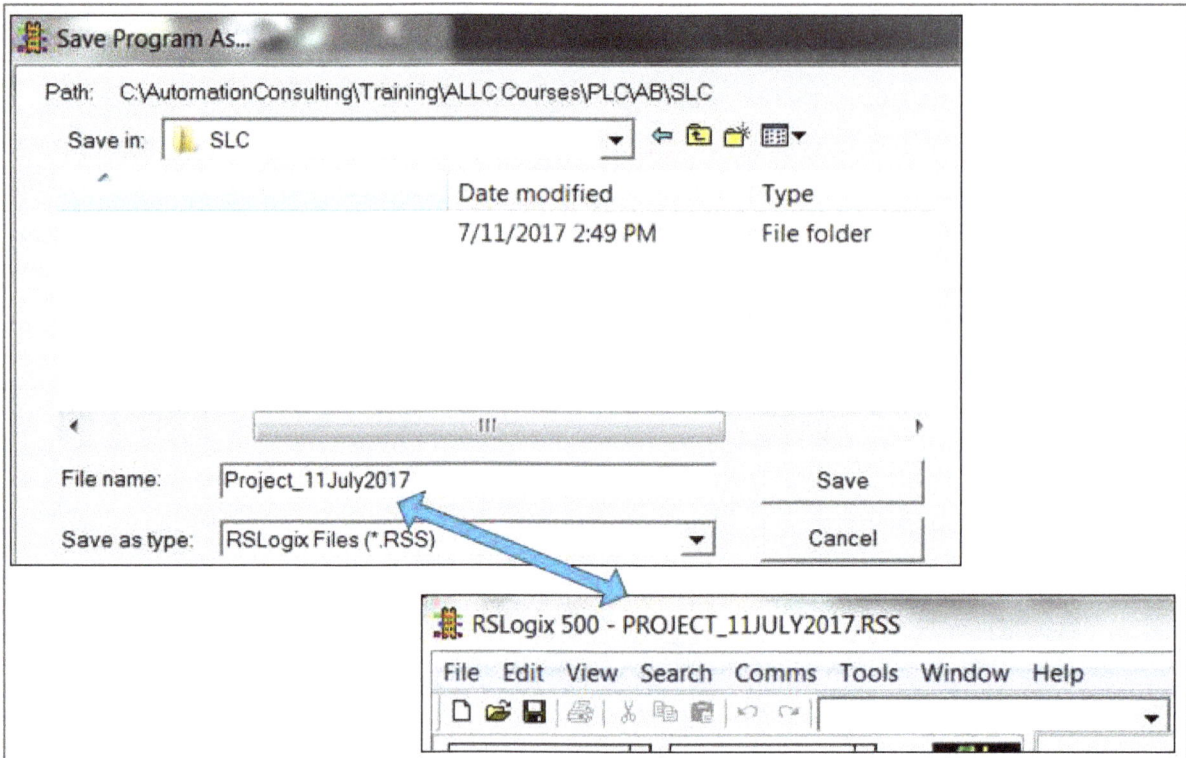

Initially, the project will be saved under the name of the processor. Selecting "Save As" will allow you to save with a date and a different title.

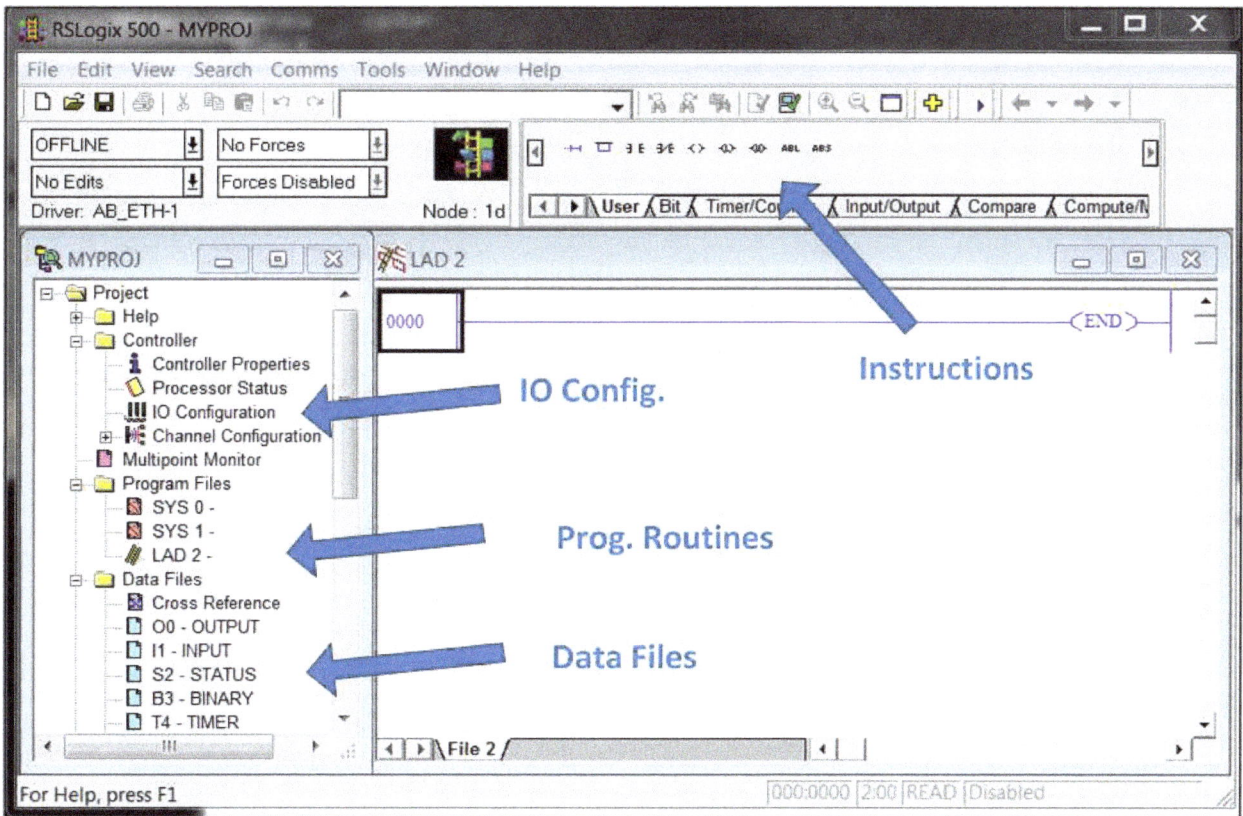

After creating the project, the programming environment will appear as shown above. The area on the left is where most of the items you will be adding to the project are located; the first task you will want to accomplish is to configure the I/O.

Hardware Configuration

Double-clicking the I/O Configuration icon will open a selection window containing all of the different types of I/O cards available. This is also where you select your rack size.

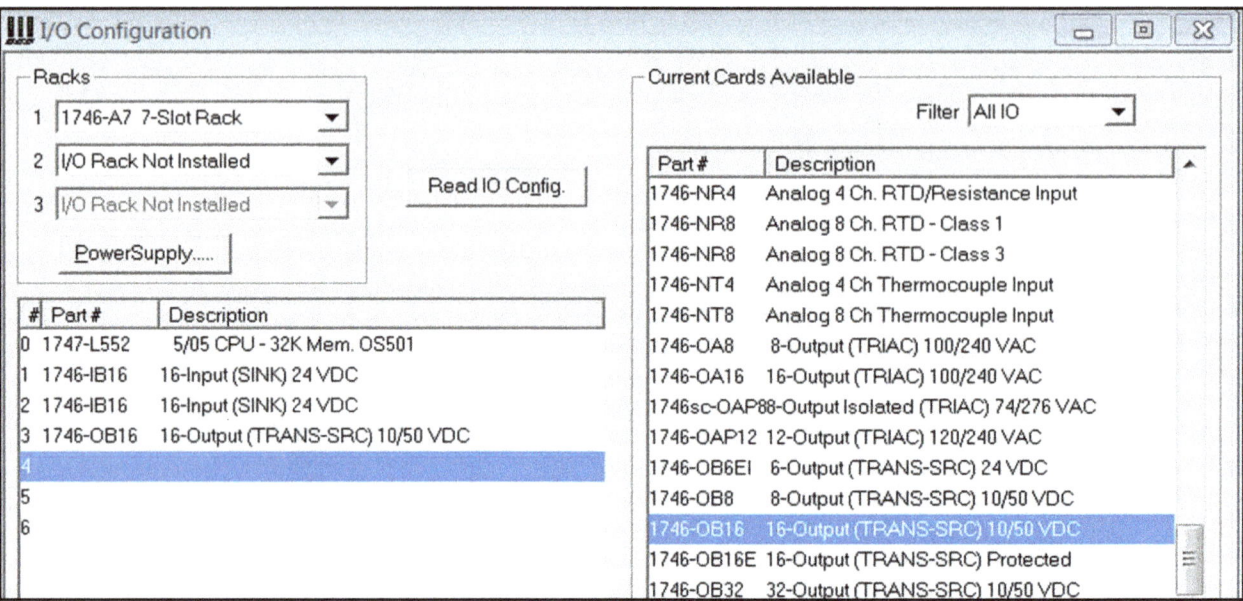

SLC processors are always located in slot 0 on the left side of the rack next to the power supply. Notice the power supply button; this must also be selected.

MicroLogix projects are configured similarly, but some may have I/O integrated with the processor. Also, some of the MicroLogix PLCs have a "rackless" design that allow I/O cards to plug in from the side, building the backplane as you go.

Writing the Program

After configuring the hardware, programming can begin. Organization of the code is done within Ladder 2, which is designated to run first. When the program is first created, LAD 2 (the main routine) is automatically created. Other subroutines are created by right-clicking the Program Files folder and selecting "New".

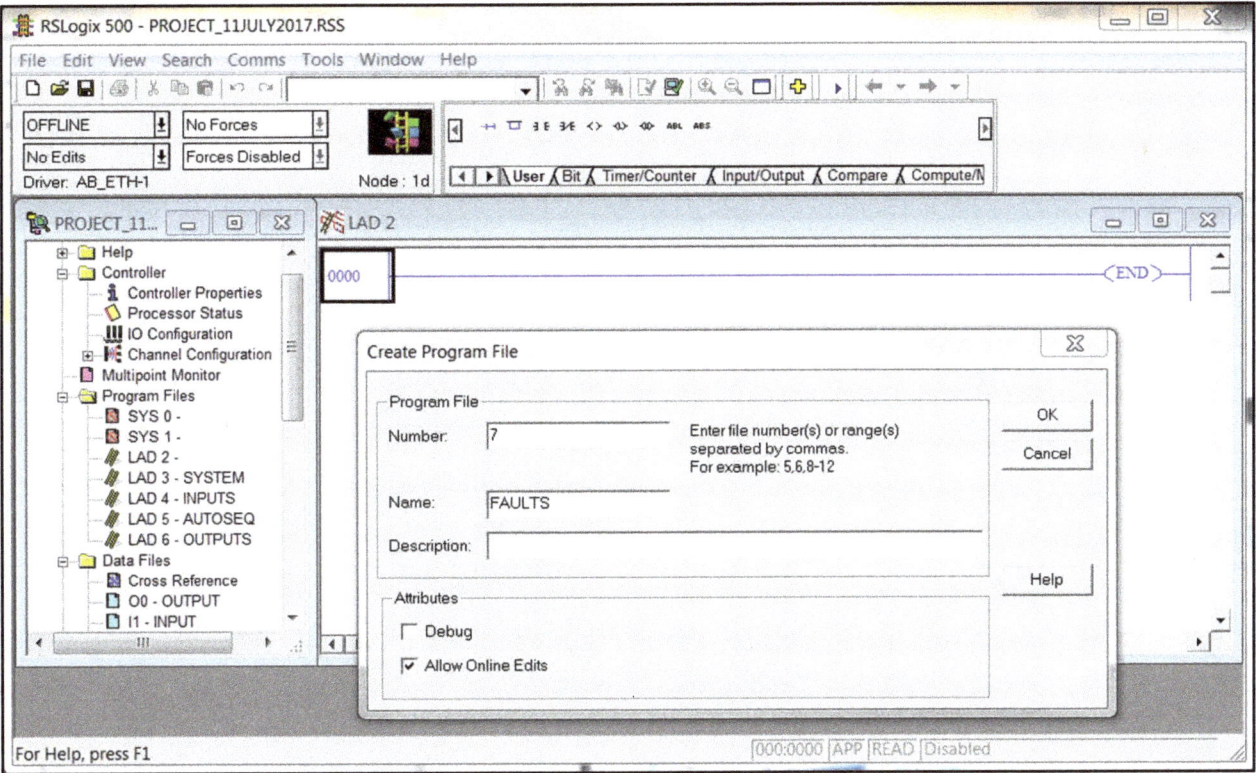

Routines are numbered from 2 to a maximum of 255. The only routine that runs automatically is LAD 2; all other routines need to be called using a JSR (Jump to Subroutine) instruction.

If you don't call the subroutine, the code inside it will not run!

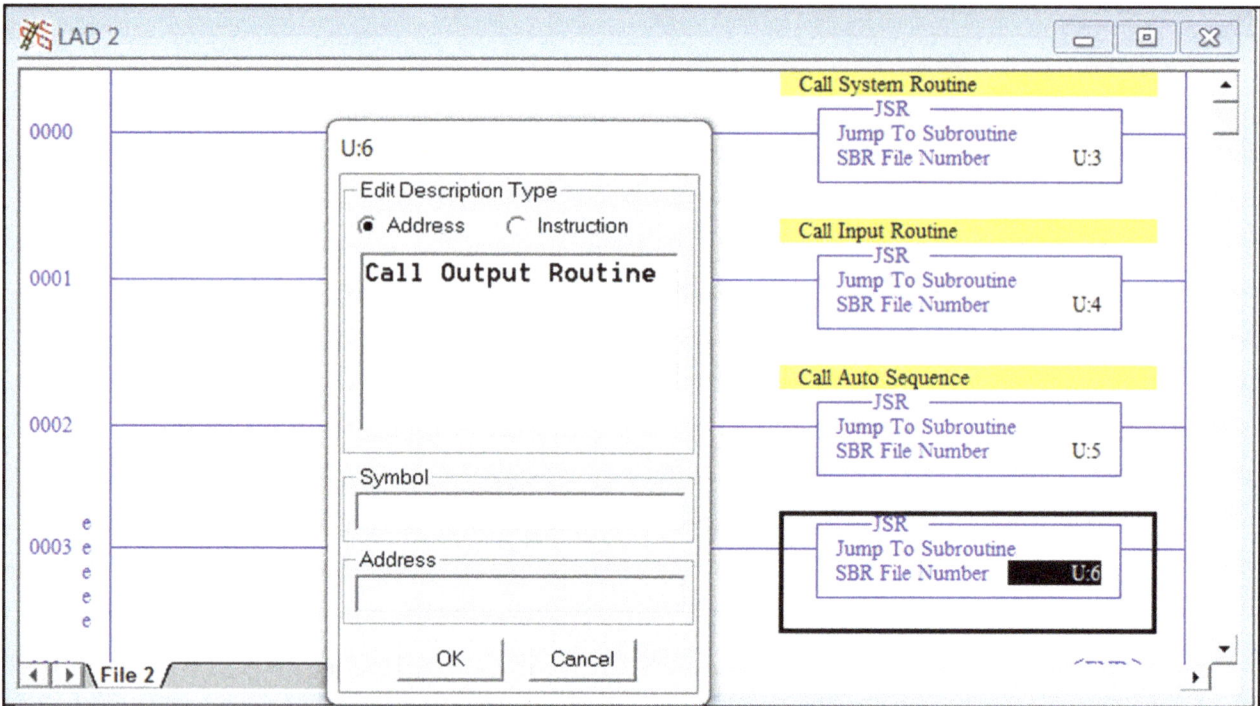

It is typical to use LAD 2 almost exclusively for subroutine calls. Note that the rungs don't have a contact in front of the JSR instruction; they are called unconditionally. This helps in organization, separating different functions of the program into sections so that they can be easily followed.

The lower case "e" on rung 0003 indicates that it is being edited. Instructions can be selected from the tabs in the "Instructions" section or a mnemonic can be typed at the beginning of the rung. In this case, "JSR" was typed with the cursor over the 0003 label and the dialog box for the instruction appeared as shown.

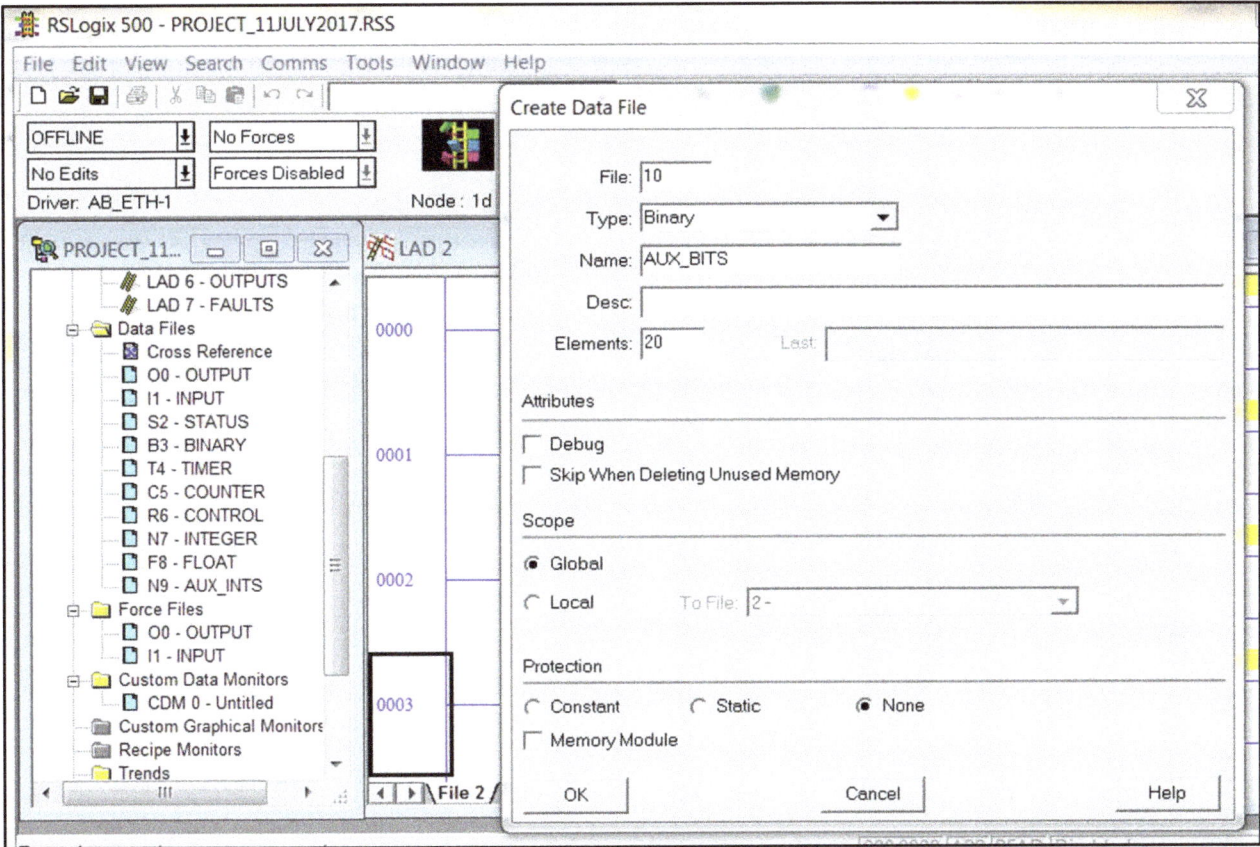

Data files can be added by right-clicking the Data Files folder and selecting "New". As described in the next section on Memory Registers, different types of data registers can be added as required. The Input and Output registers are created automatically based on hardware selection, while B3-F8 are created with the program. An important note on these registers: they only contain one element each when the program is created; you will need to go into the registers and edit the "Elements" field to size the registers. A maximum of 255 elements can be designated, but memory is subject to the limits of the processor chosen.

Allen-Bradley CompactLogix and ControlLogix Platforms

Rockwell Software – RSLogix 5000 and Studio 5000

Until version 20, Allen-Bradley's CompactLogix and ControlLogix programming software was named RSLogix 5000. As of version 20.011, (about 2014), the software was renamed Studio 5000 Logix Designer; many newer products are not supported by the older RSLogix 5000.

Name	Catalog #	Controllers	Description
Service Edition	9324-RLD000ENE	CompactLogix5370 ControlLogix5500 SoftLogix5800 Compact GuardLogix5300 GuardLogix5500	Upload/ Download and View Only
Mini Edition	9324-RLD200ENE	CompactLogix5370 Compact GuardLogix5300	Ladder Diagram (LD) is fully supported. Function Block Diagram (FBD), Sequential Function Chart (SFC), and Structured Text (ST) is <u>not</u> supported.
Lite Edition	9324-RLD250ENE	CompactLogix5370 Compact GuardLogix5300	Ladder Diagram (LD), Function Block Diagram (FBD), Sequential Function Chart (SFC), and Structured Text (ST) <u>is</u> supported.
Standard Edition	9324-RLD300ENE	CompactLogix5370 ControlLogix5500 SoftLogix5800 Compact GuardLogix5300 GuardLogix5500	Ladder Diagram (LD) is fully supported. Function Block Diagram (FBD), Sequential Function Chart (SFC), and Structured Text (ST) is <u>not</u> supported.
Full Edition	9324-RLD600ENE	CompactLogix5370 ControlLogix5500 SoftLogix5800 Compact GuardLogix5300 GuardLogix5500	Ladder Diagram (LD), Function Block Diagram (FBD), Sequential Function Chart (SFC), and Structured Text (ST) <u>is</u> supported.
Professional Edition	9324-RLD700NXENE	CompactLogix5370 ControlLogix5500 SoftLogix5800 GuardLogix5500	Ladder Diagram (LD), Function Block Diagram (FBD), Sequential Function Chart (SFC), and Structured Text (ST) is supported and it includes RSNetWorx for ControlNet, DeviceNet, EtherNet/IP (9357-CNETL3, 9357-DNETL3, 9357-

			ENETL3 individually or 9357-ANETL3 combined), and RSLogix Emulate 5000.

Latest version as of June 2017 is Version 30.

1769 CompactLogix 5370 Controllers

Software:

RSLogix 5000 or Studio 5000, Standard, Mini or Lite Editions

Models:

5370 L3: Max I/O points 960, 3MB User Memory, Max Motion Position Axes 16. Uses 1769 Compact I/O.

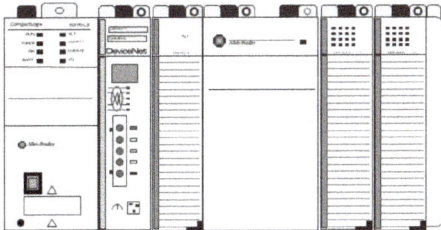

5370 L2: Max I/O points 160, 1MB User Memory, Max Motion Position Axes 4. Uses 1769 Compact I/O.

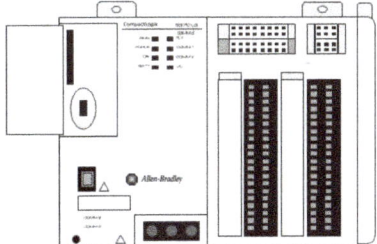

5370 L1: Max I/O points 96, 1MB User Memory, Max Motion Position Axes 2. Uses 1734 Point I/O. Has embedded power supply and I/O.

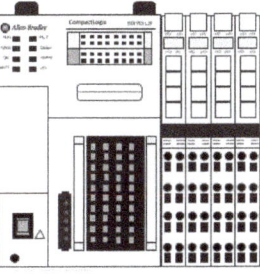

Compact GuardLogix 5370: Up to 960 I/O points, Standard 1, 2 or 3MB, Safety 0.5, 1 or 1.5 MB. 4, 8 or 16 Motion Position Axes. Uses 1769 Compact I/O. Up to 30 expansion Modules.

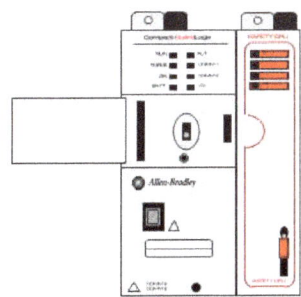

5069 CompactLogix 5380 Controllers

Software:

Studio 5000 Logix Designer, Standard, Mini or Lite Editions

Models:

5069-L3XX These controllers enable high-speed I/O, motion control, dual configurable Ethernet ports that allow dual IP addresses, enhanced diagnostics and troubleshooting. Integrated motion on Ethernet/IP is available up to 32 axes depending on the model.

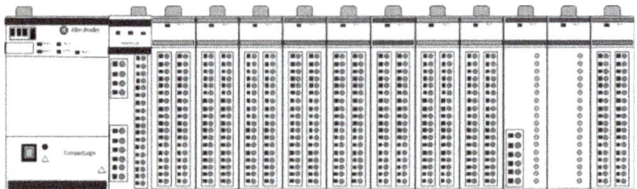

Catalog Number	Application Memory	I/O Expansion	Ethernet Nodes	Motion Axes
5069-L306ER	0.6 MB	8	16	0
5069-L310ER	1 MB	8	24	0
5069-L320ER	2 MB	16	40	0
5069-L330ER	3 MB	31	50	0
5069-L340ER	4 MB	31	55	0
5069-L310ER-NSE	1MB	8	24	0
5069-L306ERM	0.6 MB	8	16	2
5069-L310ERM	1 MB	8	24	4
5069-L320ERM	2 MB	16	40	8
5069-L330ERM	3 MB	31	50	16
5069-L340ERM	4 MB	31	55	20
5069-L350ERM	5MB	31	60	24
5069-L380ERM	8MB	31	70	28
5069-L3100ERM	10MB	31	80	32

5069 CompactLogix 5480 Controllers

Software:

Studio 5000 Logix Designer, Version 30. Available in 2017.

Models:

5069-L4XX Logix based real-time controller that runs on Windows 10 IOT Enterprise. Enables high speed I/O, motion control, Device-level Ring/Linear topologies. Includes three GbE Ethernet/IP ports; two are configurable and one is a dedicated commercial OS network interface. Includes integrated DisplayPort for high-definition monitor connectivity, enhanced security features and two USB 3.0 ports for OS peripheral and expanded data storage capability. Supports up to 31 local Bulletin 5069 Compact I/O modules.

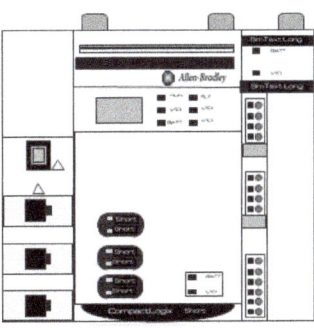

1768 CompactLogix L4x and L4xS Controllers

Software:

Studio 5000 Logix Designer, Standard, Mini or Lite Editions

Models:

1768-L4x These controllers combine both a 1768 backplane and a 1769 backplane. The 1768 backplane supports the 1768 controller, the 1768 power supply and a maximum of four 1768 modules. The 1769 backplane supports up to 16 1769 modules. All controllers have Ethernet/IP and RS-232 Port.

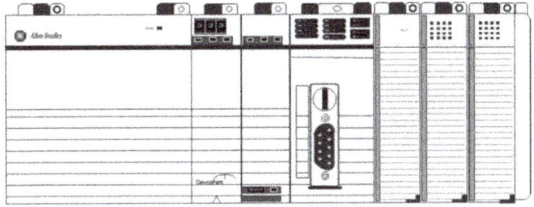

Catalog Number	User Memory
1768-L43	2 MB
1768-L43S	2MB + 0.5 MB Safety
1768-L45	3MB
1768-L45S	3MB + 1MB Safety

ControlLogix 5570 and 5580 Controllers

Software:

Studio 5000 Logix Designer, Ladder Logic, Function Block Diagram, Structured Text and Sequential Function Chart.

Models:

1756-L7X, 1756-L8X Rack-based system for larger projects. Rack sizes are 4, 7, 10, 13 and 17 slots.

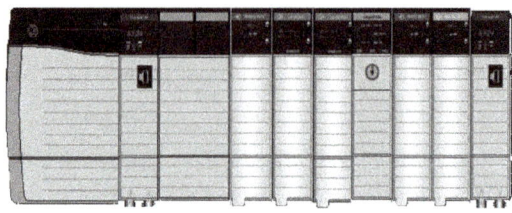

The 1756-L6x series is not supported after the release of Studio 5000 Logix Designer; version 20 of RSLogix5000 is the last firmware revision.

Catalog Number	User Memory	Catalog Number	User Memory
1756-L71	2 MB	1756-L81	3 MB
1756-L72	4 MB	1756-L82	5 MB
1756-L73	8 MB	1756-L83	10 MB
1756-L74	16 MB	1756-L84	20 MB
1756-L75	32 MB	1756-L85	40 MB

The L7X series has a USB Port, while the L8X series has an additional 1GB Ethernet port. An alphanumeric display shows processor status and name. Supports Integrated Motion on Ethernet/IP, full controller redundancy and Removal Insertion Under Power.

GuardLogix Safety Controllers (1756-LXX-S) provide safety and integrated motion in the same chassis, while Extreme Environment Controllers (1756-LXX-XT) allow operation from -25 to 70 C, or -13 to 158 F.

CompactLogix and ControlLogix Instructions

Basic Instructions

Mnemonic	Name	Purpose
XIC	Examine If Closed, Examine On	Examines a bit for an ON condition
XIO	Examine If Open, Examine Off	Examines a bit for an OFF condition
OTE	Output Energize	Turns a bit ON or OFF
OTL	Output Latch	Sets a bit ON when executed, the bit retains its state until unlatched or the register is cleared
OTU	Output Unlatch	Resets a bit OFF when executed
OSR	One-Shot Rising	Triggers a one time event on leading edge of signal, ON for one scan
OSF	One-Shot Falling	Triggers a one time event on falling edge of signal, ON for one scan
ONS	One-Shot (Rising)	Same as OSR
TON	Timer On-Delay	Counts timebase intervals when the energizing instruction is TRUE
TOF	Timer Off-Delay	Counts timebase intervals when the energizing instruction is FALSE
RTO	Retentive Timer	Counts timebase intervals when the energizing instruction is TRUE, retains accumulated value when energizing instruction is FALSE
CTU	Count Up	Increments the accumulated value at each false to true transition and retains the accumulated value when the instruction goes false or when power cycle occurs
CTD	Count Down	Decrements the accumulated value at each false to true transition and retains the accumulated value when the instruction goes false or when power cycle occurs
HSC	High-speed Counter	Counts high-speed pulses from a fixed controller high-speed input
RES	Reset	Resets the accumulator value and status bits of a timer or counter. *Do not use with TOF Timers!

Comparison Instructions

Mnemonic	Name	Purpose
EQU	Equal	Test whether two values are equal
NEQ	Not Equal	Test whether two values are not equal
LES	Less Than	Test whether one value is less than another value
LEQ	Less Than or Equal	Test whether one value is less than or equal to another value
GRT	Greater Than	Test whether one value is greater than another value
GEQ	Greater Than or Equal	Test whether one value is greater than or equal to another value
MEQ	Masked Equal	Test portions of two values to determine whether they are equal through a mask
LIM	Limit Test	Test whether one value is within the range of two other values

Math Instructions

Mnemonic	Name	Purpose
ADD	Add	Adds source A to source B and stores the result in the destination
SUB	Subtract	Subtracts source B from source A and places the result in the destination
MUL	Multiply	Multiplies source A by source B and places the result in the destination
DIV	Divide	Divides source A by source B and places the result in the destination
DDV	Double Divide	Divides the contents of the math register by the source and stores the result in the destination and the math register
CLR	Clear	Sets all bits of a word to zero
SQR	Square Root	Calculates the square root of the source and places the result in the destination
SCP	Scale with Parameters	Produces a scaled output value that has a linear relationship between the input and scaled values
SCL	Scale Data	Multiplies the source by a specific rate, adds to an offset value, and stores the result in the destination

Mnemonic	Name	Purpose
RMP	Ramp	Provides the ability to create linear acceleration, deceleration and "S" curve ramp output data waveforms
ABS	Absolute	Calculates the absolute (positive) value of the source and places the result in the destination
CPT	Compute	Evaluates an expression and places the result in the destination
SWP	Swap	Swaps the low and high bytes of a specified number of words in a bit, integer, ASCII or string file
COS	Cosine	Takes the cosine of a number and stores the result in the destination
SIN	Sine	Takes the sine of a number and stores the result in the destination
TAN	Tangent	Takes the tangent of a number and stores the result in the destination
ASN	Arc Sine	Takes the arc sine of a number and stores the result (in radians) in the destination
ACS	Arc Cosine	Takes the arc cosine of a number and stores the result (in radians) in the destination
ATN	Arc Tangent	Takes the arc tangent of a number and stores the result (in radians) in the destination
LN	Natural Log	Takes the natural log of the value in the source and stores the result in the destination
LOG	Log to the base 10	Takes the log base 10 of the value in the source and stores the result in the destination

Data Handling Instructions

Mnemonic	Name	Purpose
TOD	Convert to BCD	Converts the integer source value to BCD format and stores it in the destination
FRD	Convert from BCD	Converts the BCD source value to an integer and stores it in the destination
DEG	Convert from Radians to Degrees	Converts radians (source) to degrees and stores the result in the destination
RAD	Convert from Degrees to Radians	Converts degrees (source) to radians and stores the result in the destination
DCD	Decode 4 to 1 of 16	Decodes a 4 bit value (0 to 15), turning on the

		corresponding bit in the 16 bit destination
ENC	Encode 1 of 16 to 4	Encodes a 16 bit source to a 4 bit value. Searches the source from the lowest to the highest bit and looks at the first set bit. The corresponding bit position is written to the destination as an integer.
COP	Copy File	Copies data from the source file to the destination file
FLL	Fill File	Loads a source value into each position in the destination file
MOV	Move	Copies the source value to the destination
MVM	Masked Move	Copies the source value to part of the destination
AND	And	Performs a bitwise AND operation
OR	Or	Performs a bitwise OR operation

XOR	Exclusive Or	Performs a bitwise inclusive OR operation
NOT	Not	Performs a NOT (invert) operation
NEG	Negate	Changes the sign of the source and stores it in the destination
FFL	FIFO Load	Loads a word into a FIFO (First In - First Out) stack on each false to true transition. The first word loaded is the first to be unloaded.
FFU	FIFO Unload	Unloads a word from a FIFO (First In - First Out) stack on each false to true transition. The first word loaded is the first to be unloaded.
LFL	LIFO Load	Loads a word into a LIFO (Last In - First Out) stack on each false to true transition. The last word loaded is the first to be unloaded.
LFU	LIFO Unload	Unloads a word from a LIFO (Last In - First Out) stack on each false to true transition. The last word loaded is the first to be unloaded.

Program Flow Instructions

Mnemonic	Name	Purpose
JMP, LBL	Jump to Label and Label	Jump forward or backward to a specified "Label" instruction
JSR	Jump to Subroutine	Jump to a designated subroutine or ladder
SBR	Subroutine label	Designates the start of a subroutine or ladder
RET	Return from subroutine	Returns from a subroutine to the point from which it was called
MCR	Master Control Reset	Turn off all non-retentive outputs in a section of ladder
TND	Temporary End	Mark a temporary end that halts program execution
SUS	Suspend	Identifies specific conditions for program debugging and system troubleshooting
IIM	Immediate Input with Mask	Program an immediate input update using a mask
IOM	Immediate Output with Mask	Program an immediate output update using a mask
REF	Refresh	Interrupt the program scan to update the I/O and service communications

Application Specific Instructions

Mnemonic	Name	Purpose
BSL, BSR	Bit Shift Left or Right	Loads a bit of data into a bit array, shifts the pattern through the array and unloads the last bit of data in the array. BSL shifts data to the left, while BSR shifts it to the right
SQO, SQC	Sequencer Output and Sequencer Compare	Controls sequential machine operations by transferring 16 bit data through a mask to image addresses
SQL	Sequencer Load	Captures referenced conditions by manually stepping the machine through its operating sequences
TDF	Compute Time Difference	Calculates the number of 10 microsecond ticks between any two captured time stamps
FBC	File Bit Compare	Compares bits between two different files

DDT	Diagnostic Detect	Used to monitor machine or process operations to detect malfunctions
RPC	Read Program Checksum	Copies the program checksum from processor memory or from the memory module into the data table

ASCII Instructions

Mnemonic	Name	Purpose
ABL	Test ASCII Buffer for Line	Determine the number of characters in the buffer up to and including the user configured end of line characters
ACB	Number of ASCII Characters in Buffer	Determine the total number of characters in the buffer
ACI	String to Integer	Convert a string to an integer value
ACL	ASCII Clear Buffer	Clear the receive and /or transmit buffers
ACN	String Concatenate	Combine two strings into one
AEX	String Extract	Extract a portion of a sring to create a new string
AHL	ASCII Handshake Lines	Set or reset modem handshake lines
AIC	Integer to String	Convert an integer value into a string
ARD	ASCII Read Characters	Read characters from the input buffer and place them into a string
ARL	ASCII Read Line	Read one line of characters from the input buffer and place them into a string
ASC	String Search	Search a string
ASR	ASCII String Compare	Compare two strings
AWA	ASCII Write with Append	Write a string with user configured characters appended
AWT	ASCII Write	Write a string

Block Transfer and PID Instructions

Mnemonic	Name	Purpose
BTR	Block Transfer Read	Receive data from a remote device
BTW	Block Transfer Write	Send data to a remote device
PID	Proportional Integral Derivative	Controls physical properties such as temperature, pressure, liquid level or flow rate using closed process loops

Interrupt Routine Instructions

Mnemonic	Name	Purpose
	User Fault Routine	Provides the option of preventing a processor shutdown
STI	Selectable Timed Interrupt	Allows you to interrupt the scan of the main program file automatically, on a periodic basis, to scan a specified subroutine file
STD	Selectable Timed Disable	Disables STIs from occurring
STE	Selectable Timed Enable	Enables STIs to occur
STS	Selectable Timed Start	Sets or changes the file number or setpoint frequency of the STI routine
DII	Discrete Input Interrupt	Allows the processor to execute a subroutine when the input pattern of a discrete input card matches a compare value that you programmed.
ISR	I/O Interrupt	Allows a specialty I/O module to interrupt the normal processor operating cycle in order to scan a specific subroutine file
IID	I/O Interrupt Disable	Disables I/O Interrupts from occurring
IIE	I/O Interrupt Enable	Enables I/O Interrupts to occur
RPI	Reset Pending Interrupt	Aborts a pending I/O Interrupt
INT	Interrupt Subroutine	Optional instruction to identify interrupt subroutines

Communication Instructions

Mnemonic	Name	Purpose
SVC	Service Communications	Interrupts the program scan to execute the service communication portion of the operating cycle
MSG	Message Read/Write	Transmits data from one node to another on the network
CEM	ControlNet Explicit Message	Transmits CIP generic commands to other ControlNet nodes via the 1747-SCNR
DEM	DeviceNet Explicit Message	Transmits CIP generic commands to other DeviceNet nodes via the 1747-SDN
EEM	EtherNet/IP Explicit Message	Transmits CIP generic commands to other Ethernet/IP nodes via channel 1

For detailed information see Allen-Bradley Instruction Set Reference Manual 1747-rm001_-en-p

Starting and Editing a Project with RSLogix 5000

  Whether using RSLogix 5000 (version 19 and earlier) or Studio 5000 Logix Designer (version 20.011 and later), the software is installed in the Rockwell Software folder. There may also be an icon on the desktop as a shortcut.

Double-clicking on one of the icons will open the software environment. It is important to know ahead of time the firmware of your PLC controller, since these two versions can only be used to create projects within their scope. If an existing project is opened, the correct version will automatically be selected.

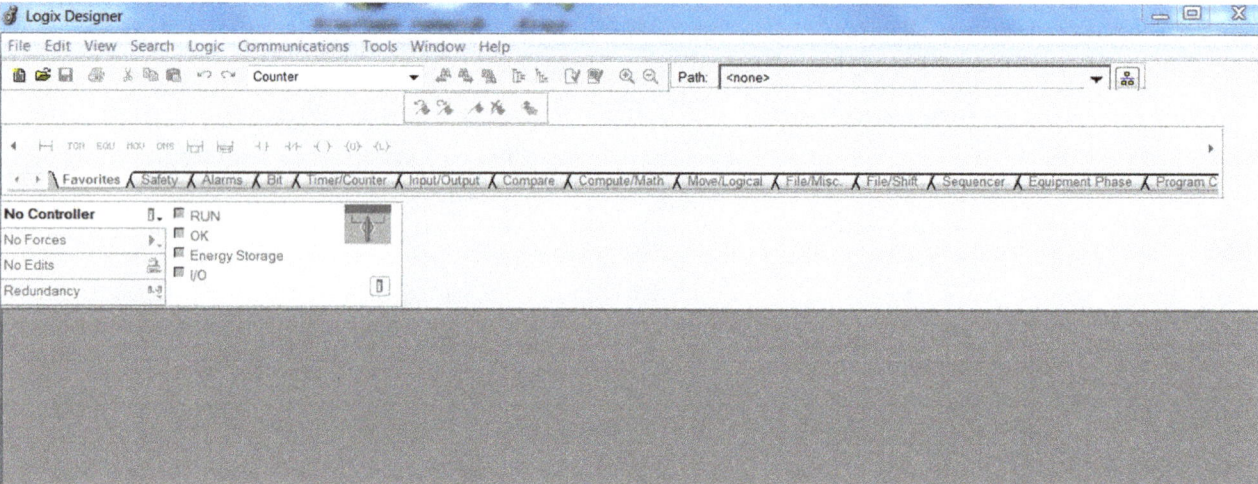

The programming environment looks like the figure shown above. This picture shows that the software selected was Studio 5000 as indicated by the "Logix Designer" title, while the image on the following page is titled "RSLogix 5000".

Creating a Project

To start a new project, a controller needs to be selected and named. This is done from the "File" tab. After selecting "New", the following screen will appear:

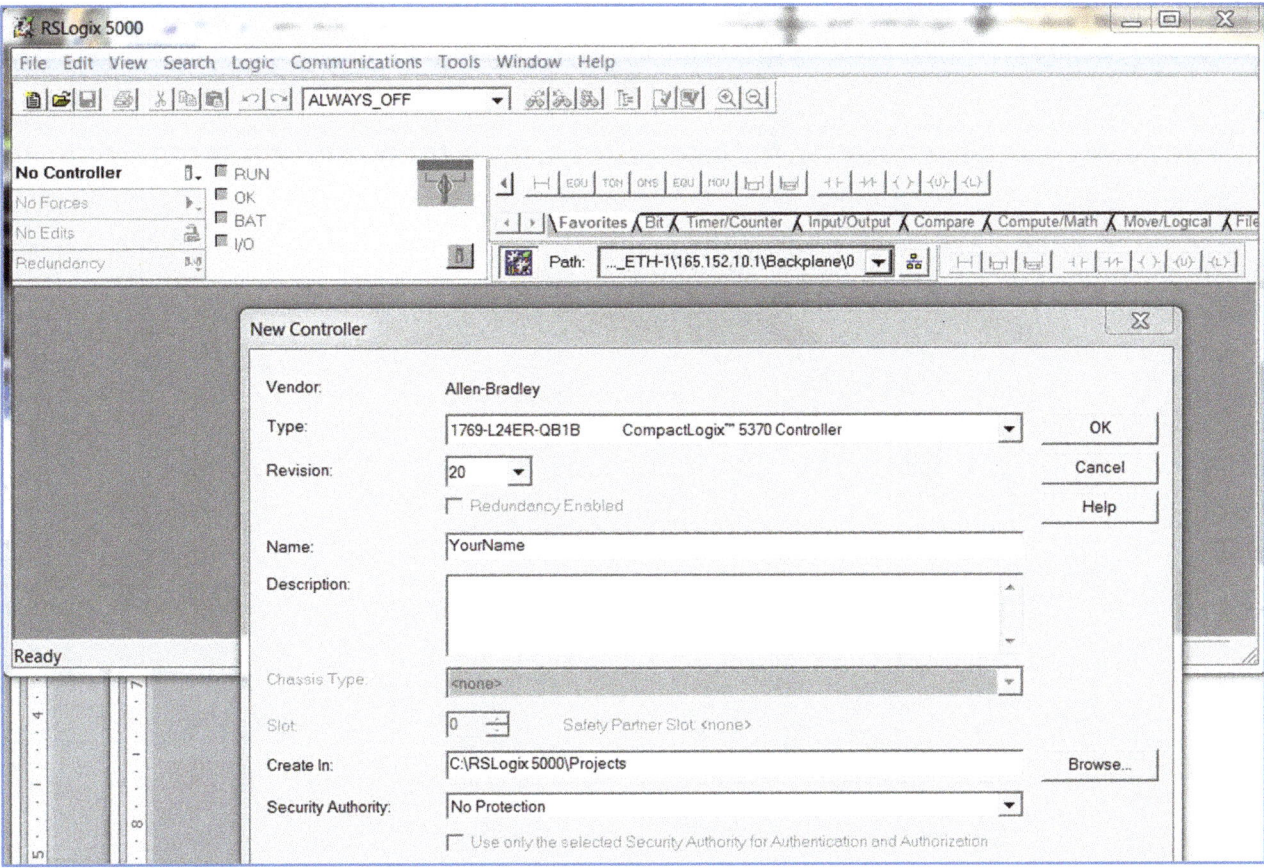

Select the processor and firmware revision and type the name you wish to use. Typically, this will be something descriptive of the machine you are controlling, such as "BearingPress03" or "Line2_Control". For this course, you may wish to use your name.

After the processor name has been entered, press "OK". This is when the project is actually created; the default location for the project is shown above in the RSLogix 5000 folder. If you wish to put it in your own designated location, press the "Browse" button.

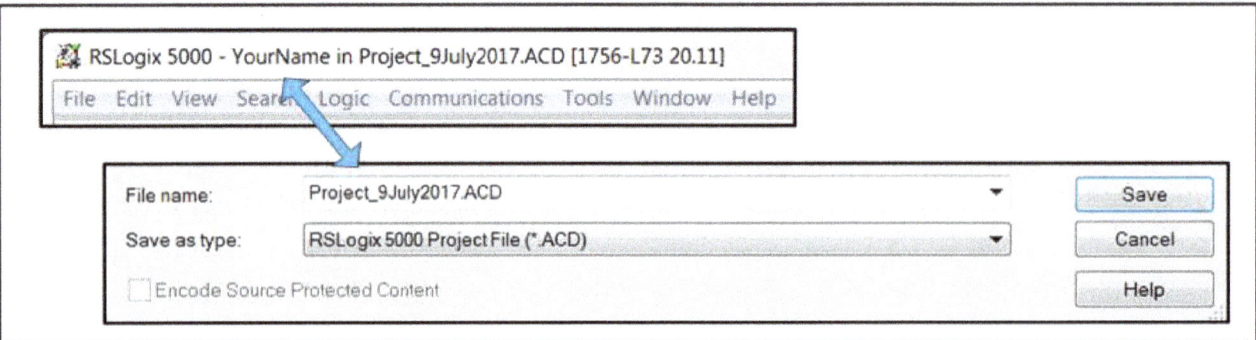

When the project is first created, it will be saved under the processor name. It is also common to save copies of the project with the date embedded, as shown above. When this is done, the header will show the processor name saved "in" the filename.

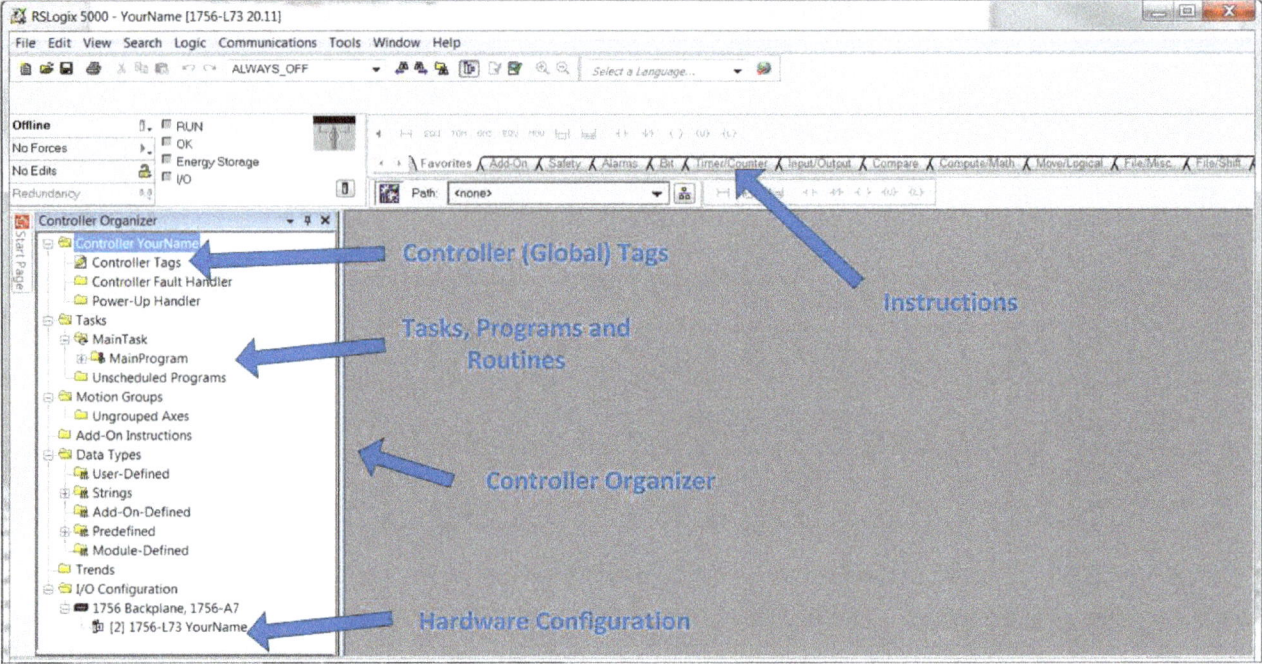

After the project has been created, you are ready to configure hardware, then start designing and programming. The figure above shows some of the different areas of the programming environment. The Controller Organizer on the left side is where you will find all of the different components of your project. This is where you can create new routines and programs, create tags, configure hardware and Ethernet communication networks, and make new data types. Above the Organizer is a status display showing whether or not you are online, in Run mode, or if the processor has a fault.

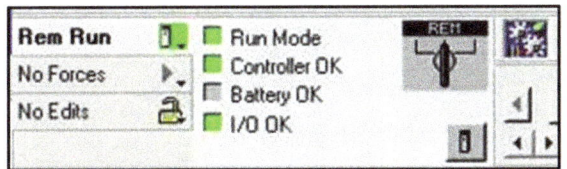

It is important to check the processor status periodically if you are editing online; if editing is done while offline, you will have to download to the processor.

While offline, items in the status area will appear gray; online status will look as shown in the figure above.

Configuring Hardware

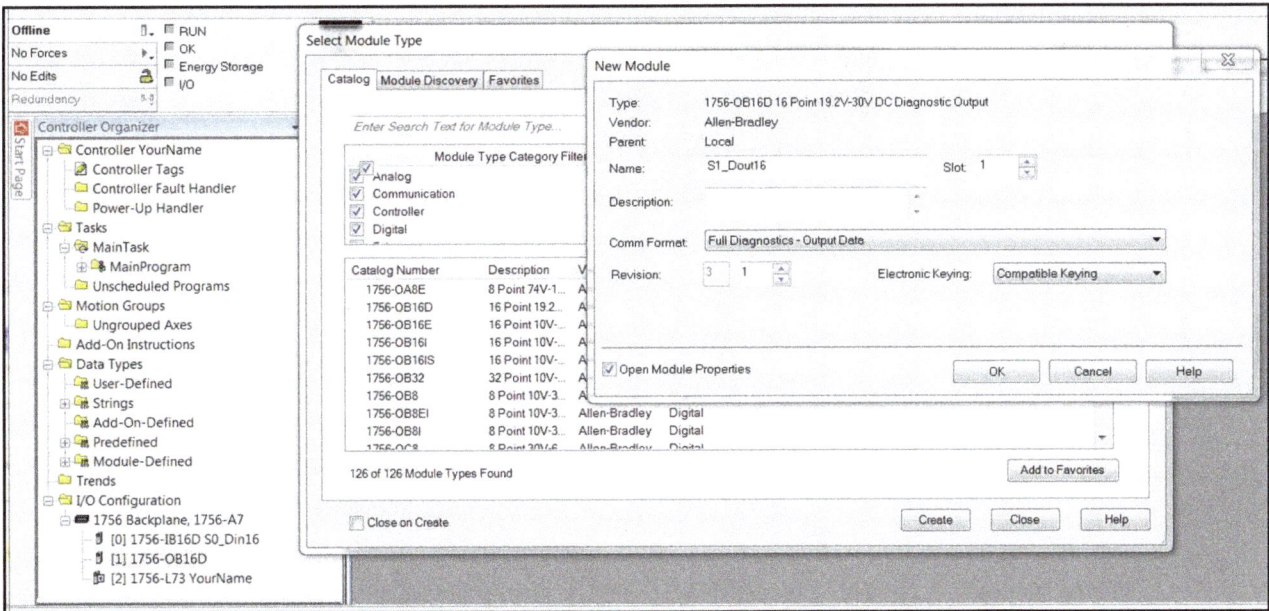

Right-clicking on the Backplane icon will allow selection of modules for your system. Note that in the above figure, the processor was placed in slot 2; the ControlLogix platform allows processors to be placed in any slot. You can even use multiple processors!

Naming your I/O cards is optional. Many different features can be configured for different types of cards; check your documentation for your particular module.

The "Select Module Type" window requires you to select the major revision number. This can't be changed after the card is configured, so it is important to ensure that the correct number is entered. The minor revision number can be changed online or offline after creation.

Electronic keying options include Exact Match, Compatible Keying (shown; the minor revision must match), and Disable Keying. Validated software must select the Exact Match option.

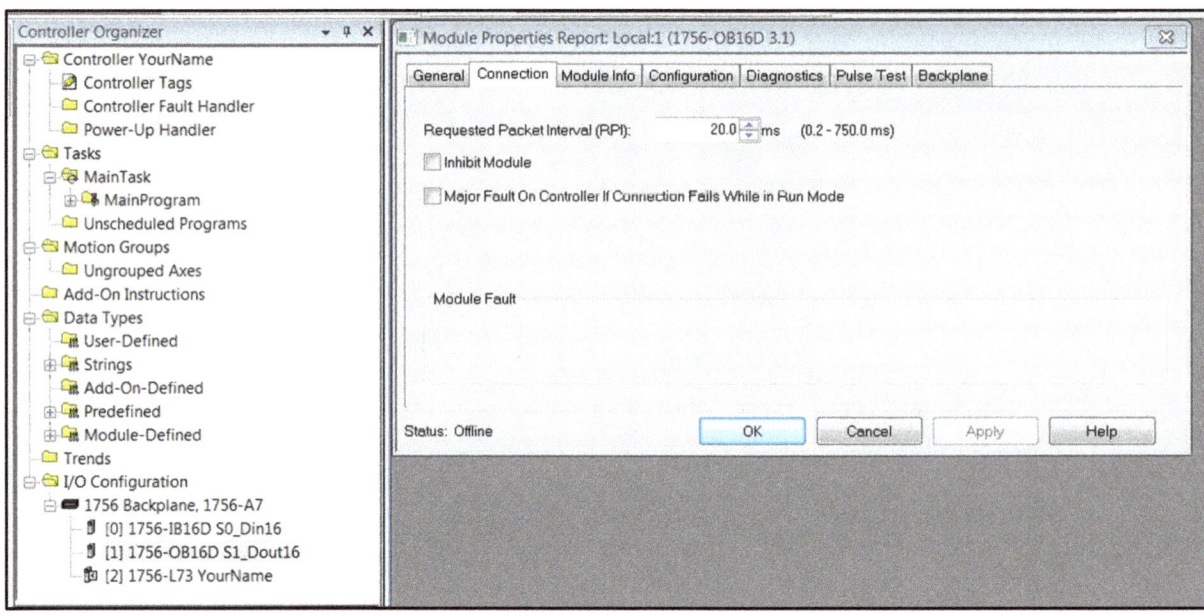

After creating the module, the properties screen will open. There are various settings that can be changed here, but all properties screens have the Requested Packet Interval (RPI) setting. The "Program Processing > Scanning" chapter of this manual has an explanation of this setting, as well as a description of how a PLC scans.

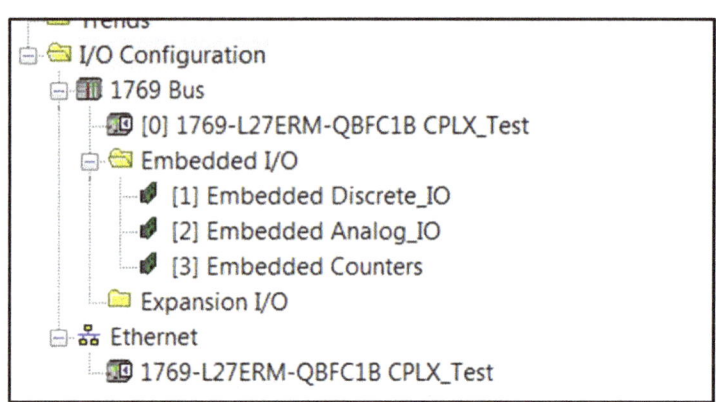

CompactLogix processors often have built-in I/O, and it may not be necessary to add modules. Right-clicking on objects in the I/O configuration and selecting "Properties" will allow them to be configured.

Writing the Program

Tasks, Programs and Routines

After configuring the hardware, programming can begin. Organization of the code is done within the Main Task, which is **Continuous**. There can only be one continuous task, but additional **Periodic** Tasks or **Event** Tasks can also be added. The Continuous Task runs as described in the "Program Processing > Scanning" chapter in this manual.

Tasks can contain multiple programs, which, in turn, contain routines. When the project is created, a "Main Program" is generated, in which is contained a "Main Routine". Programs are added by right-clicking the "Main Task" folder and selecting "New"; routines are added by right-clicking the program.

Each program contains its own Program Tags, which are local to the routines within it. More on this topic is covered in the next section on data.

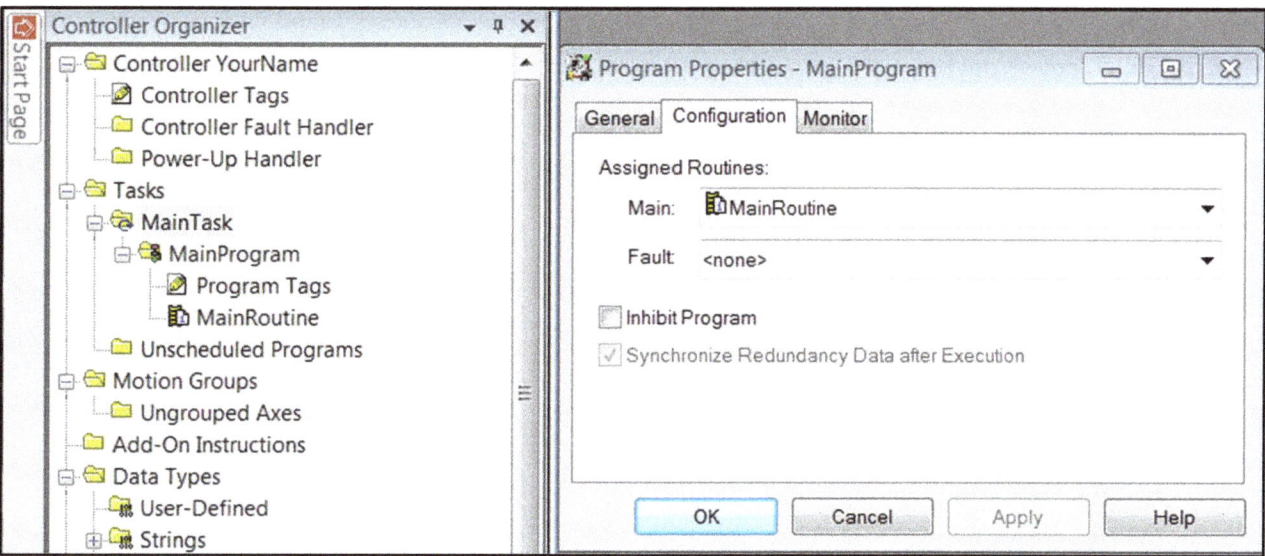

Within each program, a routine needs to be designated to run first. This is done in the dialog box after selecting "Properties" of the program as shown in the figure above. Initially, the "MainRoutine" is configured this way by default, but if you add new programs you will have to select the routine that you want to run as "Main". The name can be anything you wish.

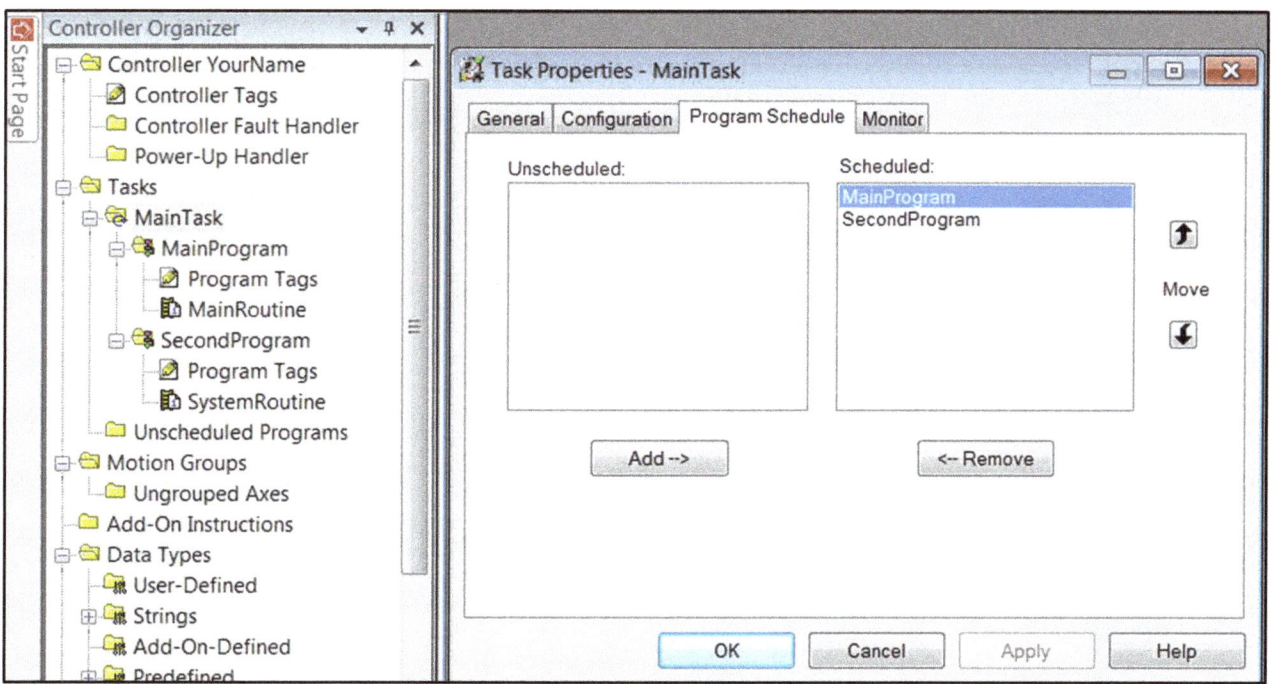

As programs are added to the MainTask, they can be *scheduled* to run in any order. This is done under the Program Schedule tab, which is found under "Properties" of the task. Programs can also be Unscheduled, meaning that they don't run at all.

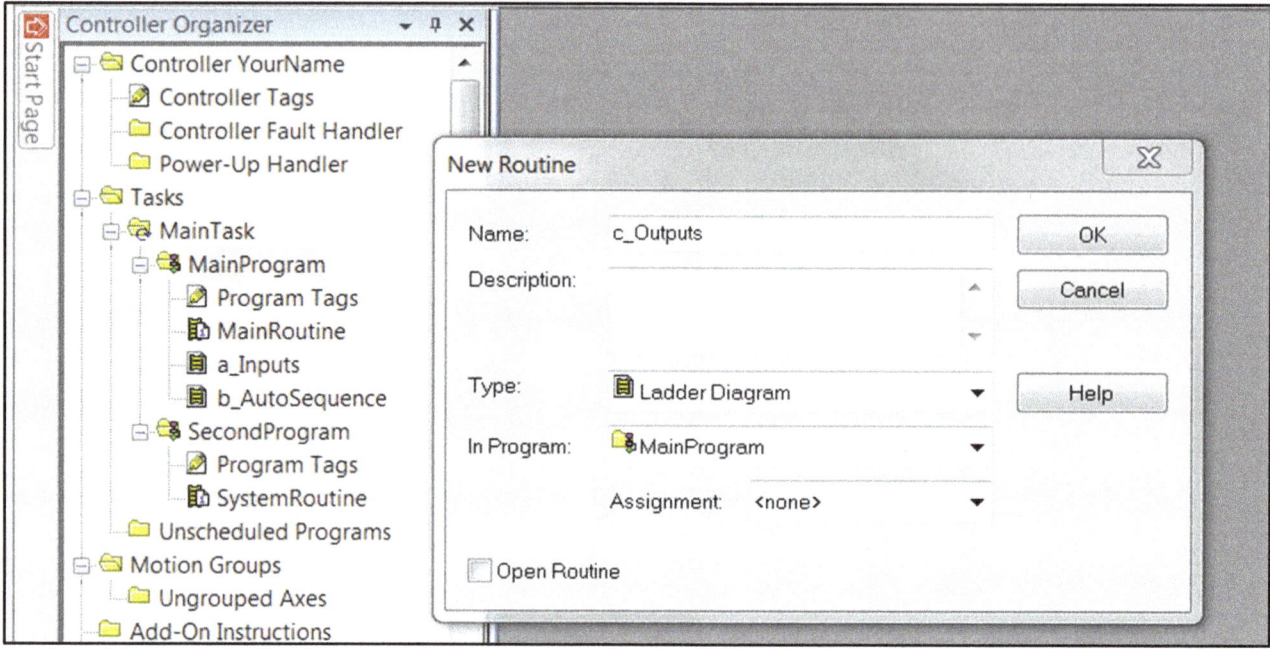

Routines are added by right-clicking the program you want to call the routine from and selecting "New". Routines must be called from the Main routine or the code will not execute. They are not listed in the order they are called; because of this, it is often a good idea to ensure that the routines are alphabetized, as shown in the illustration above, so that the programmer will easily see the order of execution.

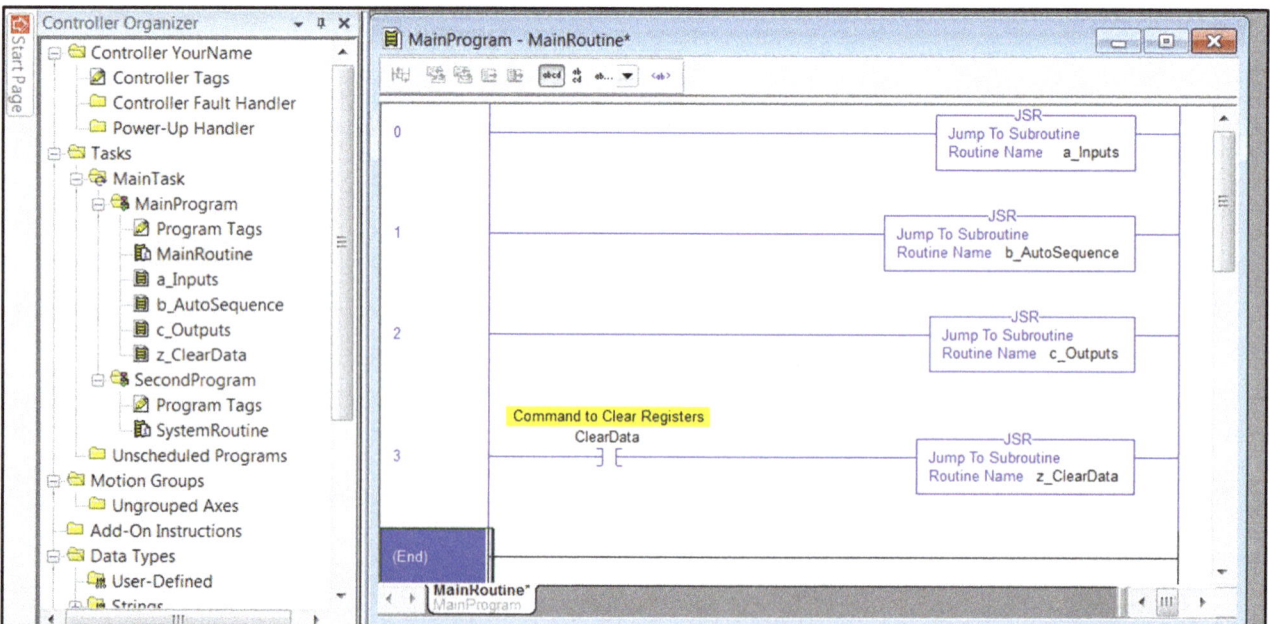

Routines are usually called unconditionally, that is, without a contact in front of the JSR (Jump to Subroutine) instruction. The figure above shows a typical Main routine with unconditional calls, along with a conditional JSR that clears data. Notice that the routines were named in alphabetical order. The MainRoutine and SystemRoutine have a small "1" and appear at the top of the list because they were configured as "Main" in the program properties.

Editing a Routine in Ladder Logic

After a routine has been created, it can be opened for editing by either double-clicking its icon in the Controller Organizer or selecting "Open" after right-clicking on it.

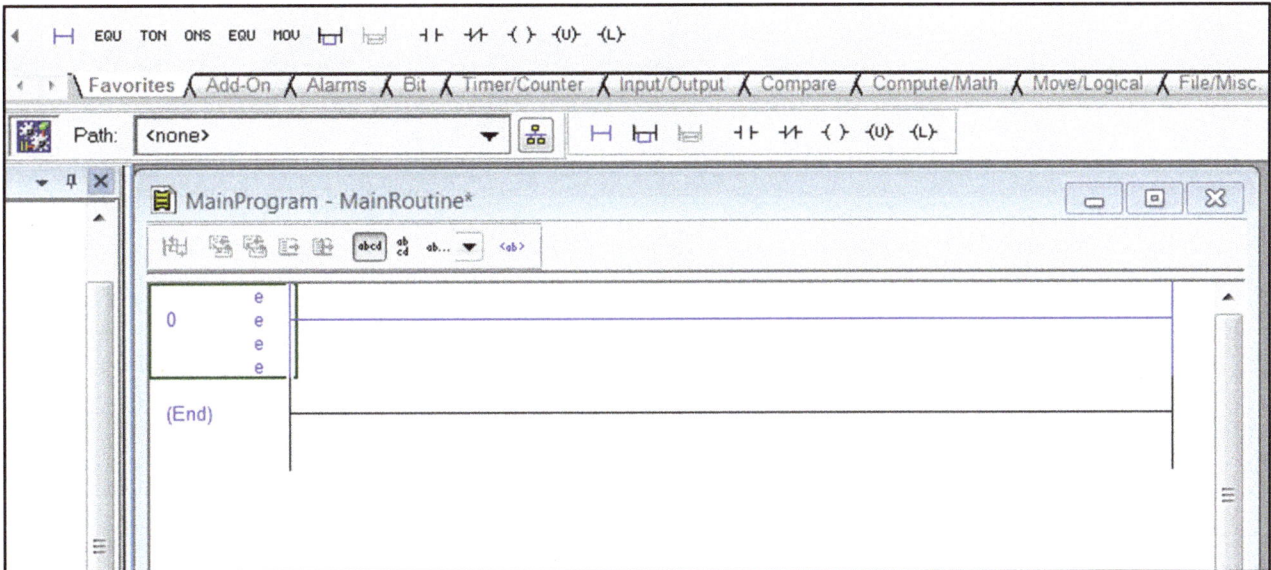

When a new routine is created, it will automatically contain an initial rung of logic as shown above. The lower case "e" means that it is in edit mode; when online, it also signifies that it is not acceptable yet. This can cause problems when new routines are created and not opened before attempting to accept the program or download it.

There are many tabs at the top of the editing window. Selecting one will open a list of **mnemonics** and icons as shown at the upper left of Figure 14. These will differ as each tab is selected.

What is a Mnemonic?

A mnemonic is an abbreviation for an instruction. Allen-Bradley's instructions use three- or four-letter mnemonics to represent instruction names; for example, a normally open contact is abbreviated as XIC, which represents "Examine If Closed". Typing these abbreviations is a shortcut for creating instructions; right-clicking on an instruction and selecting "Change Instruction Type" will allow the programmer to type a mnemonic to replace the instruction.

A list of mnemonics is included in the PLC Hardware section.

After inserting instructions, they must be given an address or name. This is done by creating or using an existing Tag.

ControlLogix and CompactLogix Data

Both the ControlLogix and CompactLogix are tag-based systems, supporting all IEC data types. User defined types (UDTs) can also be defined by the programmer. Tags may be up to 40 characters in length and can be composed of alphanumeric characters and the underscore (_) character.

Data Type	Name	Description
BOOL	Boolean	A single on or off element. A bit.
SINT	Single Integer	8 Bits, or a Byte
INT	Integer	16 Bits, or a Word
DINT	Double Integer	32 Bits, or a Double Word
REAL	Real Number, or Float	32 bit Floating Point Number

Many other complex data types are made up of these "elementary" types. Arrays are also supported up to three dimensions.

The ControlLogix platform operates on a "Producer - Consumer" object model, where the Controller and other physical objects such as I/O cards produce and consume information based on a time schedule, known as an RPI, or "Requested Packet Interval". Tags sent and received from other ControlLogix-based systems can also follow this protocol.

Older PLC platforms, such as the PLC 5 and SLC 500, used numeric addresses corresponding to registers. With these PLCs, addresses are categorized by the type of address (bit, integer or real/floating point). Additional registers for the PLC processor or program status were also used. Timers, Counters and control words for more complicated instructions also had their own registers.

Tags can be named however the programmer wishes. Tags still point to memory address registers, but the user doesn't need to know where the registers are. They do need to know what type of data a tag represents though.

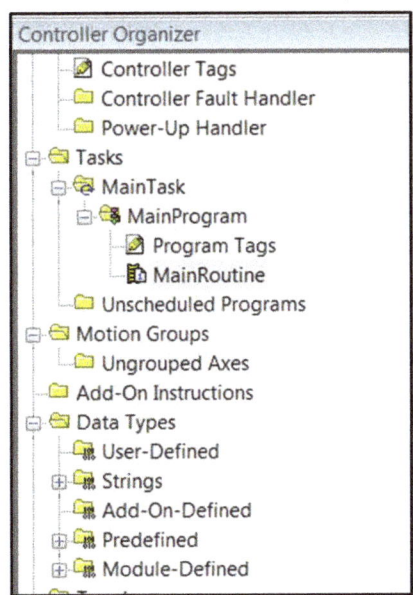

In addition to the simple data types, more complex types, such as timers and counters, also are made up of the data types listed. The figure on the left shows the section of the Controller Organizer that contains all of the different data types; the folder labeled "Predefined" contains all of Allen-Bradley's instruction-based data types. The Module-Defined folder contains all of the data structures for I/O cards, while the Strings folder contains different sized String (text) arrays. User Defined and Add-On-Defined tags are created by the programmer.

Arrays

An Array is a group of similar data types. If 10 Floating Point or REAL numbers are needed, an array can be created as shown below:

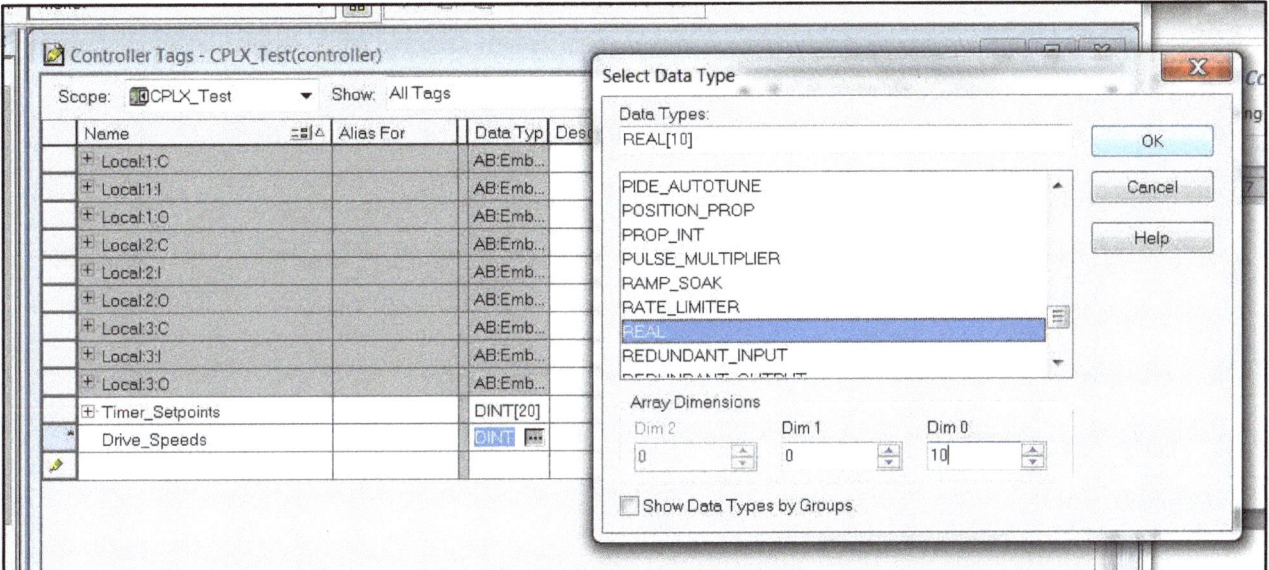

After the name of the tag is typed into the field, the data type is selected by either using a drop-down menu or by typing in the data type. Entering a number or numbers into the dimension fields will allow that number of tags to be created at once. The tag named "Drive_Speeds" will have 10 REAL numbers in it if the OK button is pressed. The tags will be named Drive_Speed[0], Drive_Speed[1] etc., up to Drive_Speed[9].

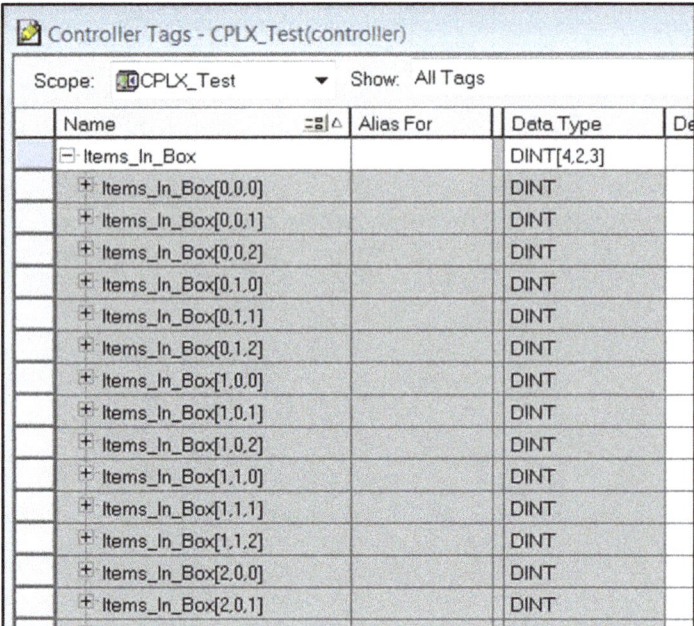

Multi-dimensional arrays can be useful for physical objects. For instance, if it was important to keep track of the number of items in a stack of boxes, a tag could be created with three dimensions, one each for level, column and row. Of course, it would then be important to remember which name represented each dimension direction!

User Defined Data Type (UDT)

A User Defined Data Type, or UDT, can be used to create groups of data that represent an object, such as a recipe, a manufacturer's product, or a device, such as a Variable Frequency Drive (VFD).

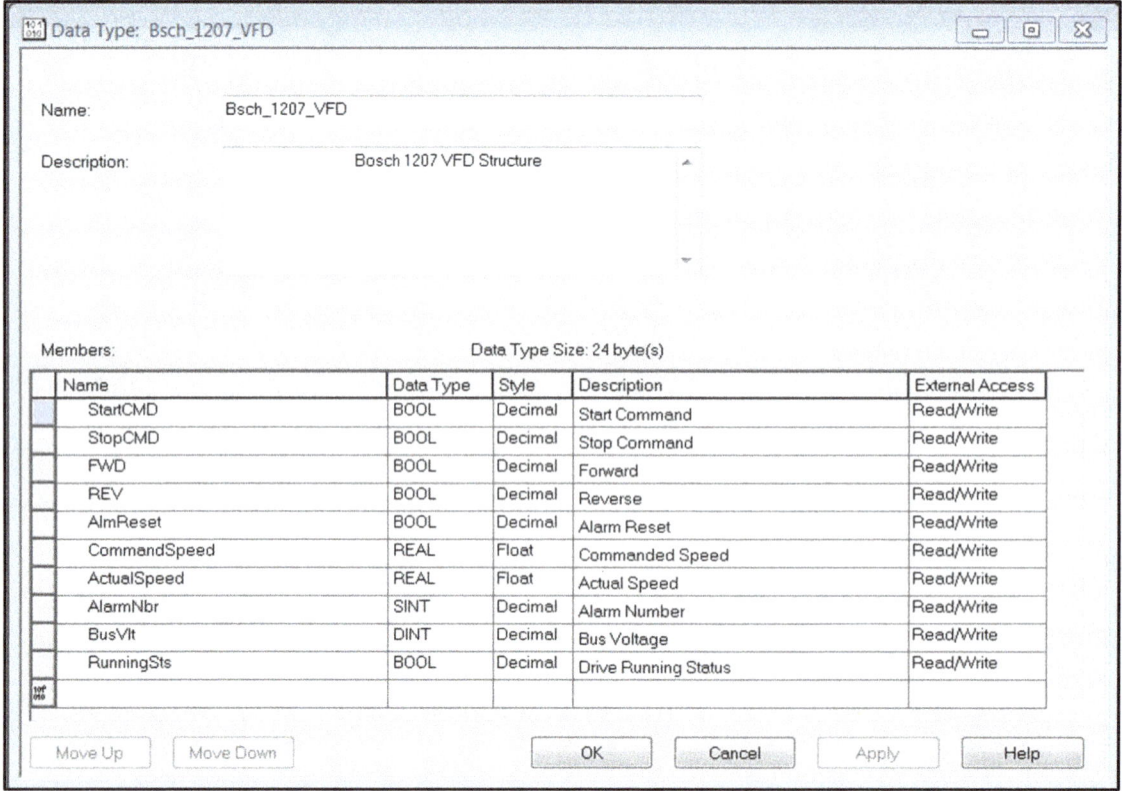

To create a UDT, right-click on the User-Defined folder in the Data Types section of the Controller organizer. Items are typed into the form very similarly to how a new tag is created.

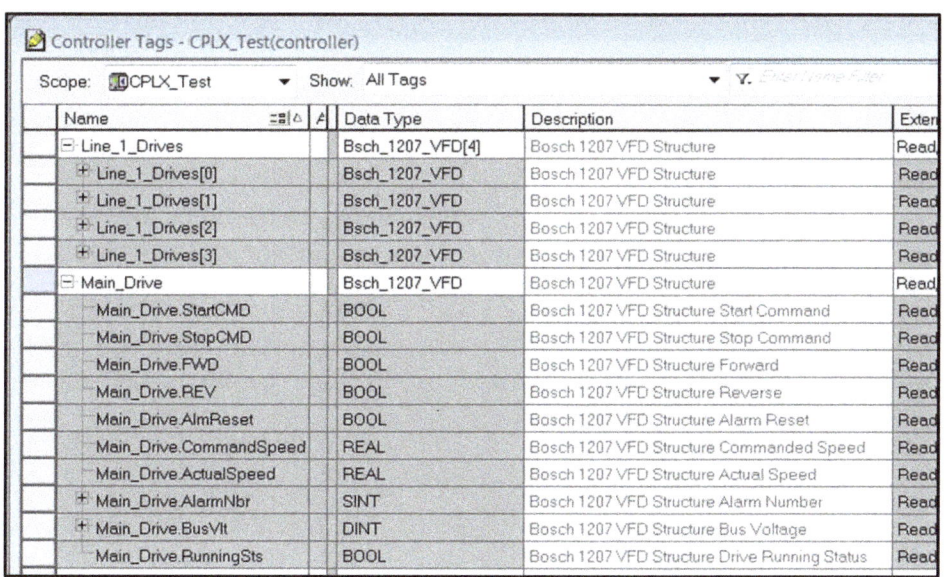

Note in the figure at left that the UDT itself is <u>not a tag</u>; instead, the newly created tag uses the UDT as its data reference. After saving the UDT, its name will be selectable as a data type. The comments are represented in the description, and UDTs can be created as arrays.

Global Tags

Rather than selecting addresses from registers as with the older Allen-Bradley platforms, a database will be created that looks something like the figure below:

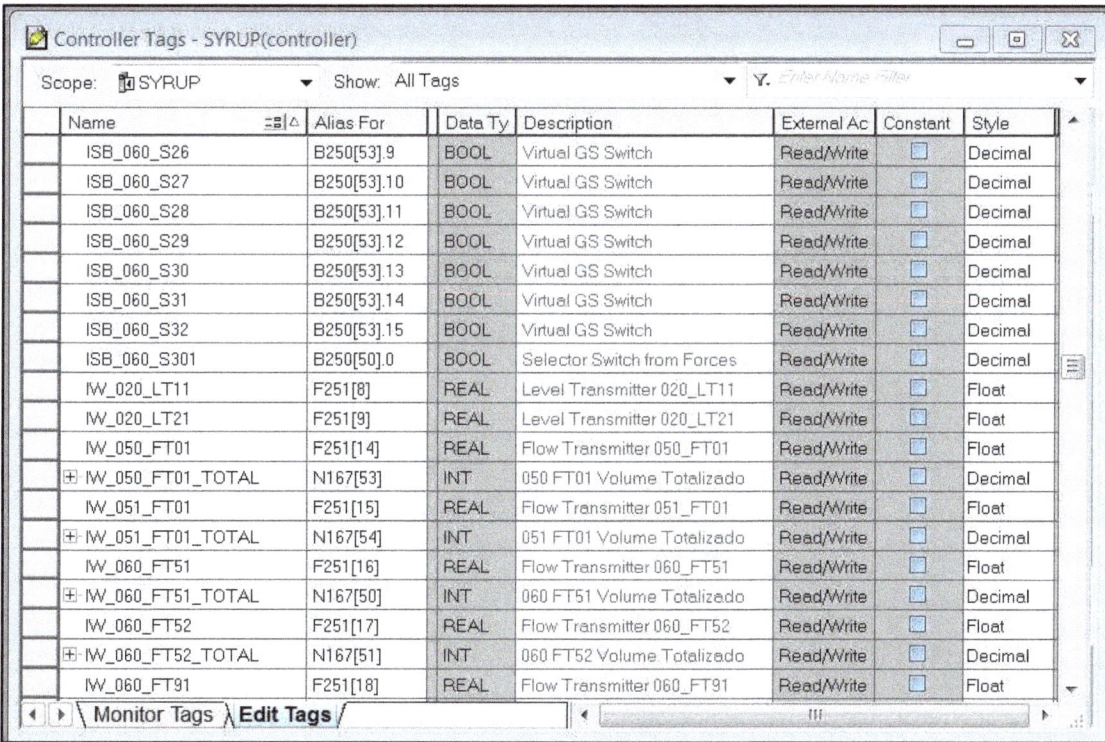

Notice that these tag names are not very descriptive and that they have an "Alias" column with register-type addresses in it. The tag names are derived from P&ID symbols, and the aliases come from another Allen-Bradley PLC platform, RSLogix500. This program was converted from a SLC500 program into this newer ControlLogix processor!

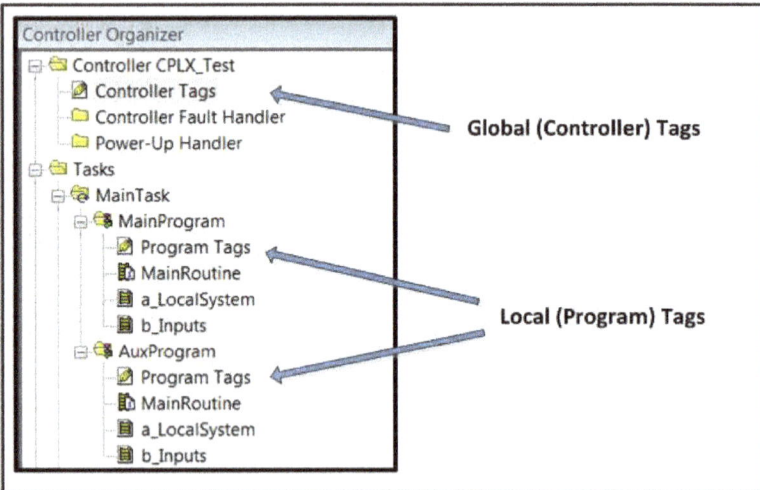

The "Scope" window at the top of the tag database dialog box shows the name SYRUP and has an icon that looks like a processor card. This indicates that the name of the processor is SYRUP and that this list of Controller tags is accessible throughout the program. Controller tags are global, which means that all programs within the PLC can access them.

Program (Local) Tags

Most tag names will be more descriptive than the previous example as shown in the figure below:

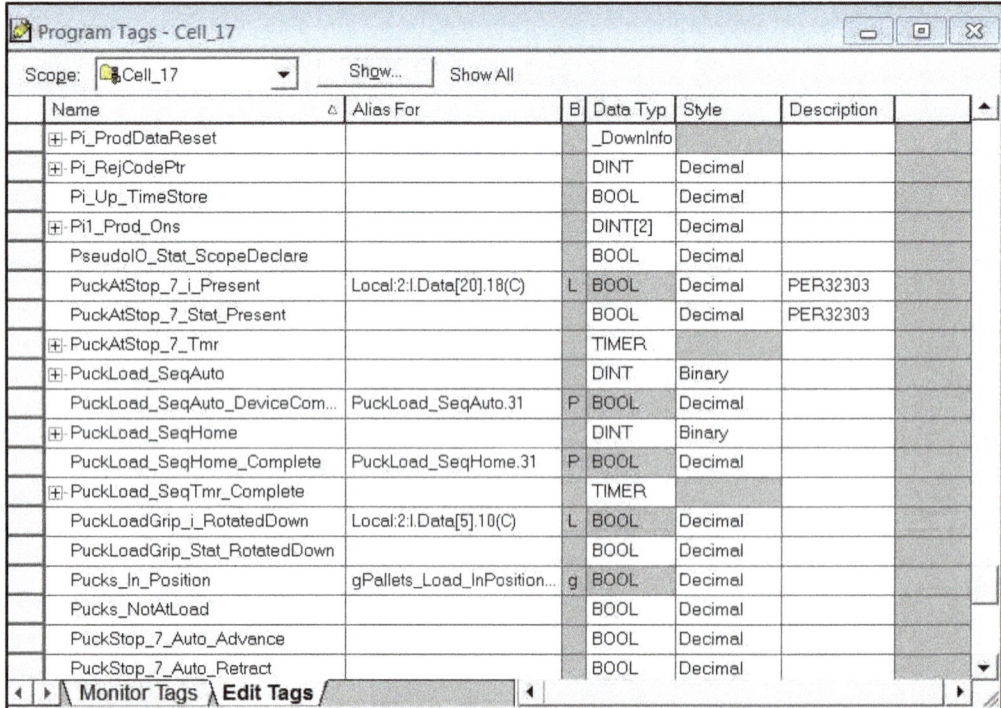

Program tags can only be seen or accessed by the particular program they are in. This particular program has the name "Cell 17"; other programs within the controller, such as those from the program "Cell 16", cannot access these tags.

Aliases

An alias is a "nickname" for a tag. In the previous figure, the tag "PuckAtStop_7_i_Present" has an address listed next to it of Local:2:I.Data[20].18. This means that the tag is connected to the physical device at that local address. Since the address is Local:2, it is accessed through slot 2 of the local I/O rack. In this case, that card is a DeviceNet card with nodes assigned to it, and Node 20 is the place that the tag is linked to. If the physical input changes state, the corresponding aliased address will do the same.

It is common to alias physical inputs and outputs to more descriptive tag names like this. To alias a tag to another, simply browse for the tag you wish to alias to in the "Alias For" column.

Add-On Instructions (AOIs)

In addition to the instructions provided by Allen-Bradley and listed in this book, it is also possible to build custom instructions.

To create an AOI, click inside the Add-On Instructions folder in the Controller Organizer and select New Add-On Instruction. A dialog box will open as shown below:

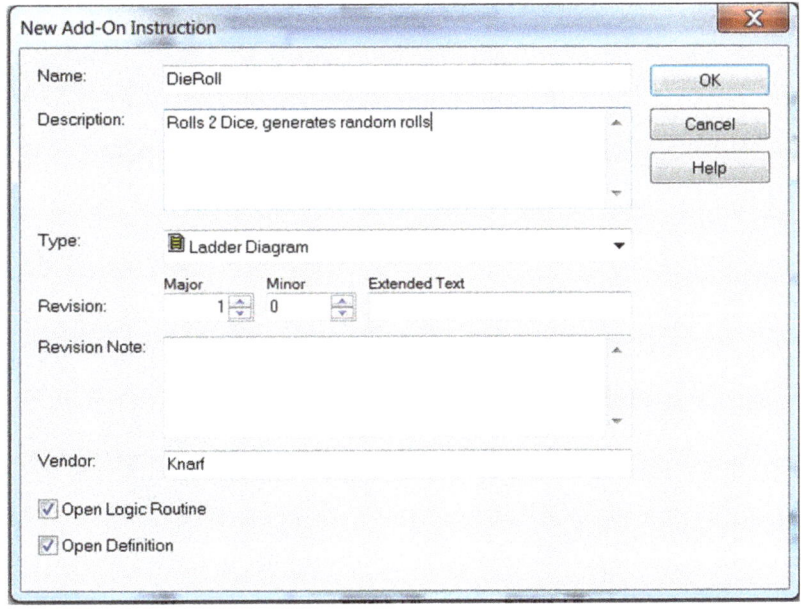

The purpose of an AOI is to create a reusable instruction that can be used in multiple projects, just like any other instruction in the tabs at the top of the editor.

This example is a fun instruction that acts as a random number generator: it rolls a set of dice.

The name of the instruction will appear as whatever you place in the Name field, so shorter names are better.

Since the boxes at the bottom of the New Add-On Instruction dialog are checked, both the logic routine and the Definition dialog will open.

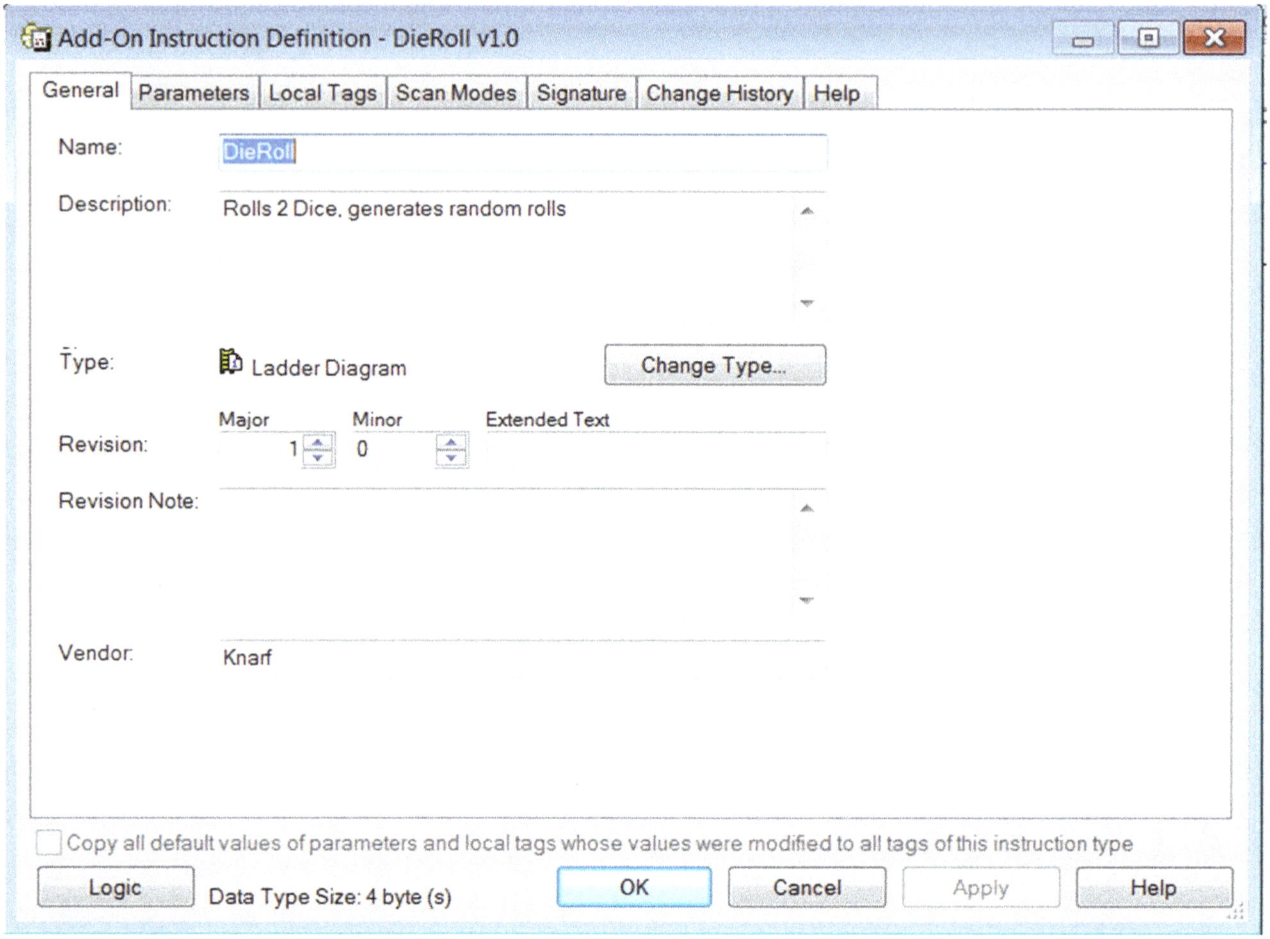

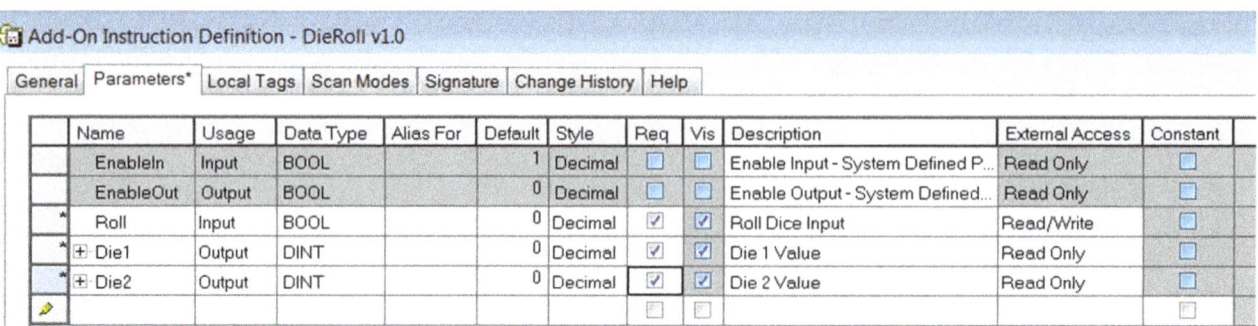

The Definition can be opened by right-clicking the name of the instruction in the Add-On Instructions folder and selecting "Open Definition" at any time. It contains a number of tabs that can be selected to configure various features of the instruction. The General tab is shown above.

The **Parameters** tab is used to configure the Input and Output variables that will be used to pass data into and out of the instruction. The Name will appear inside of the instruction if the Vis (Visibility) box is checked for that row, and the parameter is optional if the Req (Required) box is not checked.

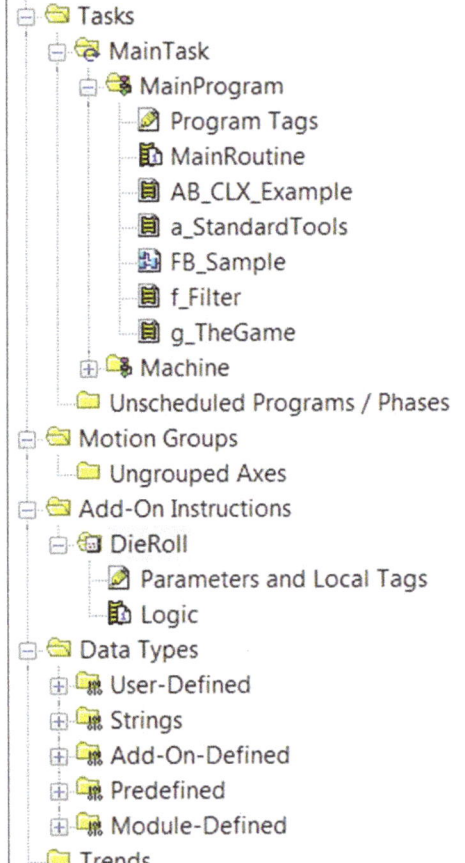

After saving the configuration, the AOI appears in the folder as shown to the left. As mentioned previously, double-clicking the object will open it for editing.

An important thing to remember when creating AOIs: they cannot be created while online! They can, however, be created in a separate offline project and then imported by right-clicking on the Add-On Instruction folder and selecting "Import Add-On Instruction".

The **Local Tags** tab will contain any internal tags that are created while building the logic. They can be filled in from the tab or be created by right-clicking in the tag field when it is entered in the logic. Double-clicking the Parameters and Local Tags icon opens this also.

The **Scan Modes** tab is used to add additional routines that will run before the Logic routine (Prescan), after the Logic Routine (Postscan) or if the EnableIn parameter is false (EnableInFalse).

The **Signature** tab is used to generate a code that uniquely identifies the instruction and seals it from modifications.

The **Change History** tab logs the time and date of modifications, and the **Help** tab allows the programmer to write descriptive information that documents the instruction and provides information to the end-user.

Roll the Dice!

The logic in this example uses two free running timers to generate numbers based on when the player presses and releases a pushbutton. Since the timer setpoints are both multiples of six, it is easy to divide the accumulator values and have an equal random chance of rolling 1-6.

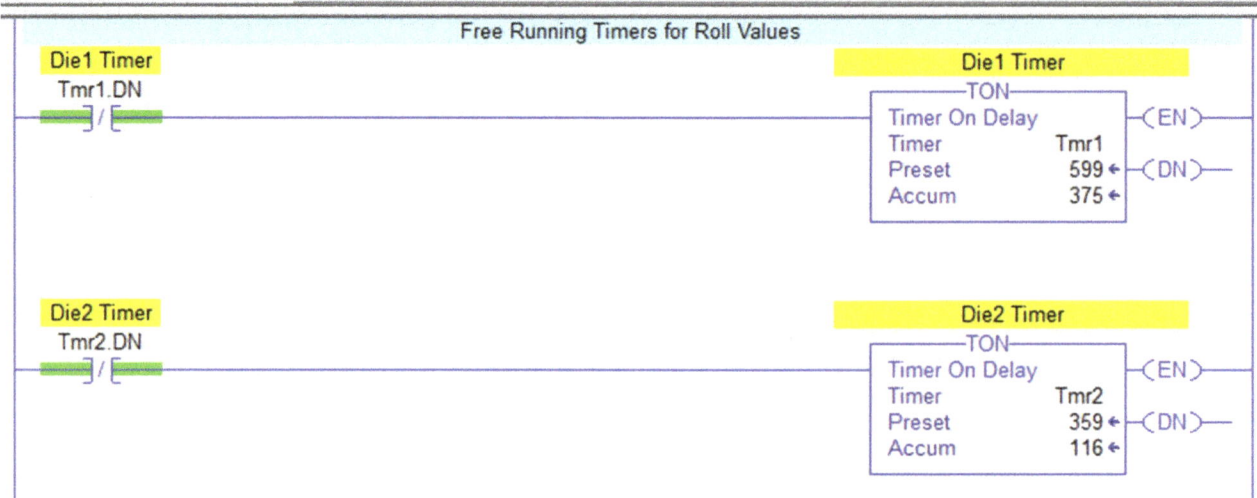

The two free running timers have different setpoints so that they run at different speeds.

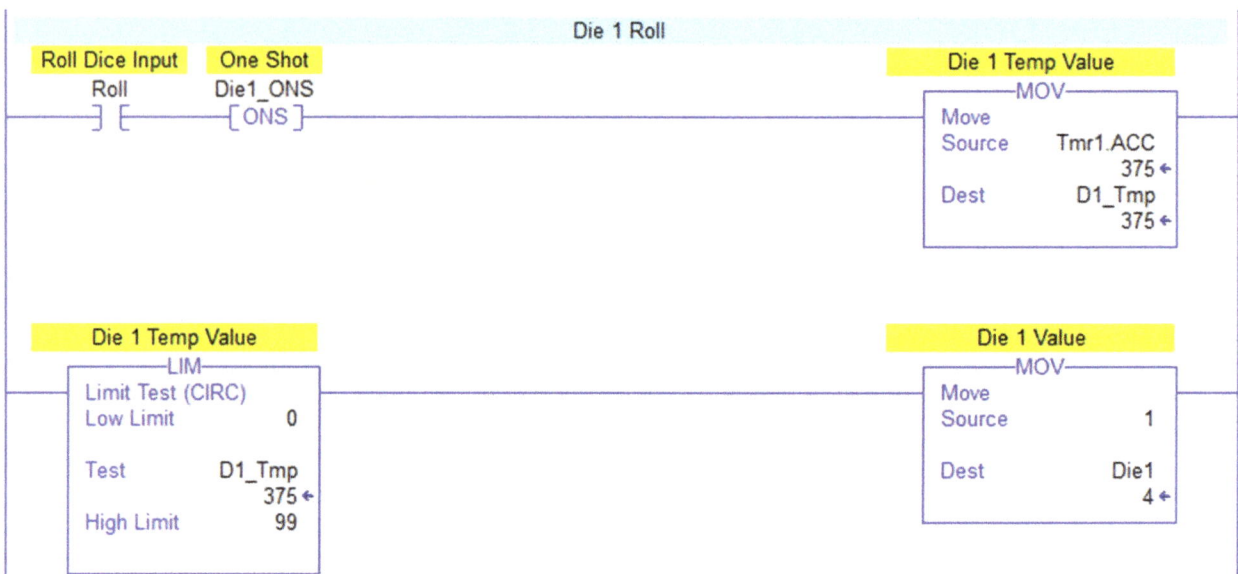

When the Roll input occurs, a one-shot captures the timer accumulator into a temporary variable. A Limit instruction is used to divide the captured value into six different ranges.

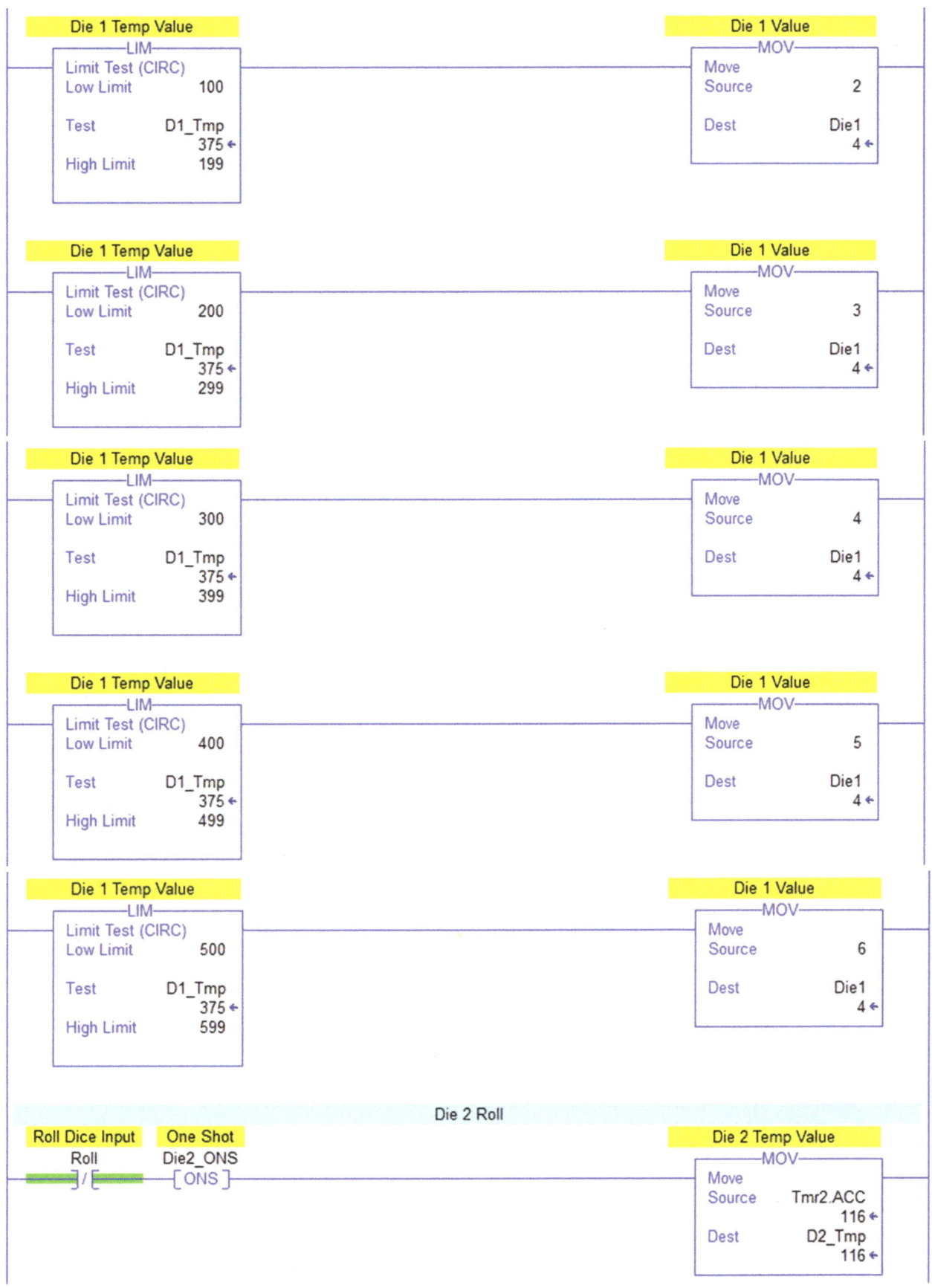

Die 2 is captured with a one-shot when the player releases the button.

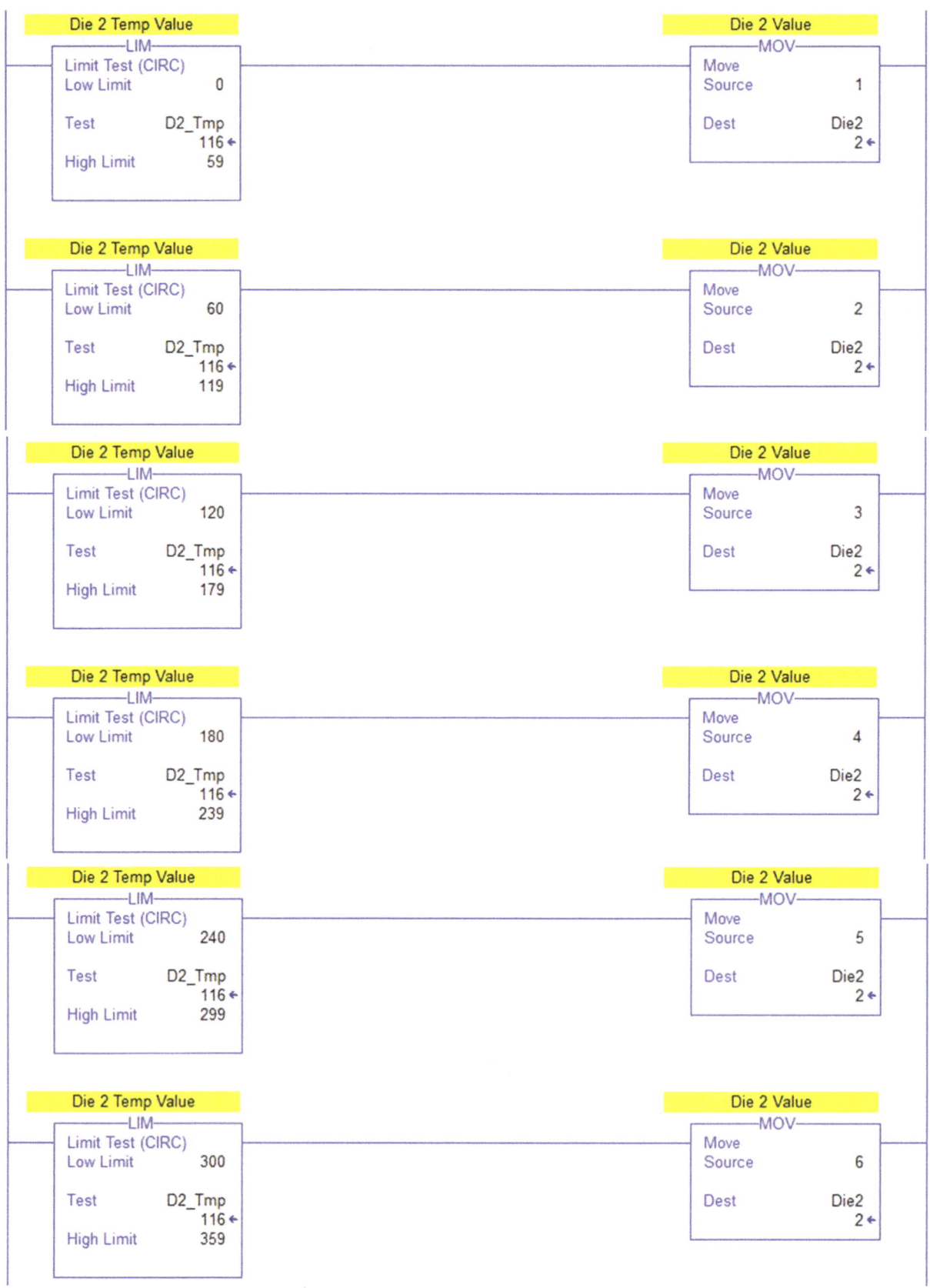

Die 2 ranges are 60 instead of 100.

After the logic is written, the AOI is ready to use. It appears in the Add-On folder as DieRoll since that was the name it was created under. When it is placed into a Rung, it appears as in the diagram below.

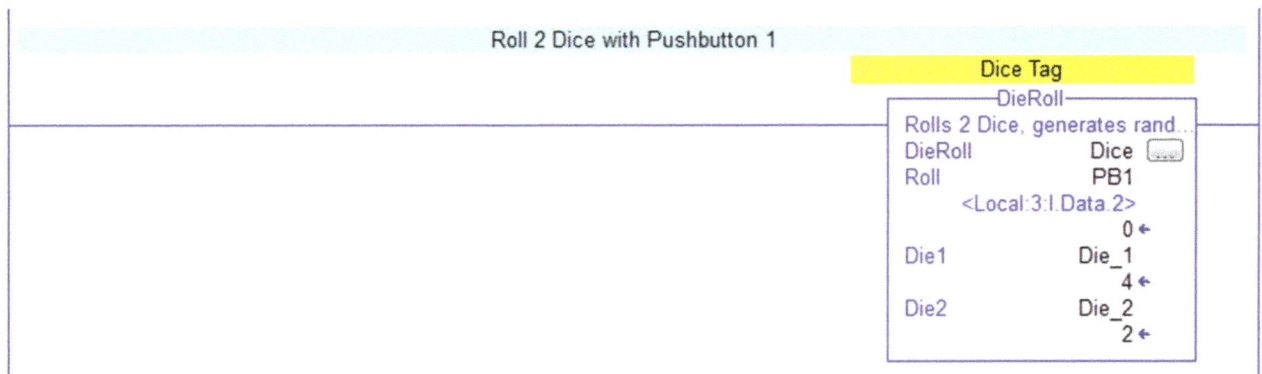

The AOI requires a tagname for each use of the instruction; in this case, it is "Dice". The tag includes the input and output variables created under the Parameters tab, as well as the EnableIn and EnableOut BOOLs.

The input assigned is a pushbutton aliased to Input 3.2 in an Allen-Bradley ControlLogix PLC, and the Die_1 and Die_2 tags are integer values.

Add-On Instructions are a powerful tool and can be used when there is a need for reusable code or it might be used in different processors. Programmers may create a library of AOIs that they use in their projects, or companies that build standardized equipment might use them for common tasks. It is also a convenient method of locking code so that other programmers can't change it or even see how it works.

Other Languages

In addition to Ladder Logic, which is the main language used by Allen-Bradley programmers, the other IEC 61131 languages are also supported to varying degrees.

ASCII Mnemonics (Instruction List)

While it is possible to view ladder logic as text, it is not common to use this feature other than for importing and exporting logic.

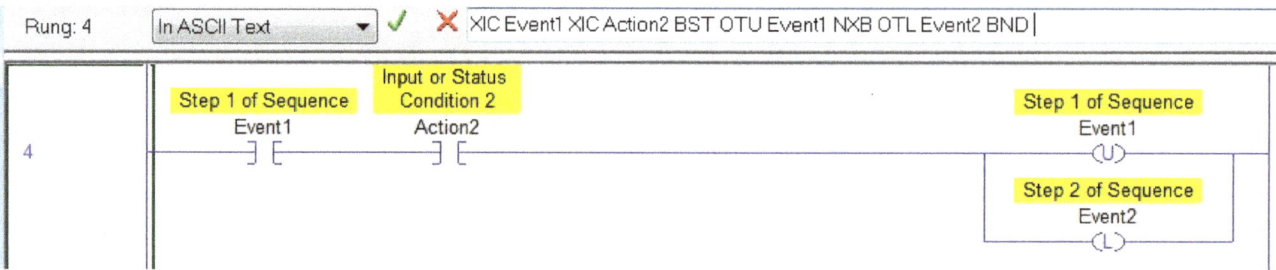

Double-clicking to the left of the rung shows the field at the top of the editor. The text description for this rung is (XIC (Examine If Closed, Normally Open Contact)) <Tagname> etc. The BST is Branch Start, the OTU and OTL are Unlatch and Latch, and the NXB is Next Branch.

Spreadsheets are often used to concatenate strings like this, substituting tags to duplicate rungs. A sample rung can first be exported as a CSV or XML file and opened in Microsoft Excel. The rung can then be disassembled and different tags substituted, making the chore of duplicating hundreds of duplicate rungs with different tags (Such as Faults, Input and Output rungs) much less tedious.

Function Block Diagram (FBD)

The programming environment can be selected as FBD when creating a new routine. FBD is part of the Standard and Lite (CompactLogix) versions of software.

Structured Text (ST)

Structured Text is not part of the standard programming software, but it can be purchased as a separate module. It is included in the Full and Professional versions.

Sequential Function Charts (SFC)

As with Structured Text, SFC can be purchased as a separate module or is included in the Full and Professional versions.

More examples of these languages can be found in the Siemens section of this book.

Communications - Allen-Bradley RSLinx

All online activities for Allen-Bradley products use a communications program called RSLinx. Before using the program to perform activities, such as uploading, downloading or going online, RSLinx communications drivers need to be configured. To open RSLinx, you can either browse for it in the Programs list on your computer or simply open the PLC program using RSLogix.

 When a program is opened, RSLinx will often open automatically as a service. The icon to the left is the RSLinx service. You can double-click on the icon to open it or browse for it in the Programs list.

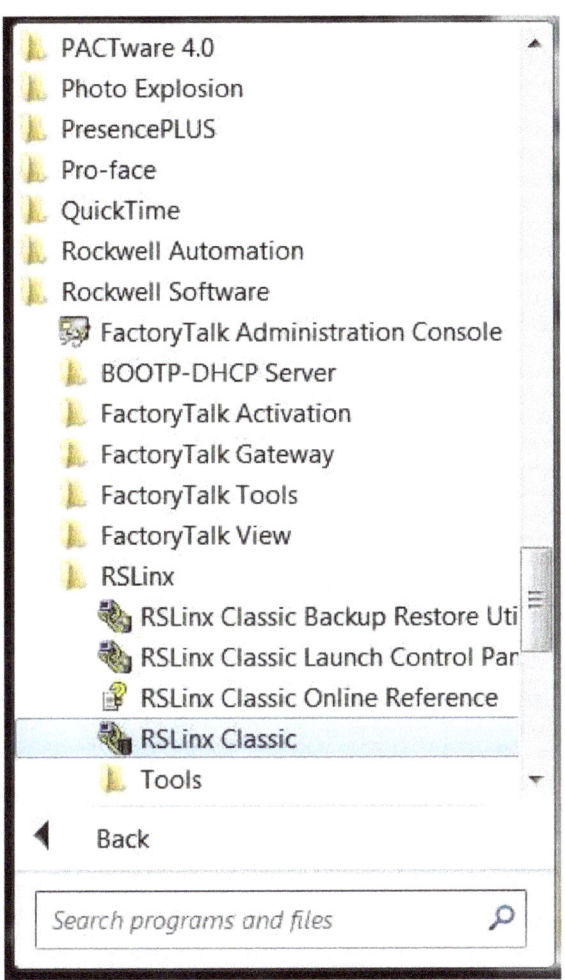

Drivers may already be configured in RSLinx. If they are not, the following procedures can be used.

Newer Allen-Bradley PLCs have several methods of communications, including USB, RS232 DF1 (Serial Communications) and Ethernet. USB communications is used with the ControlLogix L70 series of processors; the port is on the front of the processor.

Serial communications is available on all other processors. They either use a cable with a 9-pin plug on one end and a round plug on the other (MicroLogix), or a 9-pin serial cable with a null modem adapter, as shown in the RS-232 diagram in this document's Communications section. Ethernet uses a standard non-crossover Ethernet cable if using a switch (normal) or a crossover cable if attaching directly to the processor or Ethernet card. The communication ports are on the left side of the PLC for MicroLogix, on the processor for the SLC 5/05 or CompactLogix, or on the bottom of the Ethernet card if using a ControlLogix system.

When a PLC is first commissioned, it has no Ethernet address assigned to it. The Ethernet address can either be assigned using the BootP Server utility from Allen-Bradley or by downloading the program using the serial cable. If the Ethernet address has been set up in the PLC program, a serial download will also configure the Ethernet port.

To configure the serial driver:

1. Open RSLinx and select Communications > Configure Drivers.

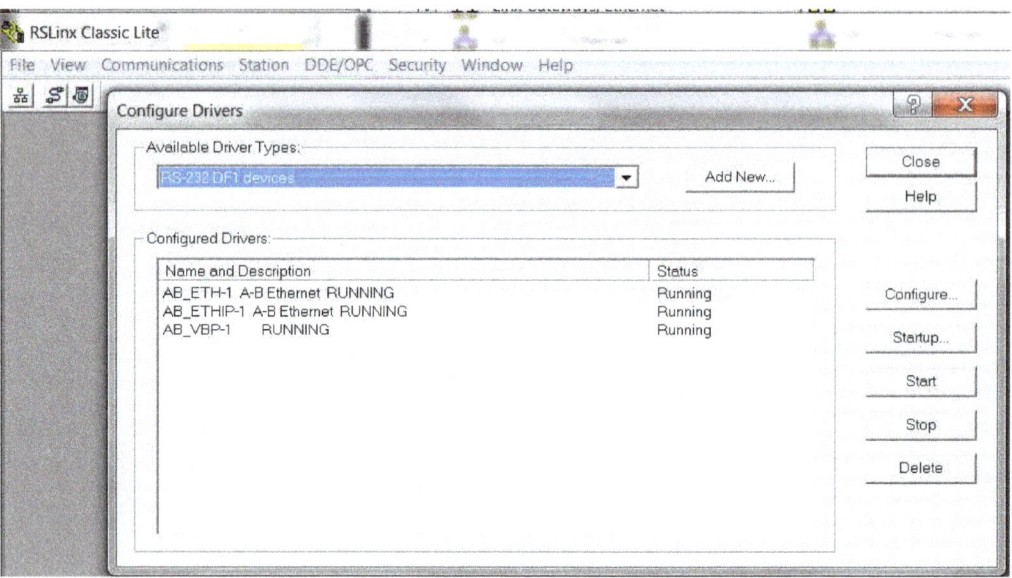

2. Select RS-232 DF1 devices in the pull-down menu under Available Driver Types. Press the "Add New" button. A dialog box will appear asking you to name the new driver. The default is AB_DF1-1; press the OK button to keep the default.

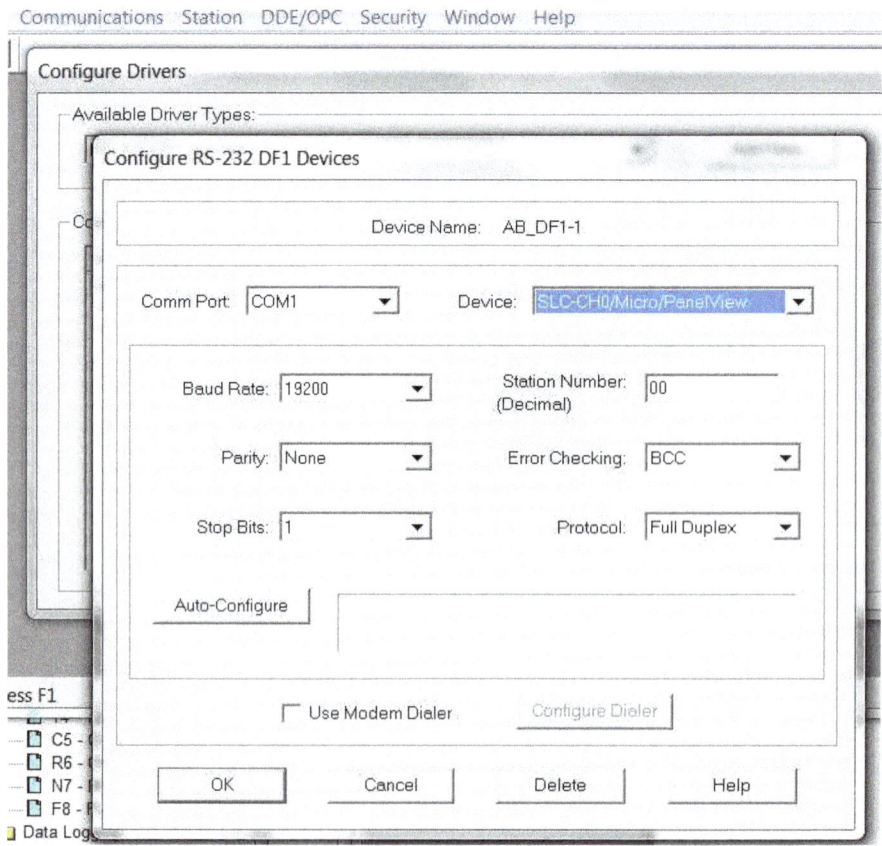

3. Under the Device pull-down, select SLC-CH0/Micro/Panelview (SLC or MicroLogix) or Logix 5550/CompactLogix. Other settings are as shown in the above diagram except for Error Checking, which should be set to CRC.

    If the cable is connected to the processor and it is powered up, you can press the Auto-Configure button. This should interrogate the PLC for its current settings and you will receive the message "Auto Configuration Successful!" This confirms that you have communications with the PLC and are ready for download.

    Because most computers no longer have a serial port, it is often necessary to use a USB to Serial adapter. Some of these adapters will assign a serial port automatically; if so, you will need to use Device Manager on your computer to discover which port is used. If you see no assigned port for the adapter, try Port 8 or higher and press Auto Configure. Even if the driver displays "Port Conflict", it will usually show the "Auto Configuration Successful!" message anyway.

    It is often necessary to use a "Null-Modem" adapter on the serial cable. This crosses pins 2 and 3 (TX and RX). This is true for ControlLogix and SLC500 processors.

To Configure the Ethernet Drivers:

There are two Ethernet drivers listed in the Available Driver Types list. The first is "Ethernet Devices" and the second is "Ethernet/IP Driver". **Ethernet Devices** allows the computer to locate processors and other devices by typing in the address; it works on all Allen-Bradley Ethernet items and also finds many compatible devices not made by Allen-Bradley.

Ethernet Devices:

1. Open RSLinx and select Communications > Configure Drivers.

2. Select Ethernet Devices in the pull-down menu under Available Driver Types. Press the "Add New" button. A dialog box will appear asking you to name the new driver. The default is AB_ETH-1; press the OK button to keep the default.

    If there is more than one group of Allen-Bradley processors and devices in your plant, it may be advisable to create more than one Ethernet Devices driver and give them different names, such as ETH_Line1, ETH_Line2 etc. This is to prevent the driver from attempting to find devices that are not present, which takes more time.

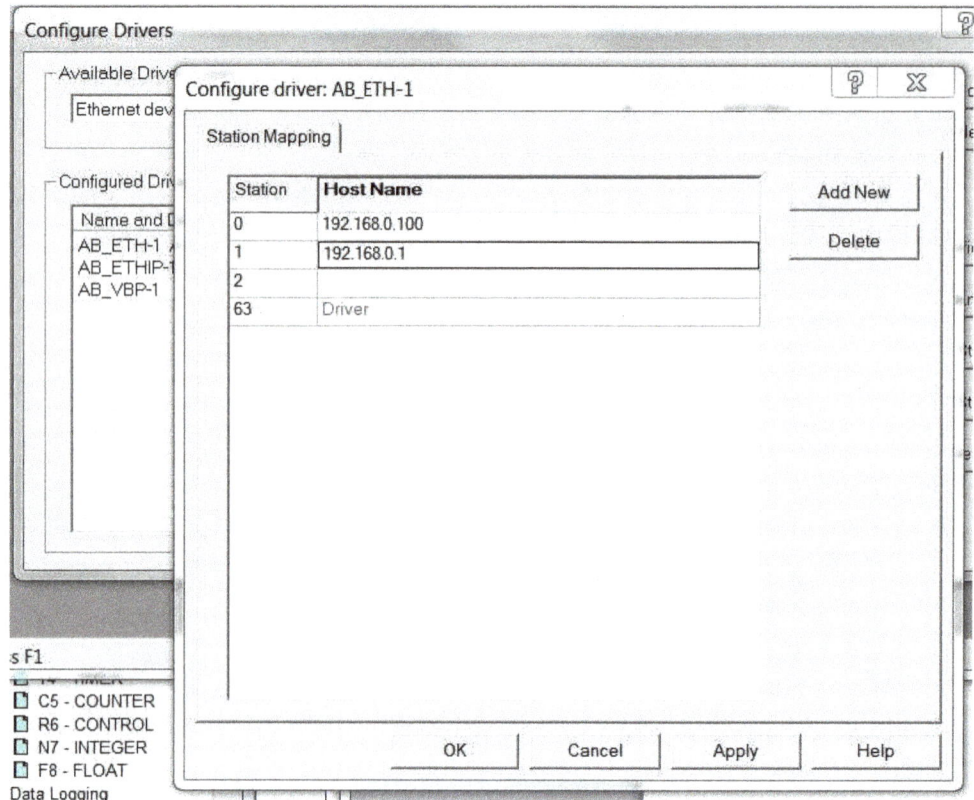

3. Type in the Ethernet addresses as shown. This assumes a computer with assigned address 192.168.0.100 and a PLC at 192.168.0.1. The first address (Station 0) is not necessary to communicate with the PLC, but it is there to show the current configuration of the computer itself. The second address (Station 1) is the address configured for the communications port in the PLC program. More PLCs and devices can be added to this list as required.

Ethernet/IP Driver:

This driver works for all Allen-Bradley CIP (Ethernet/IP) devices, which does not include the SLC 5/05 and MicroLogix. It does not require that you know the addresses of the devices.

1. Open RSLinx and select Communications > Configure Drivers.

2. Select EtherNet/IP Driver in the pull-down menu under Available Driver Types. Press the "Add New" button. A dialog box will appear asking you to name the new driver. The default is AB_ETHIP-1; press the OK button to keep the default.

3. This driver only requires that you select your Ethernet card from the list of devices on your computer. This diagram shows a wired port (192.168.4.204) and a wireless card (50.94.219.161)

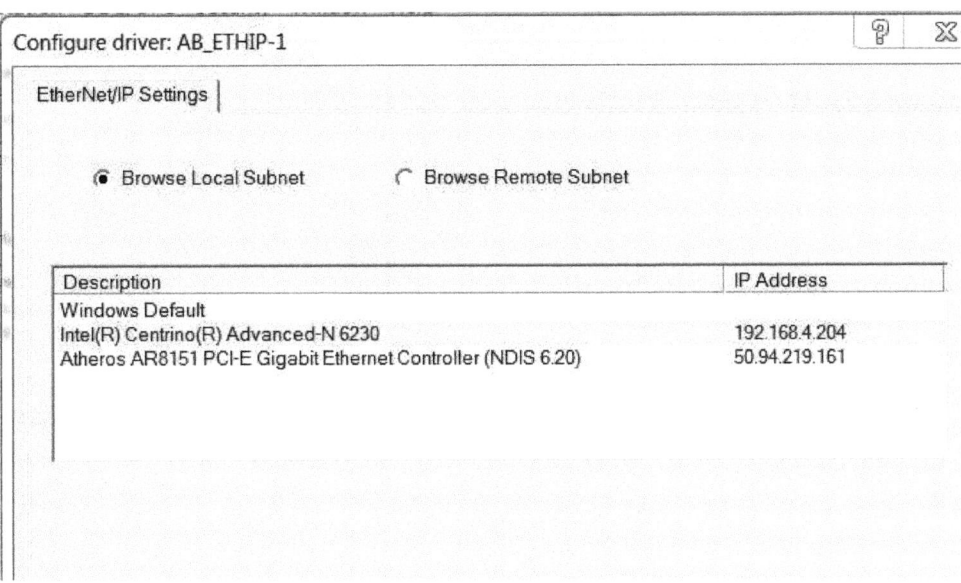

Since this driver does not maintain a list of devices, it is not necessary to create multiple drivers as with the Ethernet Devices driver. Previous connections may show up on the list when browsing in "RSWho"; just right click on them and select "remove" to eliminate them.

Siemens PLCs

Siemens was founded under the name Siemens & Halske in 1847. Werner von Siemens and Johann Georg Halske invented a device based on the telegraph that would use a needle to point to a sequence of letters, instead of using Morse code. The full name of this company was *Telegraphen-Bauenstalt von Siemens & Halske*.

The company grew and diversified in the late 1800s into making AC alternators, electric trains and light bulbs. The company was incorporated in 1897 and merged part of its activities with Schuckert & Co. in 1903 to become Siemens-Schuckert. During the 1920s and 1930s, they started to manufacture radios, television sets and electron microscopes. By 1966, Siemens and several other companies again merged to become Siemens AG.

In 1959, Siemens showed the first generation of a "Building Block System for Solid State Controls" called Simatic G. It was a hard-wired turret lathe control that automated basic machine functions. By the 1970s and 1980s, Siemens controllers were no longer based on fixed wiring but were programmable. This formed the basis of the S3, which was later improved to the S5 PLC platform.

In 1991, Siemens acquired the Industrial Systems Division of Texas Instruments Inc. based in Johnson City, Tennessee. This division was organized as Siemens Industrial Automation, Inc. and was later absorbed by Siemens Energy and Automation.

In 1994, the S7 PLC platform was released, and by 1998, it had replaced the S5. There are still many legacy remnants of the S5 platform in the S7, however, including byte-based registers, BCD timers and counters and other non-IEC-compliant features.

TIA (Totally Integrated Automation) controllers made their appearance in 2009. This is a fully IEC compliant tag-based PLC system that replaced the S7-200 with the S7-1200 and also made an S7-300 hardware-compatible processor, the S7-1500.

Siemens Terminology and Abbreviations

Siemens Terminology		
Term	Stands for:	Definition
S7	Combo PLC/HMI system	PLC and HMI platform (Step 7)
CFC	Continuous Function Chart	Optional Programming Language
CP	Communications Processor	Modules used for special communication protocols
DB	Data Block	Memory storage areas for data
DP	Decentralized Peripherals, Profibus	Mnemonic for Profibus, RS485 protocol

FB	Function Block	A Function Call (FC) with its own data block
FBD	Function Block Diagram	Standard Programming Language
FC	Function Call	Program blocks, Subroutine
FM	Function Module	Modules with special functions
GSD	Generic Station Description	Files used for ProfiBus and ProfiNet hardware descriptions.
HiGraph		Optional Programming Language
IM	Interface Module	Modules to connect remote racks
LAD	Ladder Logic Diagram	Standard Programming Language
M7	Programmable Modules	A module with processing capabilities
MMC	Micro Memory Card	A compact portable memory card
MPI	Multi Point Interface	Standard Communication Protocol
OB	Organizational Block	Blocks for user programs based upon different operating systems
OP	Operator Panel	A display panel with or without buttons
PCS	Process Control System	Software for the entire process chain
MPI	Multi Point Interface	Standard Siemens networking protocol
PG	Programming Terminal	A dedicated Siemens device (PC)
PPI	Point to point interface	Serial RS-232 communication
Profibus DP	Profibus Decentralized Peripherals	Networking Protocol used for factory automation
Profibus PA	Profibus Process Automation	Networking Protocol used for process automation
RLO	Result of Logic Operation	State of a discrete instruction after the STA or Status
SCL	Structured Control Language	Optional Programming Language, Structured Text
SFB	System Function Block	Integrated FB for CPU information
SFC	System Function Call	Integrated FC for CPU information
SM	Signal Module	Standard Input/Output Modules
Siemens Terminology Continued		
Term:	Stands For:	Definition
STA	Status	Status of a Discrete Instruction
STL	Statement List	Text Based Programming Language, Instruction List
TIA	Totally Integrated Automation	Newer PLC and HMI platform
TP	Touch Panel	Touch screen display
UDT	User-Defined Data Type	Data structures defined by the user
VAT	Variable Access Table	Tables used to monitor/modify values

Siemens Step 7 PLC Software

The programming software for the older Siemens S7 PLCs is called Step 7. Siemens offers several types of Step 7 software packages for programming the S7-300 and S7-400 PLCs. Modules can be purchased separately and added to the basic S7 package also.

Name	Controllers	Description
Step 7	S7-300, S7-400	The standard environment supporting all PLCs, networking & 3 languages (LAD, STL, FBD). Includes HMI software (WinCC)
Step 7 Professional	S7-300, S7-400	Comes with additional support for the SCL and Graph programming languages, S7PLCSIM –a software PLC simulator. Includes HMI software (WinCC)
Step 7 TIA Portal	S7-1200/1500	This is the newest version of S7. (Is NOT compatible with pre-2007 S7-300/400 CPUs.) Basic and Professional versions. Also includes WinCC, HMI programming software.

As of November 2018, the latest release of Step 7 software was 5.6, and the latest version of TIA Portal was v15.

Siemens Step 7 PLC Hardware

PLC Platform	Range of usage
S7-200	Micro PLC (Obsolete)
S7-300	Mid and low-end performance, rackless design. Up to 4 expansion racks with 8 slots each can be configured. Some processors have integrated I/O.
S7-400	High-end /medium performance, rack-based design. Multiple CPUs can be used in the same rack.

There are also a wide variety of network-based I/O devices for Profibus and Profinet in the ET200 family.

S7-300 Platform

The SIMATIC S7-300 is primarily used for mid-sized system solutions in the manufacturing industry and also serves as an all-purpose automation system for applications that need a flexible platform for central and local configurations.

The S7-300 is a rail-based system that can be expanded by adding modules to a large DIN rail using bus connector clips that plug into the back. There is no required sequence for the I/O modules, which can be freely addressed.

The power supply module for the S7-300 has all the connections on the front of the module. There are 2A, 5A, and 10A current supply models. S7-300 PS modules are optional (i.e. power supplies other than Siemens can be used to power the modules).

The main rack of the S7-300 can hold up to 8 modules in the central rack and up to 8 per expansion unit, for a total of 32. In a single-tiered configuration, this results in a maximum of 256 I/O and in multi-tiered configurations up to 1024 I/O.

Micro Memory Cards are required for the more recent S7-300 processors to operate. The older S7-300 CPUs use battery backed RAM or an optional flash module; these older models are identifiable by having a key switch for the mode select on front of the CPU.

All Siemens PLCs come with one or more communication ports on the front of the CPU. Multi Point Interface (MPI) is an RS-232 protocol typically used for programming or interfacing with an HMI, while Profibus (DP) is an RS-485 fieldbus protocol used for I/O, also including HMIs. Processors may also have two ethernet ports for ProfiNet, an industrial ethernet (IP) protocol.

Slot 2 is reserved for a CPU and Slot 3 is reserved an Interface Module (IM) on the S7-300 even if it is not present.

S7-300 CPUs

CPU	Catalog Order number	Description
CPU 312	6ES7312-1AE14-0AB0	Entry CPU for Small Applications
CPU 313C	6ES7313-5BG04-0AB0	Older Series CPU for Mid-Level Applications
CPU 313C-2 PtP	6ES7313-6BG04-0AB0	Older Series CPU for Mid-Level Applications, Peer to Peer Communications
CPU 313C-2 DP	6ES7313-6CG04-0AB0	Older Series CPU for Mid-Level Applications, Profibus Communications
CPU 314	6ES7314-1AG14-0AB0	Mid-Level Capability CPU
CPU 314C-2 PtP	6ES7314-6BH04-0AB0	Mid-Level Capability CPU with Peer to Peer Communications
CPU 314C-2 DP	6ES7314-6CH04-0AB0	Mid-Level Capability CPU with Profibus Communications
CPU 314C-2 PN/DP	6ES7314-6EH04-0AB0	Mid-Level Capability CPU with MPI/Profibus and Profinet Communications

CPU 315-2 DP	6ES7315-2AH14-0AB0	CPU with Medium to Large Program Memory and Profibus
CPU 315-2 PN/DP	6ES7315-2EH14-0AB0	CPU with Medium to Large Program Memory, MPI/Profibus and Profinet
CPU 317-2 DP	6ES7317-2AK14-0AB0	CPU with Large Program Memory
CPU 317-2 PN/DP	6ES7317-2EK14-0AB0	CPU with Large Program Memory, MPI/Profibus and Profinet
CPU 319-3 PN/DP	6ES7318-3EL01-0AB0	CPU with High Command Processing performance and Large Program Memory

A wide variety of digital and analog modules (Signal Modules, SM) with varying counts and resolutions are available. Special purpose modules for high-speed counting and PID control are available (Function Modules, FM) as well as communications cards (Communications Modules, CM).

S7-400 Platform

The Siemens S7-400 is a rack-based PLC. Racks have 4, 9, or slots and the high-speed backplane bus ensures efficient linking of central I/O modules. The S7-400 is used for data-intensive tasks, safety, redundancy, overall plant coordination and controlling lower-level systems.

The S7-400 is available in a SIPLUS version for use in extreme environmental conditions. Power supplies are required and must reside in slot 1, available in 4A, 10A and 20A. The CPU may be placed in any slot on the S7-400. The first communication port (X1) on the front of all Siemens PLCs is for the Multi Point Interface (MPI) RS-232 protocol. Multiple CPUs can be placed in a rack, along with various communication modules.

Up to 22 racks can be connected locally, providing an extremely high I/O capacity.

S7-400 CPUs

CPU	Catalog Order number	Description
CPU 412-1	6ES7412-1XJ	Entry CPU for Medium Performance Applications
CPU 412-2	6ES7412-2XJ	CPU with Medium Program Memory
CPU 412-2 PN	6ES7412-2EK	CPU with Medium Memory and Additional Profinet port
CPU 414-2	6ES7414-2XK	Medium Performance CPU with higher memory
CPU 414-3	6ES7414-3XM	Medium Performance CPU with higher memory and additional communications
CPU 414-3 PN/DP	6ES7414-3EM	Medium Performance CPU with higher memory and additional communications including Profibus
CPU 416-2	6ES7416-2XN	Entry CPU for Upper Performance Range
CPU 416F-2 DP	6ES7416-2FN	Failsafe Entry CPU for Upper Performance Range
CPU 416-3	6ES7416-3XR	Standard CPU for Upper Performance Range
CPU 416-3 PN/DP	6ES7416-3ES	Standard CPU for Upper Performance Range with Profinet
CPU 416F-3 PN/DP	6ES7416-3FS	Failsafe CPU for Upper Performance Range with Profinet
CPU 417-4	6ES7417-4XT	High End CPU for Upper Performance Range

Remote and Networked IO

ET 200SP Scalable IO, extensive diagnostics, hot swapping capability, single or multi-port connection, very user friendly and smaller than ET200S

ET 200S is discretely modular with multi-conductor connection. Multifunctional due to a wide range of modules: motor starters, safety technology, technology modules, distributed intelligence, as well as IO-Link modules

ET 200M Modular design using standard SIMATIC S7-300 modules, high-channel density with up to 64 channels per module, hot swap capable and allows for redundancy.

ET 200MP For use with the S7-1500, IO multi-channel and multi-functional with high performance and very short response times.

SIMATIC ET 200iSP Built for use in hazardous areas. Features hot swapping, redundancy and configuration changes during operation.

SIMATIC ET 200pro – Features a modular design and compact size. Multi-functional due to a wide range of modules including digital and analog I/O, safety systems, Variable Frequency Drives (VFDs) and identification systems. Also has extensive diagnostic capabilities, as well as hot swapping and permanent wiring.

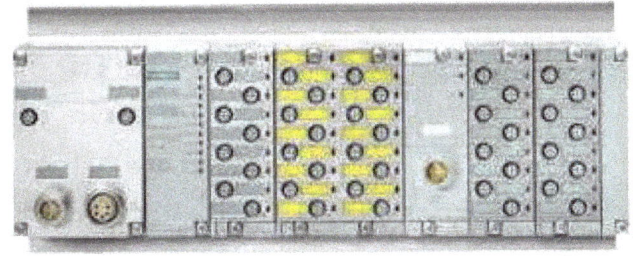

SIMATIC ET 200eco - Block IO with PROFINET connection. These low-cost, space-saving modules can be digital with up to 16 channels or analog. The electronic block can be easily replaced during operation without any interruption in bus communication or power supply.

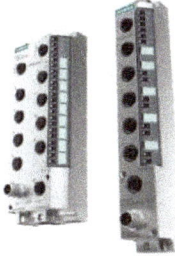

Siemens Instructions

Bit Logic Instructions

STL Mnemonic	Name	Purpose
A	Normally Open (And)	Examines a bit for an ON condition
AN	Normally Closed (And Not)	Examines a bit for an OFF condition
=	Output Coil	Turns a bit ON or OFF
(#)	Midline Output	Intermediate assigning element, Coil
S	Output Set	Sets a bit ON when executed, the bit retains its state until unlatched or the register is cleared
R	Output Reset	Resets a bit OFF when executed
RS	Reset-Set Flip Flop	Combines the Set and Reset commands into a single box. Output is a Q.
SR	Set-Reset Flip Flop	Combines the Reset and et commands into a single box. Output is a Q.
NOT	Invert Power Flow	Change the state of the RLO
FP	Positive RLO Edge Detection (One-Shot)	Triggers a one-time event on leading edge of signal, ON for one scan
FN	Negative RLO Edge Detection (One-Shot)	Triggers a one-time event on trailing edge of signal, ON for one scan
NEG	Address Negative Edge Detect (One Shot)	Compare Signal State of Address 1 with Signal State of Previous Scan (Negative Edge)

POS	Address Positive Edge Detect (One Shot)	Compare Signal State of Address 1 with Signal State of Previous Scan (Positive Edge)
SAVE	Save RLO into BR Memory	Saves the RLO to the BR bit of the Status Word
XOR	Bit Exclusive OR	Creates an RLO of "1" if the state of the bits is different

Comparison Instructions

STL Mnemonic	Name	Purpose
== I	Integer Equal	Integer Equal to Integer
<> I	Integer Not Equal	Integer Not Equal to Integer
> I	Integer Greater Than	Integer Greater than Integer
>= I	Integer Greater Than or Equal	Integer Greater than or Equal to Integer
< I	Integer Less Than	Integer Less than Integer
<= I	Integer Less Than or Equal	Integer Less than or Equal to Integer
== D	Double Integer Equal	Double Integer Equal to Double Integer
<> D	Double Integer Not Equal	Double Integer Not Equal to Double Integer
> D	Double Integer Greater Than	Double Integer Greater than Double Integer
>= D	Double Integer Greater Than or Equal	Double Integer Greater than or Equal to Double Integer
< D	Double Integer Less Than	Double Integer Less than Double Integer
== R	Real Equal	Real Equal to Real
<> R	Real Not Equal	Real Not Equal to Real
> R	Real Greater Than	Real Greater than Real
>= R	Real Greater Than or Equal	Real Greater than or Equal to Real
< R	Real Less Than	Real Less than Real
<= R	Real Less Than or Equal	Real Less than or Equal to Real

TIA Portal platforms also have an IN_RANGE and OUT_RANGE instruction for values inside or outside of a range.

Conversion Instructions

STL Mnemonic	Name	Purpose
BCD_I	BCD to Integer	Convert a Binary Coded Decimal number to an Integer
I_BCD	Integer to BCD	Convert an Integer to a Binary Coded Decimal format
BCD_DI	BCD to Double Integer	Convert a Binary Coded Decimal number to a Double Integer
DI_BCD	Double Integer to BCD	Convert a Double Integer to a Binary Coded Decimal format
I_DINT	Integer to Double Integer	Convert an Integer to a Double Integer
DI_REAL	Double Integer to Real	Convert a Double Integer to a Floating - Point number
INV_I	Ones Complement Integer	Change every bit in an Integer to its opposite state
INV_DI	Ones Complement Double Integer	Change every bit in a Double Integer to its opposite state
NEG_I	Twos Complement Integer	Change the sign of an Integer
NEG_DI	Twos Complement Double Integer	Change the sign of a Double Integer
NEG_R	Negate Floating-Point Number	Change the sign of a Floating-Point number
ROUND	Round to Double Integer	Converts a Floating-Point number to a Double Integer, converting to nearest whole number
TRUNC	Truncate Double Integer Part	Converts a Floating-Point number to a Double Integer, rounding down, uses overflow bit
CEIL	Ceiling	Converts a Floating-Point number to a Double Integer, rounding up
FLOOR	Floor	Converts a Floating-Point number to a Double Integer, rounding down

TIA Portal platforms also have a CONVERT block, which allow conversion by using a formula: a SCALE_X instruction, which maps an input to a value range, and a NORM_X function, which maps an input to a normalized range between 0 and 1.

Counter Instructions

STL Mnemonic	Name	Purpose
S_CUD	Up-Down Counter	Varies a BCD count up and down within a range. Done if not equal to zero
S_CD	Down Counter	Varies a BCD Count down within a range. Done if not equal zero.
S_CU	Up Counter	Varies a BCD Count up within a range. Done if not equal zero.
SC	Set Counter Value	Transfers the preset value to the counter's accumulator
CU	Up Counter Coil	Increments the accumulator value of the counter
CD	Down Counter Coil	Decrements the accumulator value of the counter
R	Reset	Resets the accumulator of a counter to zero

Logic Control Instructions

STL Mnemonic	Name	Purpose
JMP	Unconditional Jump	Jumps to a label destination unconditionally
JMP (JC)	Conditional Jump	Jumps to a label destination if the RLO of the previous instruction is a "1"
JMPN (JCN, JNB)	Conditional Jump Not	Jumps to a label destination if the RLO of the previous instruction is a "0" (With BR bit)
RET	Return	Conditionally exits block and returns to call point
LABEL	Jump Destination Identifier	Identifier for the destination of a jump instruction. First letter must be a letter of the alphabet, other characters can be letters or numbers.

Integer Math Instructions

Mnemonic (STL)	Name	Purpose
ADD_I (+I)	Add Integer	Adds an integer to another integer
SUB_I (-I)	Subtract Integer	Subtracts an integer from another integer
MUL_I (*I)	Multiply Integer	Multiplies an integer by another integer
DIV_I (/I)	Divide Integer	Divides an integer by another integer
ADD_DI (+DI)	Add Double Integer	Adds a double integer to another double integer

SUB_DI (-DI)	Subtract Double Integer	Subtracts a double integer from another double integer
MUL_DI (*DI)	Multiply Double Integer	Multiplies a double integer by another double integer
DIV_DI (/DI)	Divide Double Integer	Divides a double integer by another double integer
MOD_DI	Return Fraction Double Integer	Returns a double integer remainder after a DIV_DI operation

Floating Point Math Instructions

Mnemonic (STL)	Name	Purpose
ADD_R (+R)	Add Real	Adds a REAL to another REAL
SUB_R (-R)	Subtract Real	Subtracts a REAL from another REAL
MUL_R (*R)	Multiply Real	Multiplies a REAL by another REAL
DIV_R (/R)	Divide Real	Divides a REAL by another REAL
ABS	Absolute Value	Establish the absolute value (positive value) of a REAL
SQR	Square	Establish the square (a value multiplied by itself) of a REAL
SQRT	Square Root	Establish the square root of a REAL
EXP	Exponential Value	Establish the exponential value on the basis e(=2.71828...) of a REAL
LN	Natural Logarithm	Establish the natural log of a REAL
SIN	Sine Value	Establish the Sine of a REAL, where the floating-point number represents an angle in a radian measure
COS	Cosine Value	Establish the Cosine of a REAL, where the floating-point number represents an angle in a radian measure
TAN	Tangent Value	Establish the Tangent of a REAL, where the floating-point number represents an angle in a radian measure
ASIN	Arc Sine Value	Establish the Arc Sine of a REAL in the range -1 to 1, where the floating-point number represents an angle in a radian measure
ACOS	Arc Cosine Value	Establish the Arc Cosine of a REAL in the range -1 to 1, where the floating-point number represents an angle in a radian measure

| ATAN | Arc Tangent Value | Establish the Arc Tangent of a REAL, where the floating-point number represents an angle in a radian measure |

Shift Instructions

STL Mnemonic	Name	Purpose
SHR_I	Shift Right Integer	Move contents of an integer bit by bit to the right. Divides by two. Left bits are filled with zero
SHR_DI	Shift Right Double Integer	Move contents of a double integer bit by bit to the right. Divides by two. Left bits are filled with zero
SHL_W	Shift Left Word	Move contents of a word bit by bit to the left. Logic operation. Right bits are filled with zero
SHR_W	Shift Right Word	Move contents of a word bit by bit to the right. Logic operation. Left bits are filled with zero
SHL_DW	Shift Left Double Word	Move contents of a double word bit by bit to the left. Logic operation. Right bits are filled with zero
SHR_DW	Shift Right Double Word	Move contents of a double word bit by bit to the right. Logic operation. Left bits are filled with zero

Rotate Instructions

STL Mnemonic	Name	Purpose
ROL_DW	Rotate Left Double Word	Move contents of a double word bit by bit to the left. Logic operation. Right bits are filled with left bits as they are vacated
ROR_DW	Rotate Right Double Word	Move contents of a double word bit by bit to the right. Logic operation. Left bits are filled with right bits as they are vacated

Status Bit Instructions

Status bit instructions work with the contents of the status word. These bits are used extensively in STL programming and debugging.

STL Mnemonic	Name	Purpose
BR	Exception Bit Binary Result	Used in the transition from word to bit processing
CC1	Result Overflow	Floating point > 3.402823 E+38

CC0	Result Underflow	Floating point < -3.402823 E+38
OV	Exception Bit Overflow	Recognizes an overflow of the last math function executed
OS	Exception Bit Overflow Stored	Recognizes and stores a latching overflow in a math function
OR	Exception Bit Unordered	Recognizes if a value in a math function is an invalid floating-point number
STA	Status Bit	Stores the value of an addressed bit
RLO	Result of Logic Operation	Stores the value of a logic operation string or comparison instruction
/FC	First Check Bit	Controls a logic operation string along with the RLO

Timer Instructions

Step 7 software allows for two different forms for each of the timers listed. **Coil Timers** allow for only the address and setpoint to show, though other parameters are accessible programmatically. **Box Timers** show the reset input and allow accumulated time (.ET, elapsed time) to be passed out to a data location in either BCD or decimal format. It is important to remember that Siemens S5 timers, which these are, are BCD.

X	X	Base			Digit				Digit				Digit		
.7	.6	.5	.4	.3	.2	.1	.0	.7	.6	.5	.4	.3	.2	.1	.0
x	x	0	1	1	0	0	1	1	0	0	1	1	0	0	1

S5T (S5 Timer) Data Format

The structure of the S5 Timer is as shown above. The fields labeled Digit are the three BCD values for the preset, and the two bits labeled Base are the time base. The two bits with the x are not used. The maximum value that can be held in the Digit field is 999, which is multiplied by the time base, listed below:

00 – 10ms
01 – 100ms
10 – 1 Second
11 – 10 Seconds

The value shown in the data format is then 999 x 100ms, or 99.9 seconds. It is not necessary to specify the time base when entering the setpoint for a timer, but the characters S5T# must be placed before the value to specify the data type as in "S5T#3s" for a 3-second timer. Siemens Timers also time downwards, the timer is done when it reaches 0.

Mnemonics for the timers listed below are in the order <u>Coil Timer, Box Timer.</u>

STL Mnemonic	Name	Purpose
SD, S_ODT	On-Delay Timer	Timer runs when a positive signal is applied to the input, and is done when time expires at 0. Reset by removing the input signal.
SS, S_ODTS	Retentive On-Delay Timer	Timer runs when a positive signal is applied to the input, and is done when time expires at 0. Timer continues even when input signal is removed.
SF, S_OFFDT	Off-Delay Timer	Timer is done when a signal is applied to the input. Timer starts when the signal is removed from the input, done bit is turned off when time expires at 0.
SP, S_PULSE	Pulse Timer	Timer runs when a positive signal is applied to the input, and is done immediately. Done bit is turned off when time expires at 0. Reset by removing the input signal.
SE, S_PEXT	Extended Pulse Timer	Timer runs when a positive signal is applied to the input, and is done immediately. Done bit is turned off when time expires at 0. Pulse length remains the same even if input signal is removed.

Word Logic Instructions

STL Mnemonic	Name	Purpose
WAND_W	AND Word	Perform a bit by bit AND operation on the contents of IN1 and IN2, or the contents of Accumulator 1 (STANDARD) and Accumulator 2 (ACCU 2) in STL
WOR_W	OR Word	Perform a bit by bit OR operation on the contents of IN1 and IN2, or the contents of Accumulator 1 (STANDARD) and Accumulator 2 (ACCU 2) in STL
WXOR_W	Exclusive OR Word	Perform a bit by bit Exclusive OR operation on the contents of IN1 and IN2, or the contents of Accumulator 1 (STANDARD) and Accumulator 2 (ACCU 2) in STL

WAND_DW	AND Double Word	Perform a bit by bit AND operation on the contents of IN1 and IN2, or the contents of Accumulator 1 (STANDARD) and Accumulator 2 (ACCU 2) in STL
WOR_DW	OR Double Word	Perform a bit by bit OR operation on the contents of IN1 and IN2, or the contents of Accumulator 1 (STANDARD) and Accumulator 2 (ACCU 2) in STL
WXOR_DW	Exclusive OR Double Word	Perform a bit by bit Exclusive OR operation on the contents of IN1 and IN2, or the contents of Accumulator 1 (STANDARD) and Accumulator 2 (ACCU 2) in STL

Miscellaneous/Other Instructions

STL Mnemonic	Name	Purpose
OPN	Open Data Block	Transfers the number of a shared (DB) or Instance (DI) Data block into the DB or DI register. Subsequent DB and DI commands access the corresponding block.
CALL	Call Block	Calls a routine of FC, FB, SFC or SFB type with parameters
MOVE (L, T)	Assign a Value (Load and Transfer)	Copy a value of BYTE, WORD or DWORD size objects
MCR	Master Control Relay	Master Control Relay On, Off, Activate and Deactivate commands. Turns off outputs in a zone
MOVE (L, T)	Assign a Value (Load and Transfer)	Copy a value of BYTE, WORD or DWORD size objects

This is an overview of the Siemens instruction set for Step 7 and TIA Portal and does not include all instructions. For a complete listing see the Reference Manuals for Ladder and Statement List programming.

Starting and Editing a Project with Step 7

Step 7 programming always begins by opening Simatic Manager. While Allen-Bradley software is self-contained (except for the communications software RSLinx), Step 7 consists of various different programs, all opened from Simatic Manager. Usually the manager will be placed on the desktop as a shortcut.

From the Manager, some of the components that can be accessed are listed below:

Hardware Manager – Configures the CPU, I/O modules and communications ports of the PLC.

Editor – Allows modification of the program's code in Ladder, Statement List or FBD (Function Block Diagram).

Symbol Table – Allows for names to be attached to I/O and memory addresses, as well as components, such as blocks and functions.

Variable Table (VAT) – Allows the contents of I/O or memory registers to be changed.

Creating the Project

To start a new project, select "New" under the File selection in the Manager. When a project is created, several files and directories are created in a folder under the name of the project. The default location for these files is C:\Program Files\Siemens\Step7\S7Proj\, but the project can be saved to wherever the programmer wishes. After the project is saved, the software remembers the path to all saved projects.

Because there are many files and directories for a single project, it is often desirable to create a single project file for transport. Step 7 allows the programmer to zip the file by using the **Archive** utility, also under the File tab. When reopening the file, the **Retrieve** selection is used.

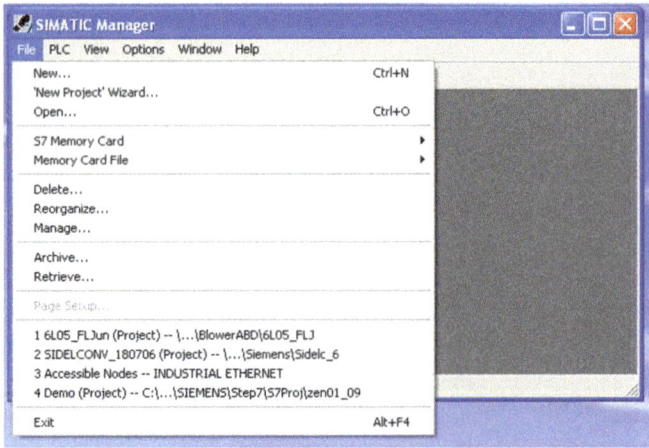

Several other selections are available under the file tab such as **Delete**, which removes the files and references to them, **Manage**, which allows a file that was not Archived to be found, and **Reorganize**, which compacts the files into unused spaces created when files are created and deleted, much like defragmenting a hard drive. Files can also be accessed or saved to and from the memory card in the CPU from this list.

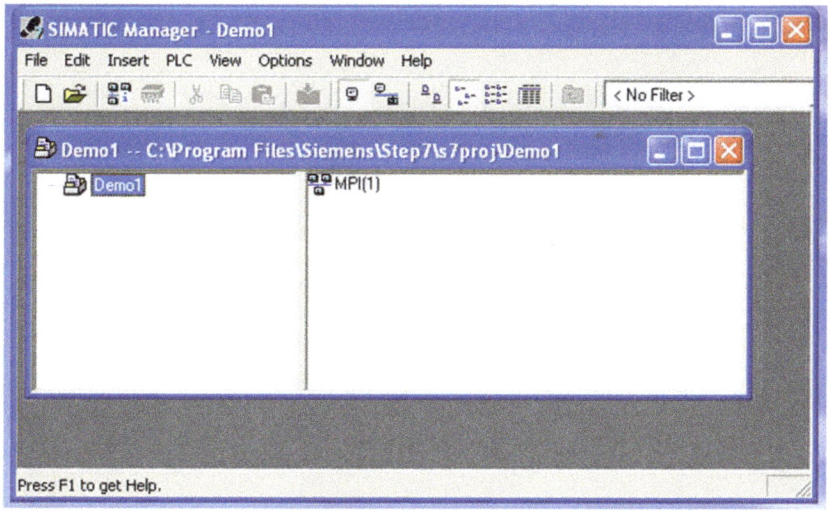

After a project has been created, the hardware must be selected and configured. Before this, the programmer must insert a "Station". This may be an S7-300, S7-400, or even a PC-based processor or operator interface. In this example, the programmer has created a project named Demo1, and an S7-300 station will be inserted.

Note that upon creating the project, an MPI (Multi-Port Interface) icon was also created. This is Siemens' form of serial communications.

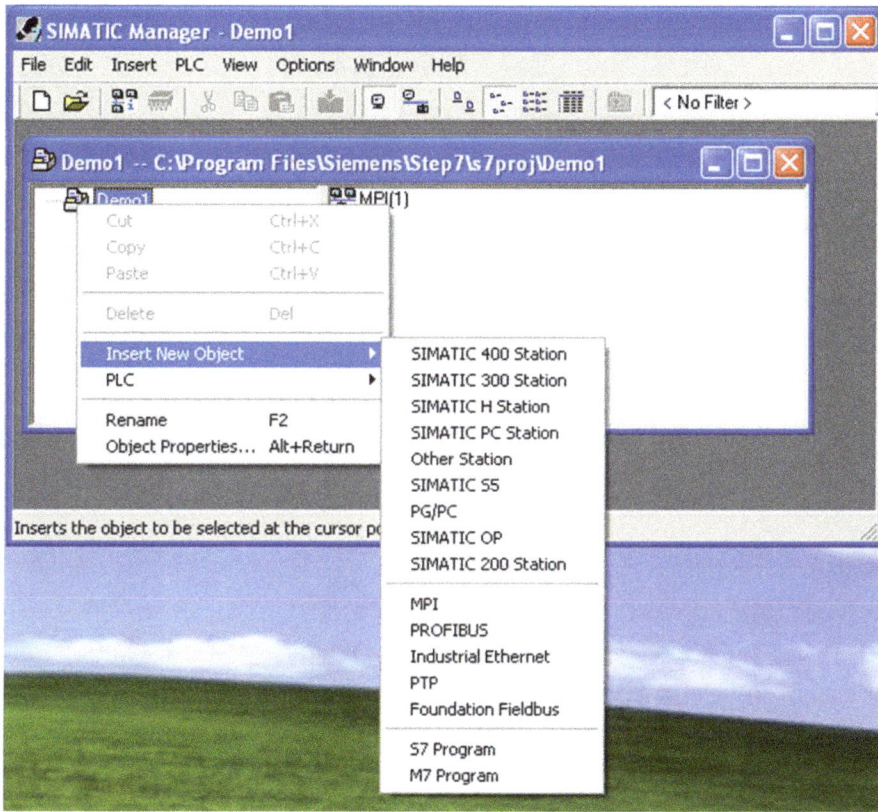

There are many different types of stations or files that can be placed into the project. This layout allows the different components within a project to easily reference each other. The 300 and 400 platforms, PC-based PLCs, redundant PLCs and even HMIs can be included in the same project. Multiple stations or processors can be placed into a single project, as well as HMIs. The HMI software, WinCC, is included in the Step7 software.

Configuring Hardware

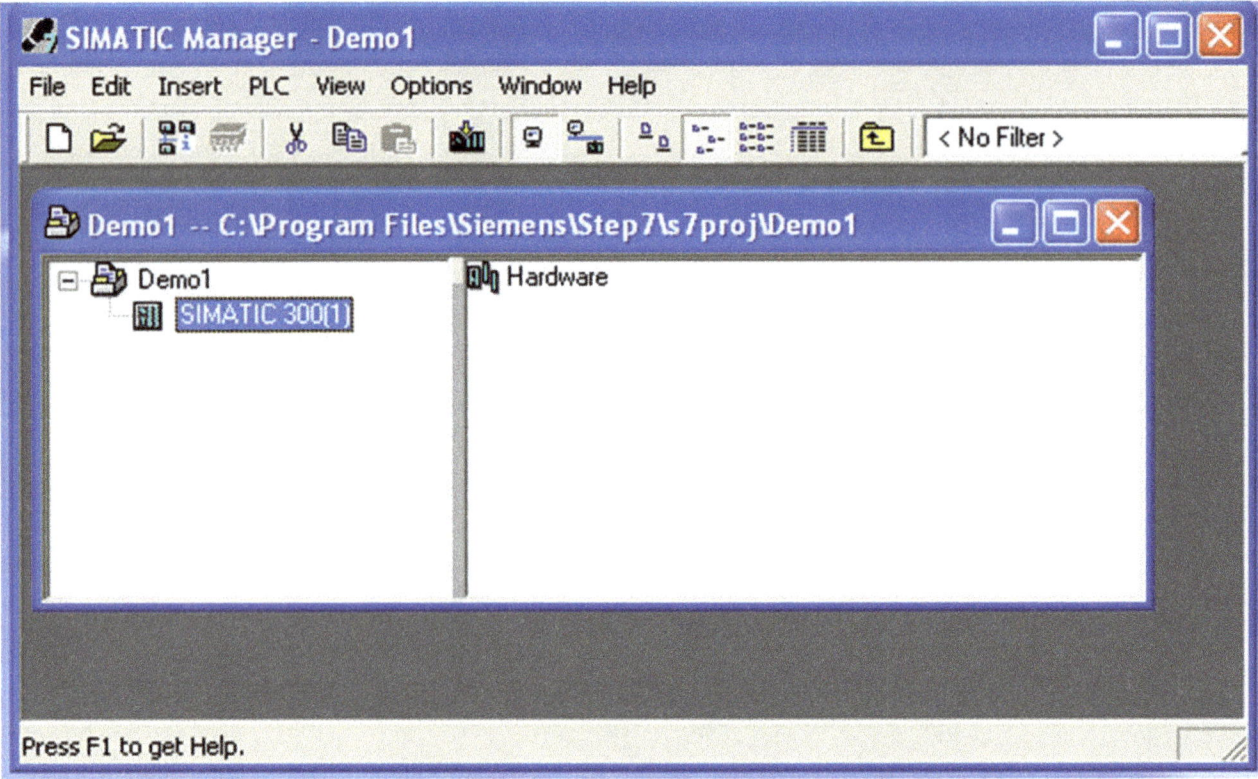

After inserting the station, a hardware icon is created. Double-clicking this icon will open the dialog for creating the hardware configuration for the project. Notice that this is a separate application than the Manager. The Station can be named whatever the programmer wishes.

Opening the hardware configuration allows the programmer to select hardware items for the platform chosen. In the case of the S7-300, the rail must first be selected from the Rack-300 folder. This creates the spaces for all of the different components to be placed into.

Folders in the hardware configuration are labeled by the types of modules contained in them. Digital and analog modules are contained in the SM (Signal Module) folder, CPUs in the CPU folder, special purpose modules, such as High-Speed Counters and PID, in the FM (Function Module) folder. Communications modules are in the CM folder, and there are many folders for Profibus and Profinet remote I/O devices and drives.

Devices that are not made by Siemens can be imported as GSD files and should be obtained from the manufacturer of the device.

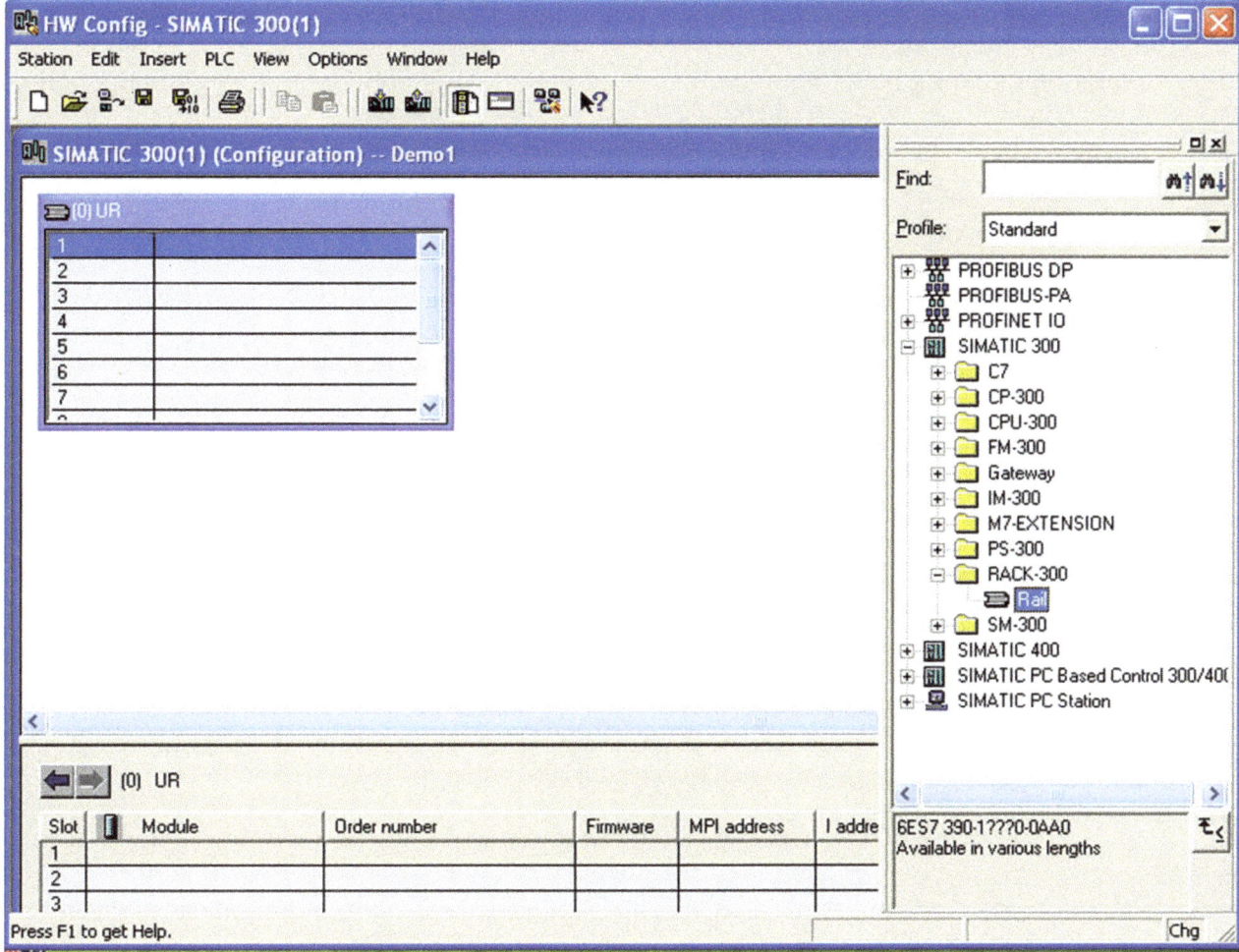

Only power supplies can be placed in the first slot, and the processor must reside in slot 2. Slot 3 is reserved for Interface Modules (IMs), which allow expansion racks to be connected. A single S7-300 processor can control up to four expansion racks with up to 32 total I/O modules.

I/O is assigned at a byte level, with each module assigned input (I) or output (Q) registers. While these addresses have default locations based on their slot positions, the I/O assignments can also be chosen by the programmer.

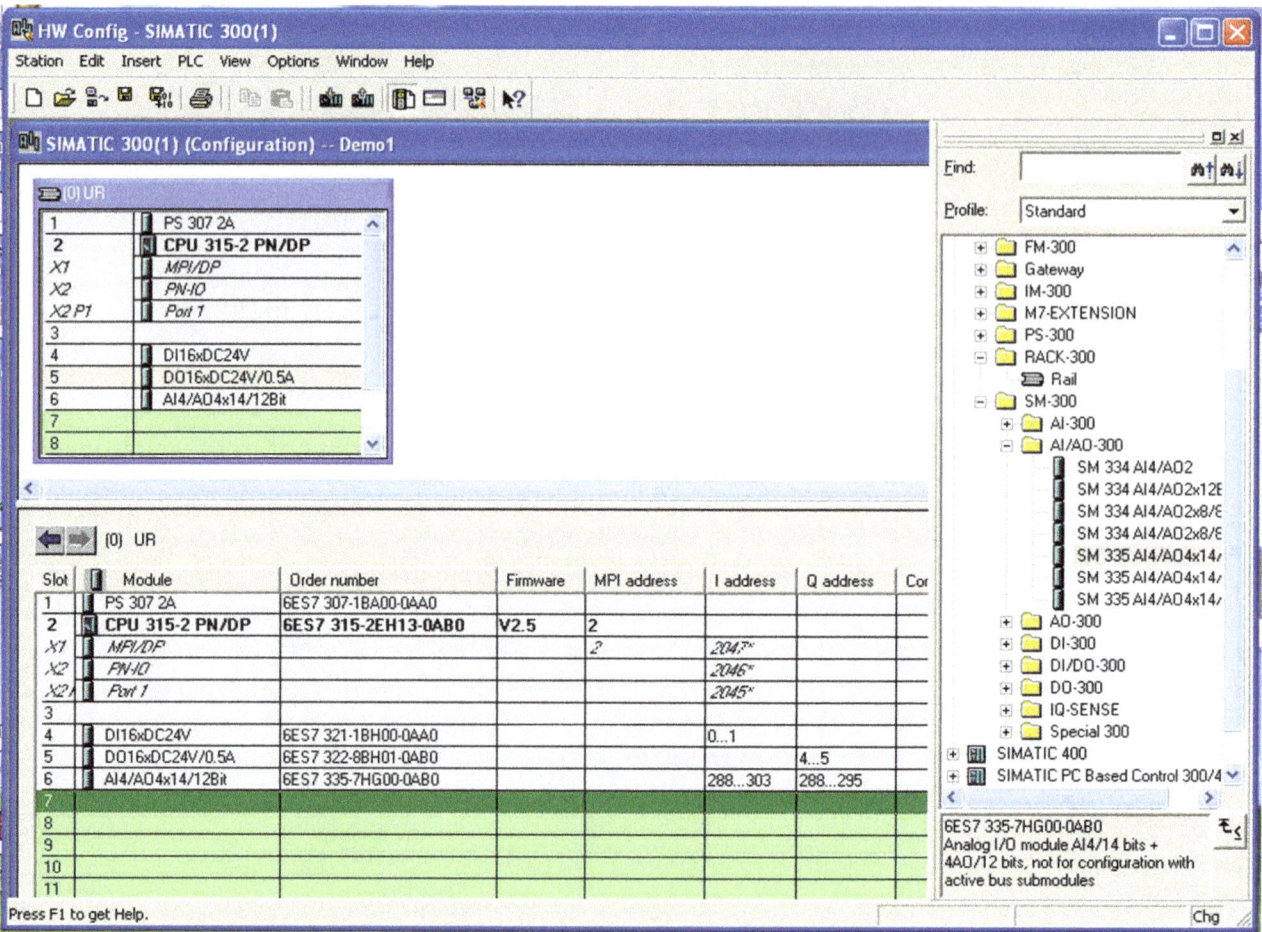

Digital and analog I/O modules are selected from the SM (Signal Module) folder. Detailed information about each module is shown in the small window below the catalog. Information includes the full part number of the module, its firmware revision and other details.

The address of the I/O is also shown in columns to the right of the slots. Double-clicking the module description brings up a dialog, which allows the addresses to be changed if desired. Some cards also have other parameters that can be modified or adjusted.

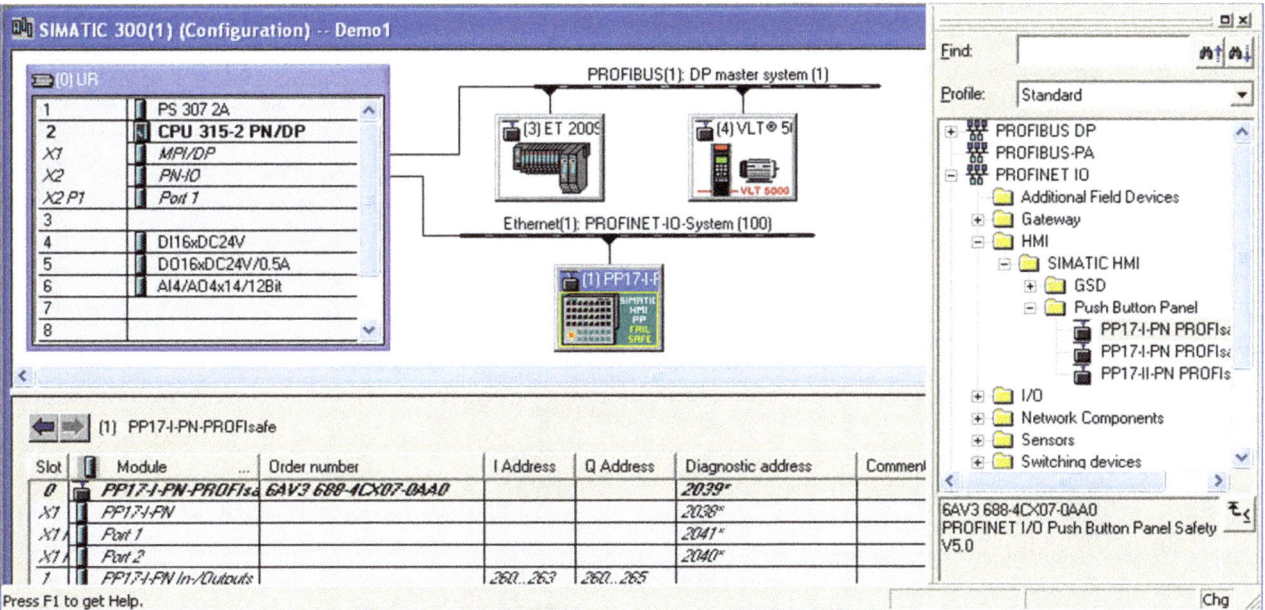

Profibus or Profinet devices can also be configured from the Hardware Manager.

An additional configuration tool called NetPro is included for further modification of Profinet devices.

Clicking on the ports of the CPU brings up dialog boxes for configuring communications. As you can see in the diagram below, it takes several clicks and windows to set up the communications. Double-click the MPI/DP Port, select Profibus, set the address (the CPU is usually 2), click New, and the name of the network can be set. The default of PROFIBUS(1) is usually fine. The network will then appear in the configuration and devices can be added from the folders.

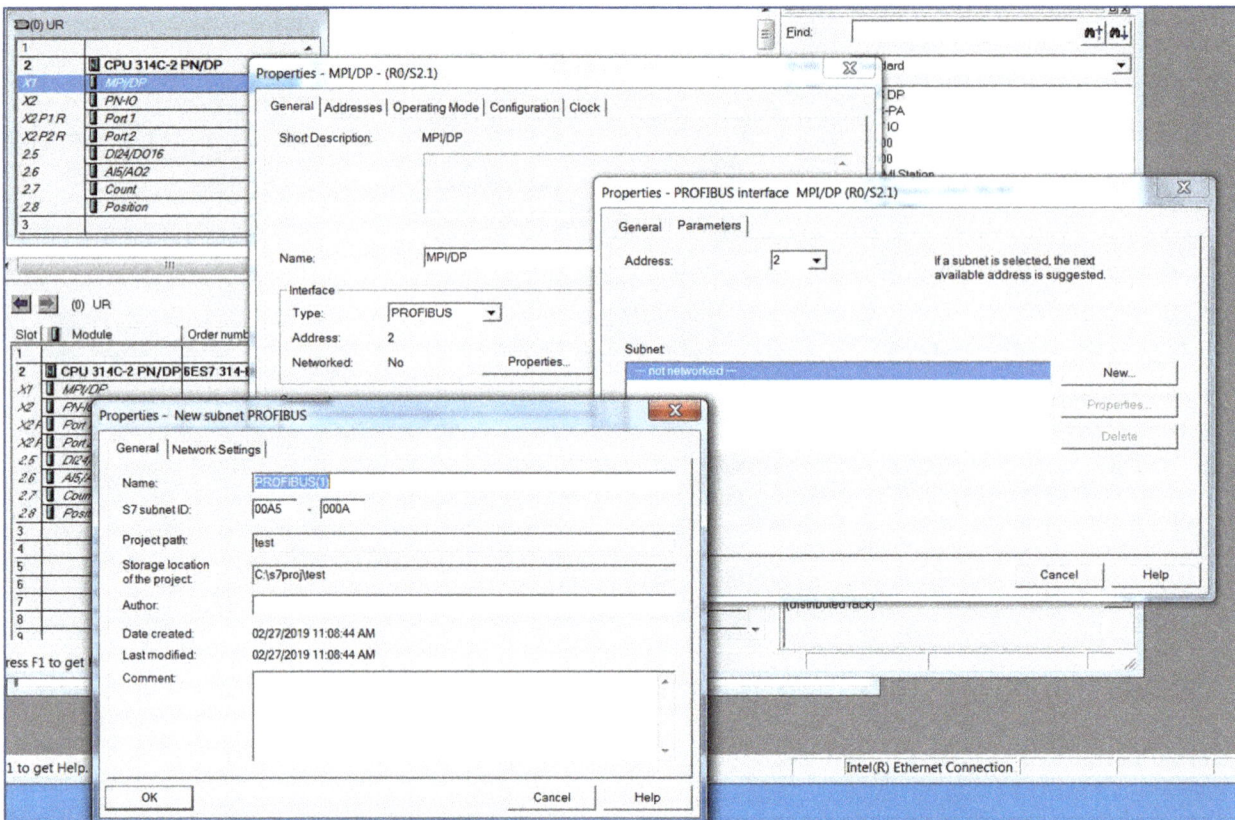

Profinet communications are configured much the same way. In addition to naming the network, the ethernet address and subnet mask needs to be entered. Further configuration is done with NetPro.

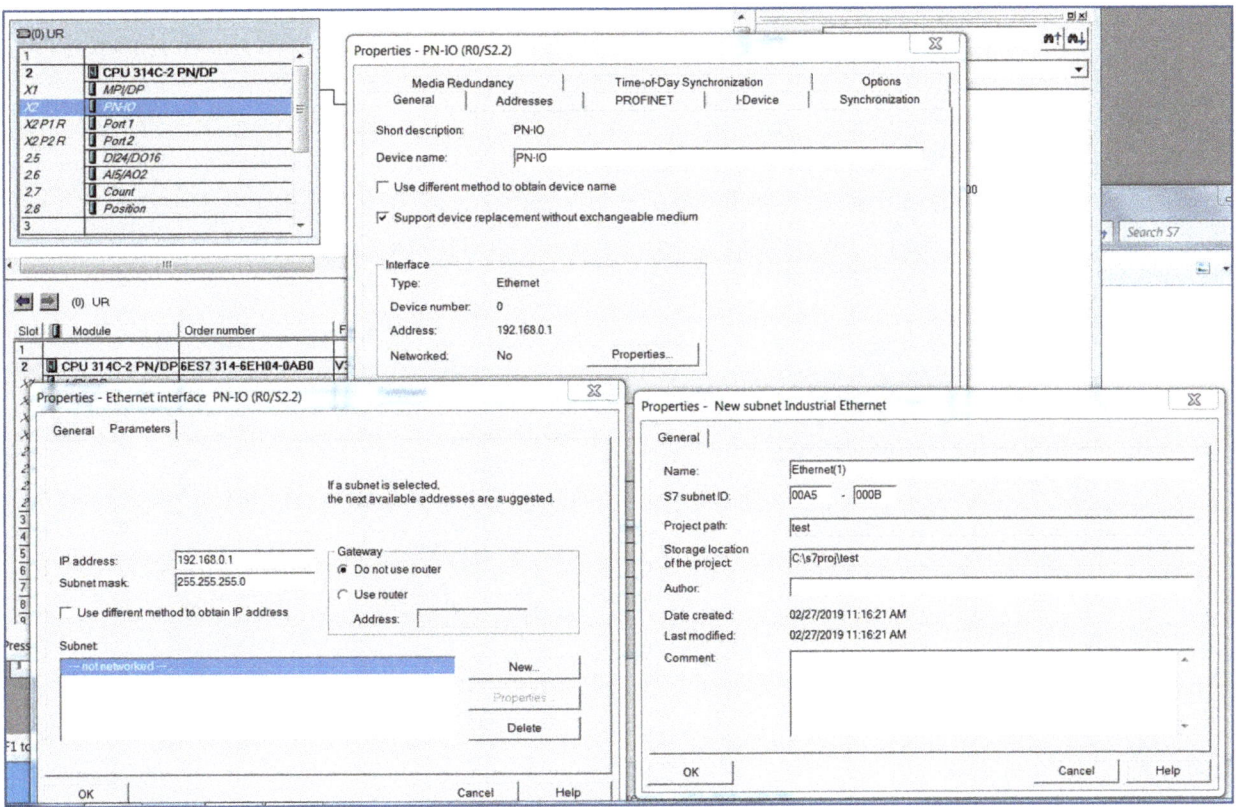

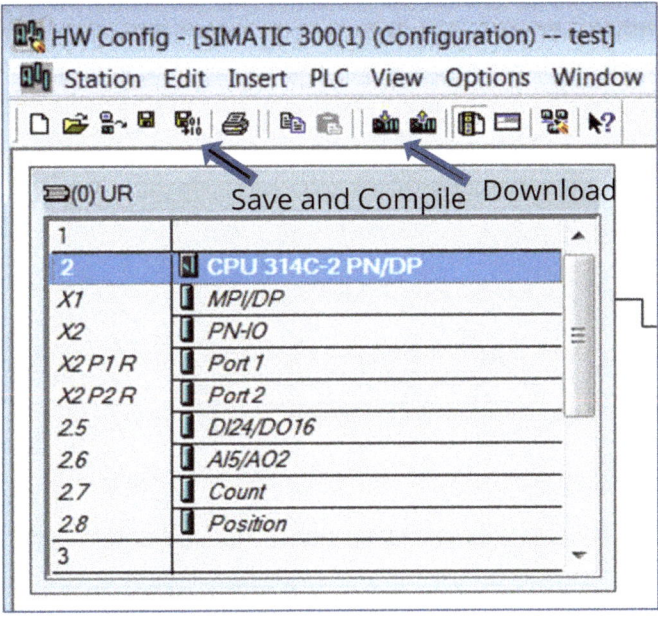

After the hardware has been configured, it needs to be saved, compiled and downloaded. This can be done using the icons shown or from the Station and PLC tabs.

After completion of the hardware configuration, it is best to close the hardware editor. Also, do not Save, Compile and Download with other editing windows online with the processor; this has been known to cause problems!

Writing the Program

After the hardware has been configured, the programmer can begin work on the program itself. When the project is created, a folder named S7 Program appears containing two sub-folders named Sources and Blocks. The Blocks folder contains all of the program elements that are to be downloaded into the processor, along with some files that are present only on the programming computer (PG).

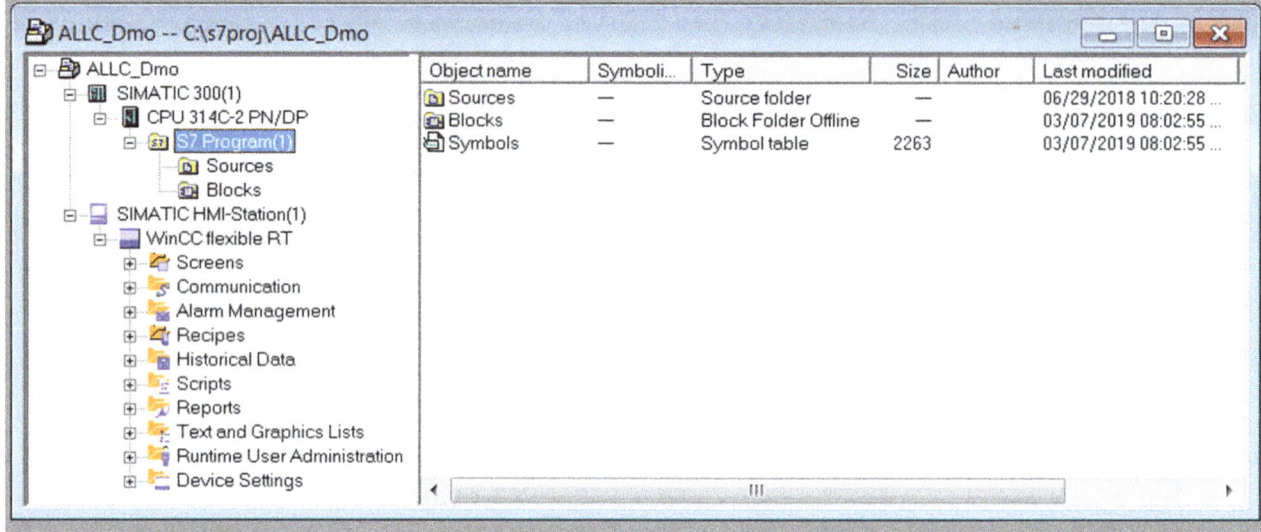

Siemens Blocks

Elements of an S7 program are called Blocks. These blocks, with the exception of Data Blocks, have local variables that can be used to pass input and output parameters, temporary variables that can be used internally, and variables specific to the type of block being used. There are several different types of blocks that make up a program:

Organization Blocks (OBs)

Special purpose blocks that contain code by the programmer. They cannot be called but are instead called automatically based on time or events. Below some of these are listed:

OB1- The originating block that runs continuously. Every platform has a routine that runs first and runs continuously; this is it for Siemens. All FC and FB routines are called and originate from here.

OB10-17- Time of Day Interrupts. These blocks run based on the time entered into the hardware configuration. Different processors have a different number of these available. Useful for collecting and logging data at a specific time, such as shift change.

OB20-23- Time Delay Interrupts. These blocks run based on setting up a system block that defines a delayed response to an event. Also useful for data logging and ensuring that events do not happen at the same time. Delay time is specified in SFC32 (System Function Call)

OB30-38- Cyclic Interrupts. These blocks run on a periodic schedule defined in the hardware configuration. Useful for events that do not need to be scanned continuously in OB1, such as PID instructions.

OB40-47- Hardware interrupts. These blocks run based on hardware events set up on I/O modules capable of generating them. Sometimes used for alarms and exceeding limits.

OB55-57- DPV1 Interrupts. These blocks run based on Profibus events (DP). Events, such as reading or writing a record or receiving an interrupt request from a DP Slave, trigger them.

OB60- Multicomputing interrupt. Used in synchronous operation of several PLCs.

OB61-64- Synchronous cycle interrupts. Used to coordinate short and equal-length process reaction times on Profibus (DP).

OB70, 72- Redundancy error interrupts. Used in H Series (redundant) PLC processors and related I/O.

OB80-87- Asynchronous error interrupts. These blocks run based on different hardware and communication network error events, includes watchdog error.

OB90- Background Organization Block. If you have specified a minimum scan time longer than the actual scan time, the CPU still has processing time available at the end of the cyclic program. This time is used to execute the background OB if desired.

OB100-102- Startup Organization Blocks. These blocks run once upon startup. Different types of startups are defined, Warm, Cold and Hot. Only warm startups are available with the S7-300 platform, so OB100 is commonly used.

OB121- Programming Error. Runs on errors such as calling a block that has not been downloaded and pointers to data areas that do not exist.

OB122- I/O Access Error. Runs when a read error occurs while accessing data on an I/O module.

Functions (FCs)

A logic block without permanent memory. This is the equivalent of a subroutine in other platforms, where permanent local variables are not required. Typically, most code is organized in these blocks, which are called from OB1 or other Functions or Function Blocks.

Temporary variables (TEMP) can be declared in these blocks, but as soon as the scan leaves the routine or function, the temporary memory is overwritten by other temporary variables in other blocks. Input and Output variables can also be declared to pass parameters into or out of the blocks, making them reusable.

Function Blocks (FBs)

A logic block with permanent memory. These blocks are subroutines like FCs, which must be called; however, they also have a field called STAT for static variables. When Function Blocks are called, a Data Block has to be assigned to it; this type of DB is called an **Instance Data Block**. Each block call needs to have its own instance data block. The instance data block can be created automatically and contains Input, Output and Static variables.

Function Blocks also have temporary variables, which are typically used for intermediate calculations as placeholders.

Data Blocks (DBs)

An organized block of data that can contain mixed data types, UDTs, arrays and others. There are two types of Data Blocks: **Global**, where the data is entered by the programmer and available to all blocks and routines, and **Instance**, which are created automatically when calling a Function Block.

Data block elements can be called by direct addressing, where the element's block number and byte offset is named, such as "DB6.dbw26". They can also be called by symbolic addressing, such as "Drive2.ConveyorSpeed". Element names and data types are typed directly into the block's definition, and the Data Block's name is defined when creating it or in the Symbol Table. An exception to direct addressing is when a data block is Optimized. When this is done, the data block is compacted, and only symbolic addressing can be used. Optimized data blocks take up less space in the processor and can be accessed faster but cannot be accessed by HMIs because of the lack of direct addressing capability. Optimizing data blocks can only be done in the newer TIA Portal software and only on S7-1500 processors.

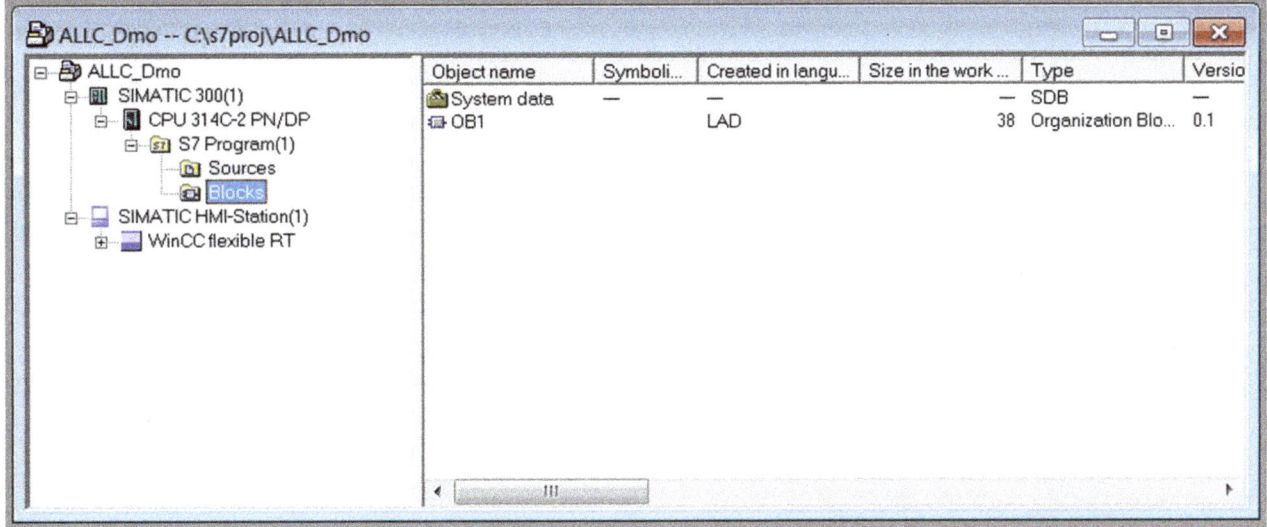

The Main block OB1 is automatically placed in the blocks folder. Other blocks are created as needed by selecting them from the Insert menu tab in the manager or by right-clicking on the Blocks folder and selecting "Insert".

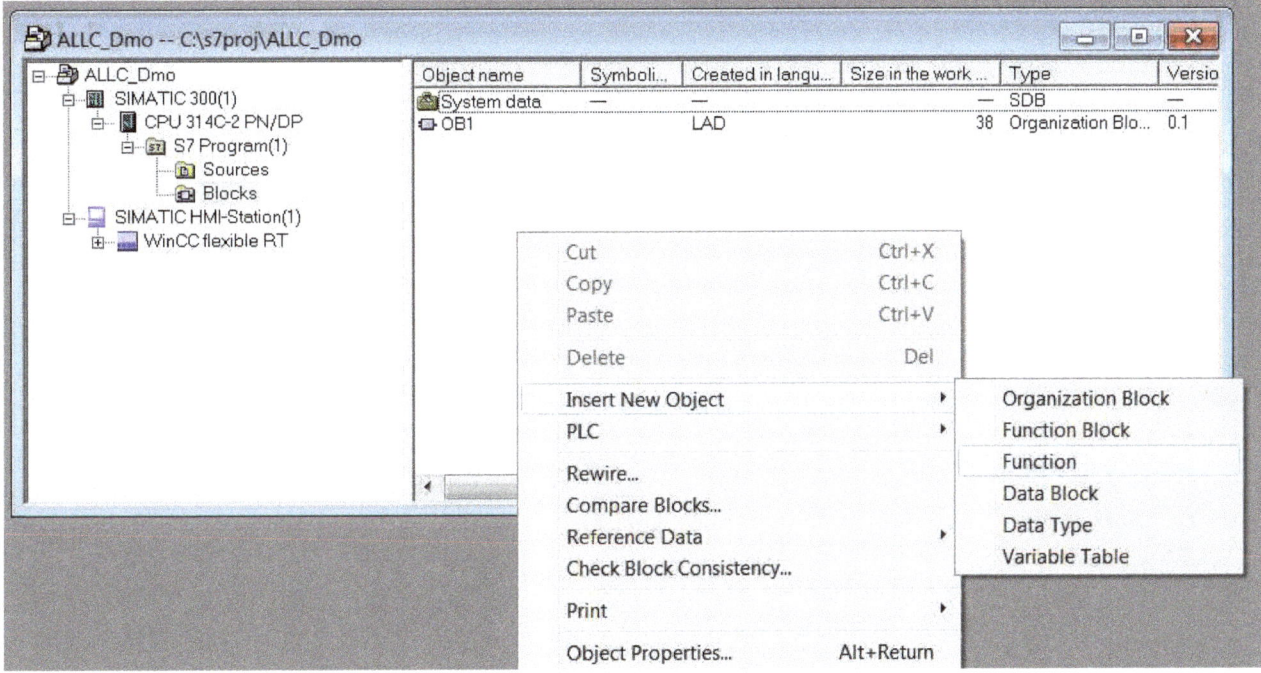

When a new block is inserted, a dialog appears allowing a name to be applied to the block. The name appears in the Symbol Table automatically. The number applied to Functions, Function Blocks and Data Blocks is not significant, but Organization Block numbers are specific to their purpose.

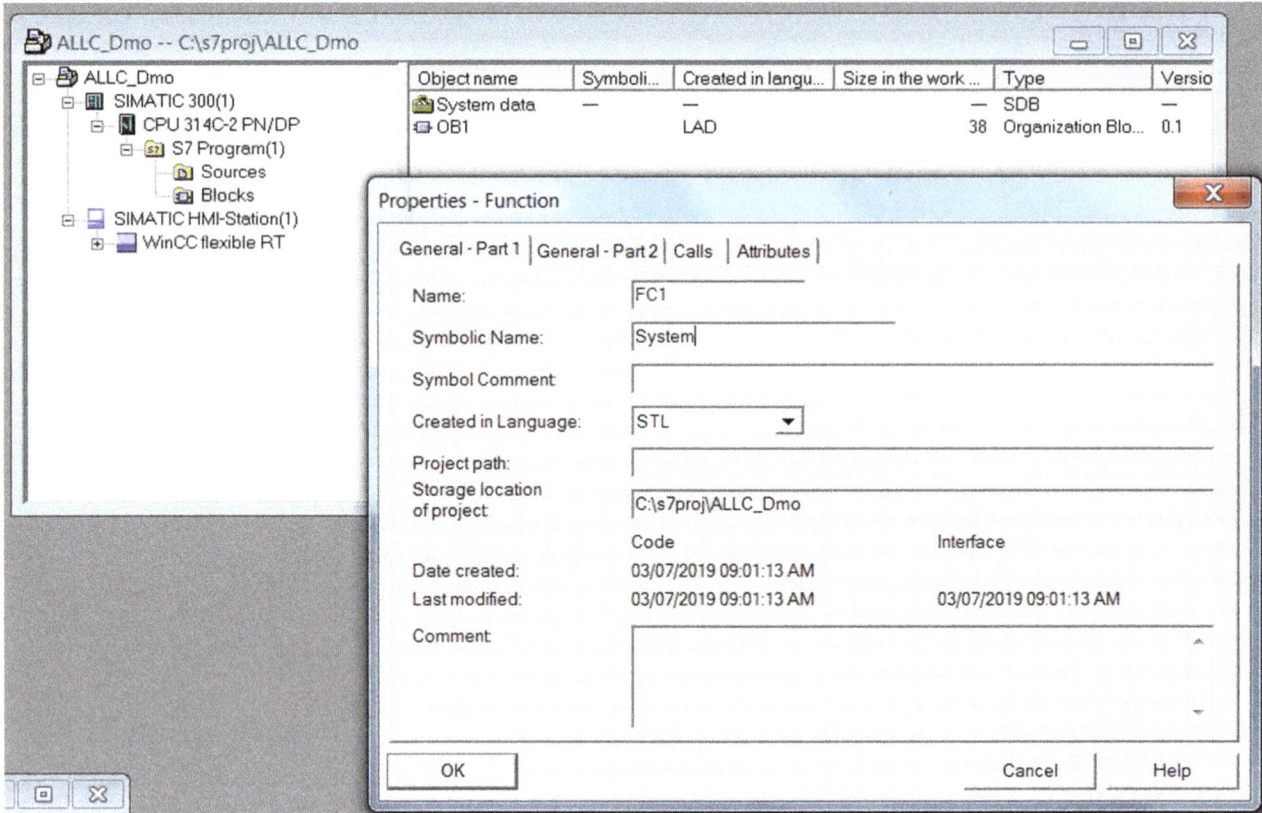

The programming language is also specified when creating the block but can be changed from the **View** menu in the editor.

After creating the block, double-clicking on the icon in the Blocks folder will open it for editing.

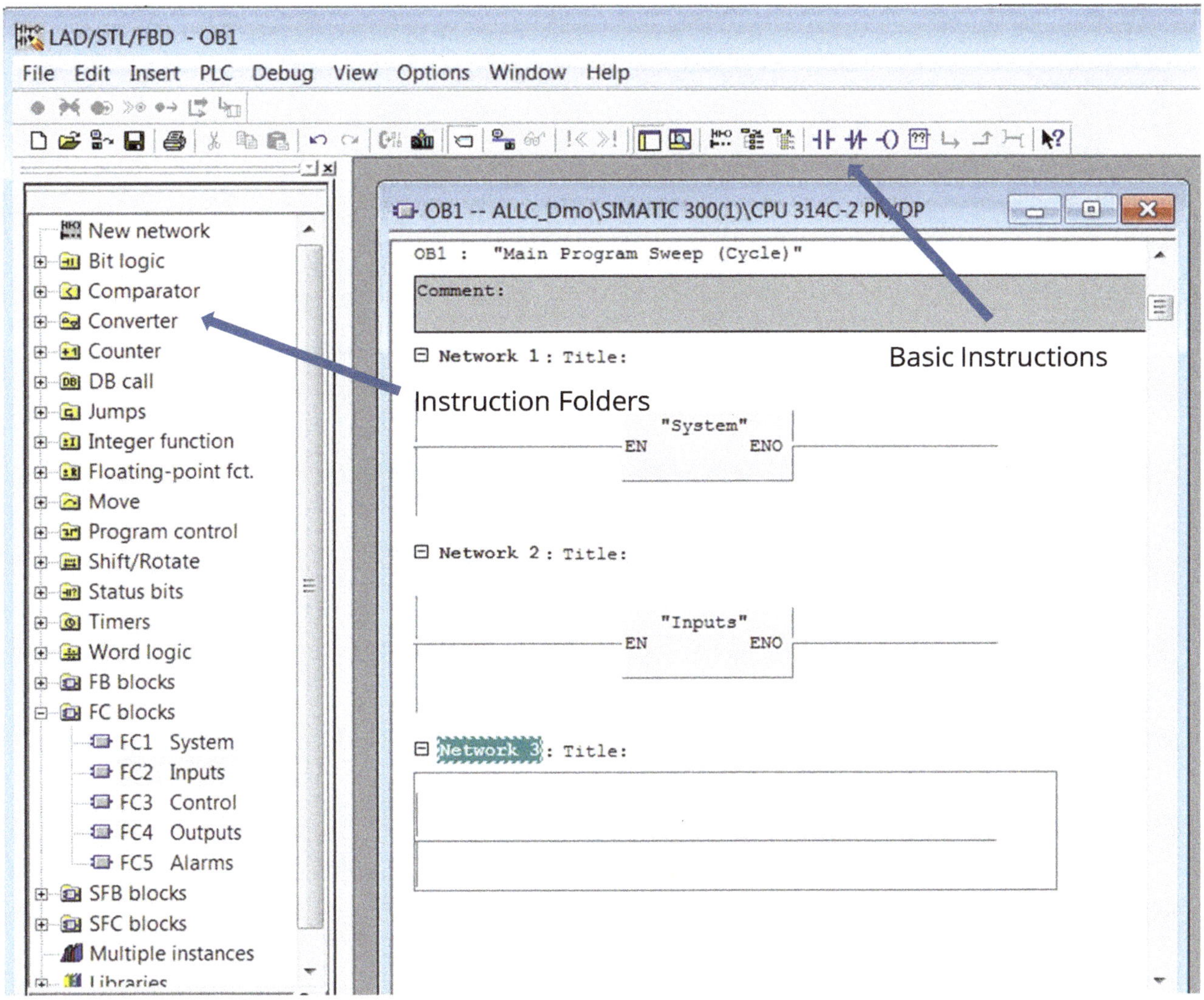

![Edit screen showing LAD/STL/FBD - OB1 with instruction folders on left and Basic Instructions toolbar at top]

Edit Screen with Calls to Functions

Instructions are inserted either by selecting them from the basic instruction icons at the top of the edit screen or from the folders on the left side. Calls to functions and function blocks are made by double-clicking or dragging them from the appropriate blocks folder in the instructions. Of course, blocks cannot be called until they are created, and if they are not called, the code inside of them does not run!

After editing the code in a block, the block should be saved and downloaded into the processor. There are several icons at the top of the edit screen that are useful for saving, downloading and viewing blocks online, as shown in the figure above.

Unlike many other PLC platforms, Siemens blocks are downloaded individually into the processor. This can make it difficult to know whether the blocks have been downloaded or not or whether the saved block is the same as the block in the CPU.

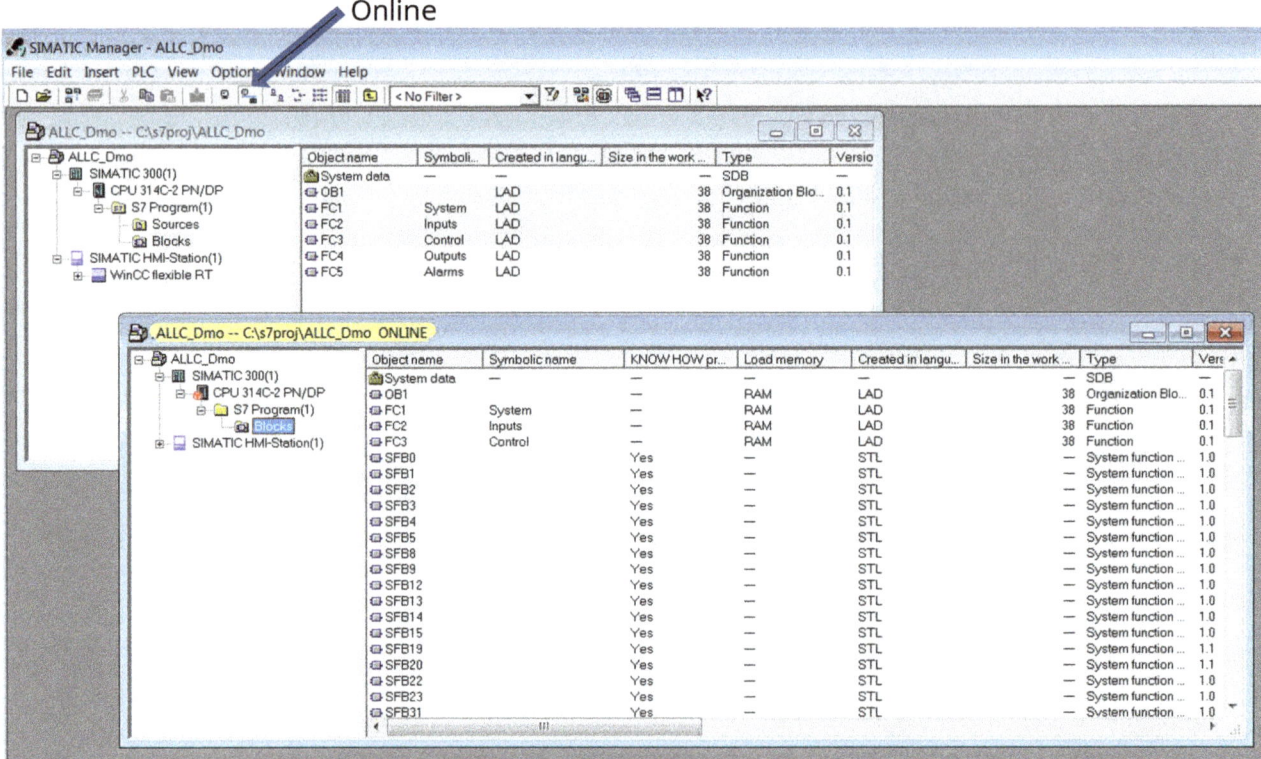

In the Simatic Manager, there is an icon that allows the blocks that are in the processor to be viewed. Notice that there are additional blocks besides the ones created by the programmer; these are system blocks and will be discussed later. In the image above, the offline view shows that there are six blocks in the offline view, but only four of them have been downloaded into the PLC. It is very important that before downloading OB1 with the function calls to the blocks that all called blocks have been downloaded! The CPU will fault if the blocks are called without being present. This is the main purpose of the online view: to determine block presence.

Symbols can be assigned to addresses as they are entered into the program or they can be entered directly into the Symbol Table all at once.

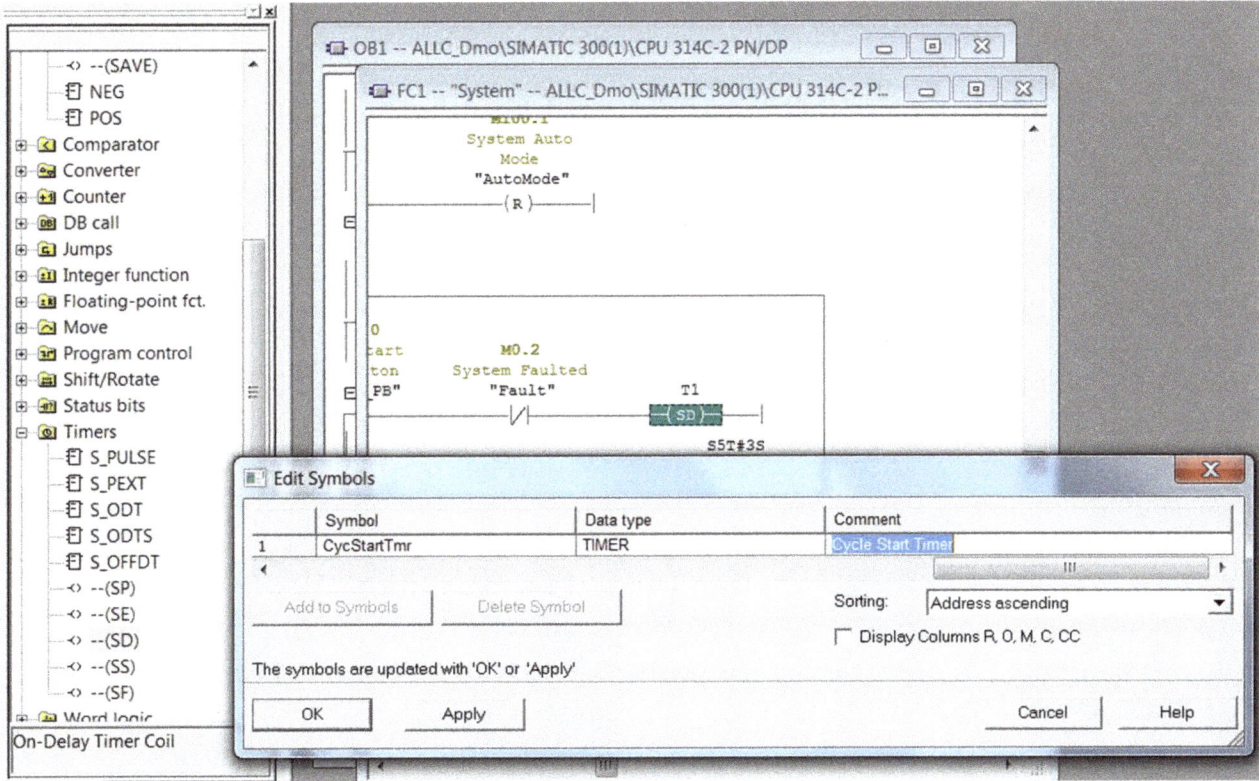

Often, I/O addresses are assigned first, along with system control addresses, such as Auto and Manual Modes, Auto Cycle, Fault, and other addresses that apply to the entire system. It can be easier to type these directly into the Symbol Table.

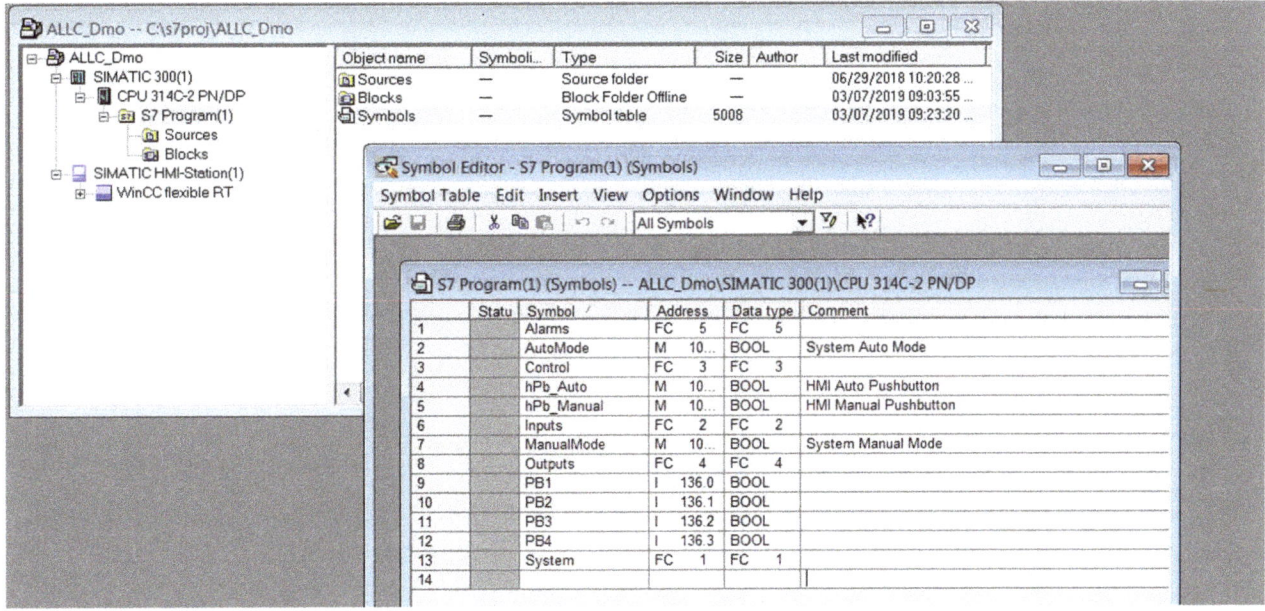

Many instructions' addresses will be assigned from global data blocks rather than being assigned to marker memory. These will <u>not</u> appear in the Symbol Table.

System Functions (SFBs and SFCs)

System Functions (SFCs) and System Function Blocks (SFBs) are already in the processor when a program is downloaded. These are just like regular FCs and FBs; however, they are created by Siemens to accomplish various tasks. They can be called by selecting them from the SFB Blocks and SFC Blocks folders on the left of the edit screen.

Other FB and FC blocks created by Siemens are selectable from the library folders in the same section, but these must be downloaded like regular blocks.

There are other objects that are not downloaded to the processor at all and are only present in the programming computer; these are listed in the following section.

Siemens Data

An important thing to remember about all Siemens data: on the S7-300 and S7-400 processors and on the newer S7-1200 and S7-1500 processors, is that everything is byte-based. This means that all addressing of bits, words, Reals, Timers, Counters and other structures are in eight-bit sections. To directly access data, especially in the S7 300 and 400s, addresses must be specified by byte number and data type. On the newer TIA Portal software and hardware, this is less critical to remember since all addressing is symbolic.

Data Type: User Defined Data Type (UDT), as defined in other sections of this book. These are also called Structures.

Variable Table (VAT): These tables are used by the programmer to monitor and change data in the processor while the program is running.

Symbol Table: Blocks and memory addresses are named here or when creating the variable or block in the software. It is not necessary to name variables and blocks, but it is best to do so.

Marker Memory: Marker memory already exists in the processor and is used as needed. Different processors will have differing amounts of Marker Memory, addressed by using the letter "M".

Marker memory, and all other addressing in Siemens, is byte-based. This means that words, double words, REALs, and all other addressed are addressed at the byte level, combining bytes as necessary.

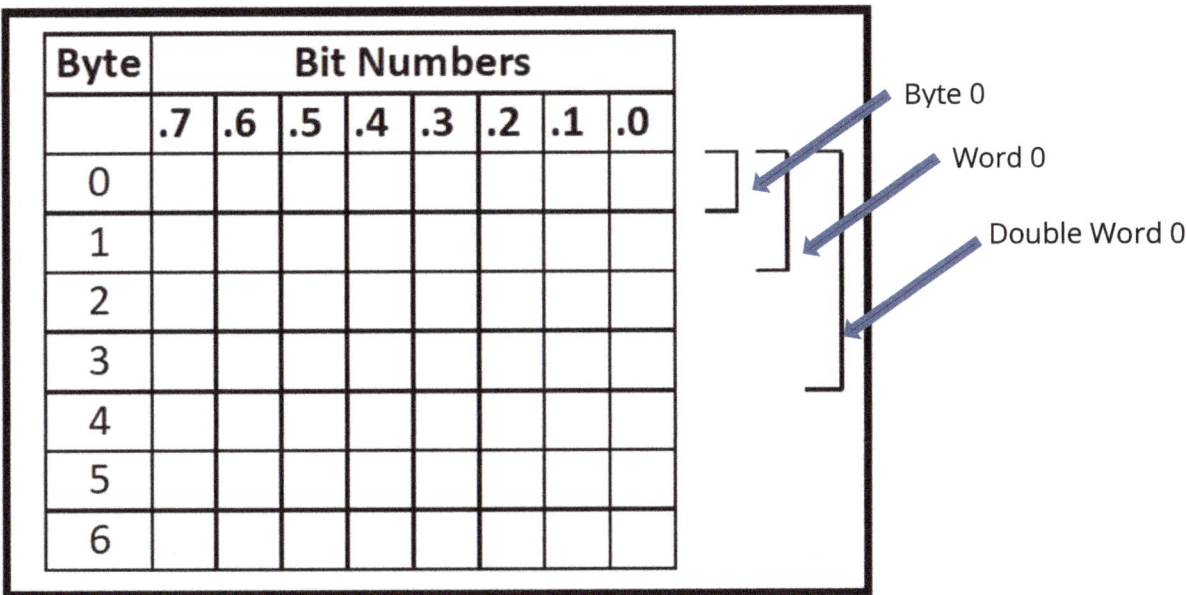

The diagram above shows that Byte 0, Word 0, and Double Word 0 actually occupy the same area in memory. Word 0 consists of Byte 0 and Byte 1, while Double Word 0 consists of Bytes 0, 1, 2 and 3. This means that Marker Memory addresses would be MB 0, MW 0 and MD 0 for this example. Because of this, Siemens programmers typically use only even numbered addresses. If MD0 is used, the next available word would be MW4, or MD4 if a double word was required. Bits are addressed in the range 0-7, as in M6.2, signifying the 3rd bit of the sixth byte.

Siemens is also very restrictive when it comes to data. Though an Integer and a Word are the same size -- 2 bytes or 16 bits -- in Siemens, they are treated differently. An integer can be used for math operations and a word can only be used for logic operations, such as AND/OR. Double words and double integers have the same restrictions.

Real numbers, of course, also occupy a double word of space. It is important when discussing data to differentiate between the data type and the space it occupies; a word can mean two different things: a type of data or two bytes.

Until a symbol is defined in the Symbol Table, data in the Marker Memory does not have a data type and can be used interchangeably. As soon as you name it, the rules about integer or word will apply, as a data type must be declared.

More on Addressing

As mentioned previously, everything in Siemens addressing is byte-based. This means that Input and Output addresses may take up multiple byte addresses for analog, so an analog input PIW272 (Process Input Word 272) actually takes up input bytes 272 and 273. The next analog input address would then be PIW 274. Analog outputs are designated as PQW (Process Output Word).

Address	Name	Type	Initial value	Actual value
0.0	CH1.MinEngineering	REAL	0.000000e+000	0.000000e+000
4.0	CH1.MaxEngineering	REAL	0.000000e+000	0.000000e+000
8.0	CH1.BiPolar	BOOL	FALSE	FALSE
10.0	CH1.Return	WORD	W#16#0	W#16#0
12.0	CH1.Scaled	REAL	0.000000e+000	0.000000e+000
16.0	CH1.Temperature_F	REAL	0.000000e+000	0.000000e+000
20.0	CH1.Delta	REAL	0.000000e+000	0.000000e+000
24.0	CH1.InRange	BOOL	FALSE	FALSE
26.0	CH2.MinEngineering	REAL	0.000000e+000	0.000000e+000
30.0	CH2.MaxEngineering	REAL	0.000000e+000	0.000000e+000
34.0	CH2.BiPolar	BOOL	FALSE	FALSE
36.0	CH2.Return	WORD	W#16#0	W#16#0
38.0	CH2.Scaled	REAL	0.000000e+000	0.000000e+000
42.0	CH2.Temperature_F	REAL	0.000000e+000	0.000000e+000
46.0	CH2.Delta	REAL	0.000000e+000	0.000000e+000
50.0	CH2.InRange	BOOL	FALSE	FALSE

Data Blocks also often need to be addressed directly. In the Data Block above, there are various data types, including BOOLs, Words and Reals, all taking up different amounts of space. The Address column at the left shows the offset within the data block. If the Data Block had the name "Channels" defined in the Symbol Table, these elements could be addressed symbolically, such as "Channels.CH1.Scaled". As can be seen in the table, the address for this element is at offset byte 12, and the next address is at byte 16. This means that the element is 4 bytes long; this makes sense since Scaled is defined as a REAL, which is 32 bits. If this was Data Block 4, its direct address would then be DB4.DBD12. DBD, or Data Block Double Word, defines the size of the data, not the data type.

The Element above Scaled is a Word data type at byte offset 10 named "Return". This means that its symbolic address is "Channels.CH1.Return", while its direct address would be DB4.DBW10. At byte offset 34, the element BiPolar is a BOOL, its symbolic address would be "Channels.CH2.BiPolar", and its direct address would be DB4.DBX34.0. Boolean elements are addressed as DBX because DBB is reserved for addressing Data Block Bytes. Note: they must also be specified to the bit level.

The Data Block in this example was actually created by using UDTs to define an element called "Channel". Two of these elements, CH1 and CH2, were defined in the **Declaration View** of the block when it was created. The view shown in the diagram is known as the **Data View**.

Statement List (STL)

When discussing Siemens programming, it is impossible to do so without detailing Statement List programming, or STL. Statement List is a version of Instruction List (IL), one of the five standard languages defined in IEC 61131; however, in Siemens programming, it is the de-facto standard for many programmers, even favored over Ladder Logic.

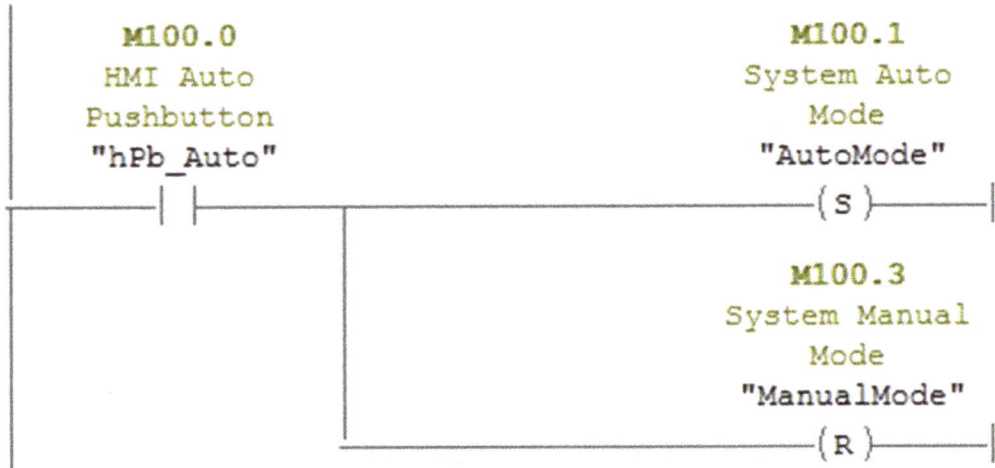

Mode Setting in Ladder

The logic above shows a simple mode control network in Ladder Logic.

The same logic is shown below written in STL:

```
Network 1: Select Auto Mode
    A    "hPb_Auto"      M100.0    -- HMI Auto Pushbutton
    S    "AutoMode"      M100.1    -- System Auto Mode
    R    "ManualMode"    M100.3    -- System Manual Mode
```

Mode Setting in Statement List

Normally open contacts are shown as A for "And", while a normally closed contact is shown as AN, for "And Not". The Set and Reset are represented as S and R just like ladder. Contacts placed in parallel are shown as O and ON, for "Or" and "Or Not".

All of the other IEC languages can be converted to Statement List, but not all STL can be converted to Ladder Logic. There are instructions that only exist in STL and not LAD or FBD. There are also sometimes NOP (No Operation) placeholders and jumps that appear when converting LAD to STL, but most STL programmers do not place the NOPs in their code, making it impossible to convert to LAD.

Statement List is more efficient than Ladder Logic but can be more difficult to read and troubleshoot for maintenance people who are not used to it.

More complex logic sometimes has to use labels or placeholders, as shown in the following examples.

Cycle Start - Ladder

The network above is a typical Cycle Start network in Ladder Logic.

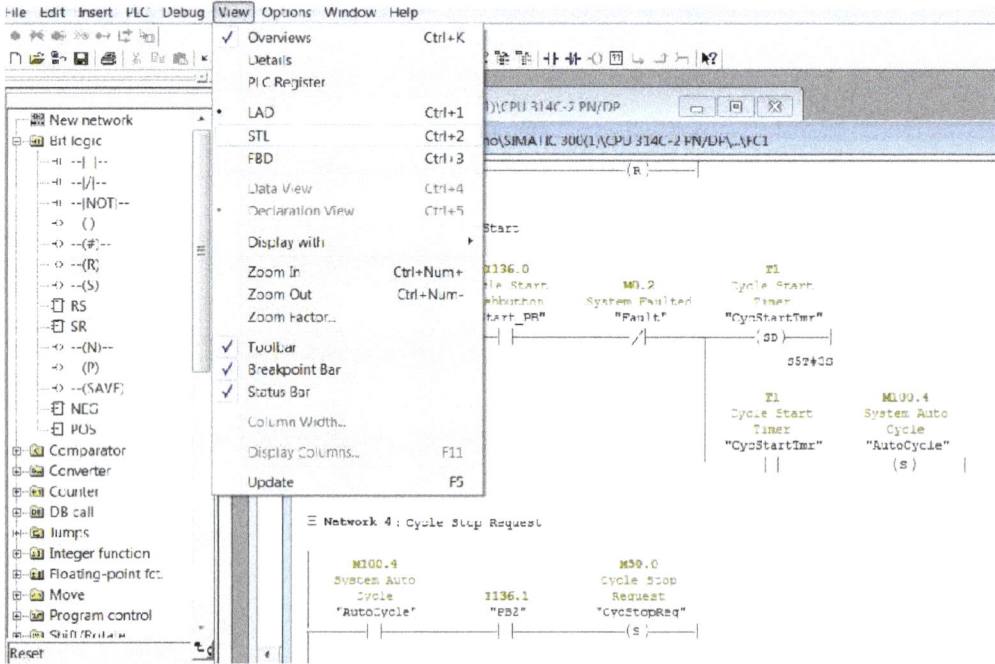

Convert to Statement List (STL)

To convert the display to STL, select the language from the View tab.

```
⊟ Network 3 : AutoCycle Start
      A      "AutoMode"           M100.1          -- System Auto Mode
      A      "Start_PB"           I136.0          -- Cycle Start Pushbutton
      AN     "Fault"              M0.2            -- System Faulted
      =      L     0.0
      A      L     0.0
      BLD    102
      L      S5T#3S
      SD     "CycStartTmr"        T1              -- Cycle Start Timer
      A      L     0.0
      A      "CycStartTmr"        T1              -- Cycle Start Timer
      S      "AutoCycle"          M100.4          -- System Auto Cycle
```

Cycle Start – Statement List (STL)

Because the logic has more than just a few instructions, a label was established at the branch point just after the NC fault contact. This point was called L0.0 and serves as a reference for the lower branch of logic. The BLD instruction is a null type program display command, which allows the STL to be converted back to a LAD display; it can be removed without affecting the operation of the program. If it is removed, the display will remain STL though, even if the display is converted to LAD.

The Load (L) instruction places a value into one of the accumulators. These are data registers in the PLC that are used to hold numbers.

Network 6 : Title			RLO	STA	STANDARD	ACCU 2
A "ManualMode"	M100.3		0	0	300	0
JNB _001			1	1	300	0
L 1						
T "Message"	MW60		1	1	300	0
_001: NOP 0						

Network 6 above shows the online view for a data movement being monitored. The columns STANDARD and ACCU 2 are the two accumulators. Note that the instruction L 1 (Load 1) has blank spaces in the cells next to both the Load (L) and Transfer (T) lines. The code is being jumped over by the JNB (Jump Not with BR Status bit). This is necessary in Statement List; if the status of the Boolean AND preceding the instruction is not addressed, the code will execute regardless of the state on Manual Mode.

All Boolean logic instructions need to take this into account; Jump instructions are used often in Statement List. The _001: label was generated automatically when converting the logic from LAD to STL. Labels are four characters long, cannot start with a number, and always have a colon at the end.

The ladder equivalent of the previous Statement List Load and Transfer is shown below.

```
Network 6 : Title:

    M100.3
 System Manual
     Mode
  "ManualMode"         MOVE
 ───┤ ├────────────EN       ENO───
                    1 ─IN         MW60
                                 Message
                                 Register
                             OUT ─ "Message"
```

There are two accumulators in S7-300 processors and three in S7-400 processors. Accumulators act as a "stack" for number manipulations, such as Moves and math operations. When a number is loaded into STANDARD (Accumulator 1), the number that was there before moves to ACCU 2 (Accumulator 2). Whatever number was in ACCU 2 previously is lost; it "falls off" of the stack.

				RLO	STA	STANDARD	ACCU 2
Network 6: Title:							
A	"ManualMode"		M100.3	1	1	300	0
JNB	_001			1	1	300	0
L	1			1	1	1	300
T	"Message"		MW60	1	1	1	300
_001: NOP	0			1	1	1	300

Before Manual Mode was true, the number 300 was in Accumulator 1 from a previous operation. After the Load (L) instruction, the 300 is moved to Accumulator 2 and is replaced by the 1.

Math operations are performed in the same way using the accumulators. The following instructions add two REAL numbers:

L 27.3
L 10.62
+R
T MD54

The double word register MD 54 would then contain the number 37.92, the result of adding the numbers. Accumulator 1 would also contain the same number, and Accumulator 2 would contain the number 10.62, the result of the last Load instruction. Math operations can be entered very quickly in Statement List.

Other Siemens Languages

All other IEC 61131 languages are supported in Siemens software; however, the software may not be included with some versions. See the *Siemens Step 7 PLC Software* description in this book for more information.

Function Block Diagram (FBD)

FBD is a standard language in all Step 7 software packages. The logic below shows the FBD equivalent of the latching Auto and Manual Mode bits shown previously.

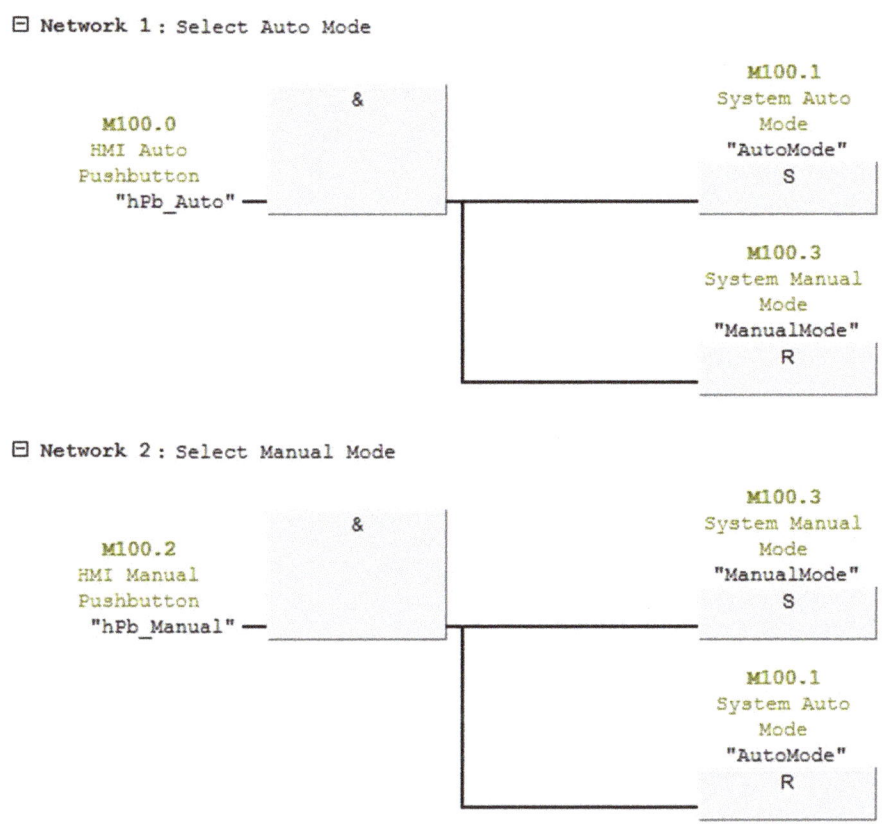

As with Statement List, FBD is selected from the View menu in the editor. All FBD can be converted to STL, but not all STL can be converted to FBD or Ladder.

Structured Control Language (SCL)

SCL is Siemens' version of Structured Text. It is a high-level language supporting Siemens objects, such as data blocks, and has templates for Siemens structures, including OBs, FBs, FCs and DBs. Like most high-level languages, it uses control structures such as IF-THEN-ELSE, CASE, FOR, WHILE and REPEAT (Looping).

SCL is a compiled language and is not included in all Siemens PLC software packages but can be purchased separately. SFC blocks are created from within the Sources folder in the Simatic Manager.

```
SCL_Example -- Advanced_Solutions\Solutions\CPU 314C-2 PN/DP
FUNCTION FC34: INT

TITLE = 'SCL Example'
VERSION: '1.0'
AUTHOR: ATrain
//KNOW_HOW_PROTECT

VAR_INPUT
  IN1: REAL ;
  IN2: REAL ;
END_VAR
VAR_OUTPUT
  OUT: REAL ;
END_VAR
BEGIN

  //Use a different formula based on the input values
  IF IN1 >= IN2 THEN
    OUT := (IN1 * 10) + 2;
  ELSE
    OUT := (IN2 * 5) + 10;
  END_IF;

  //Flag an error if the value is out of range
  //for integer values
  IF OUT > 32768 OR OUT < -32767 THEN
    FC34 := 1;
  ELSE
    FC34 := 0;
  END_IF;

END_FUNCTION
```

The code above is an example of Structured Control Language. The different colors and fields are typical of Structured Text language editors. As with most high-level languages, syntax is very important; for instance, if one of the many semicolons in the above code is not present, it will not compile.

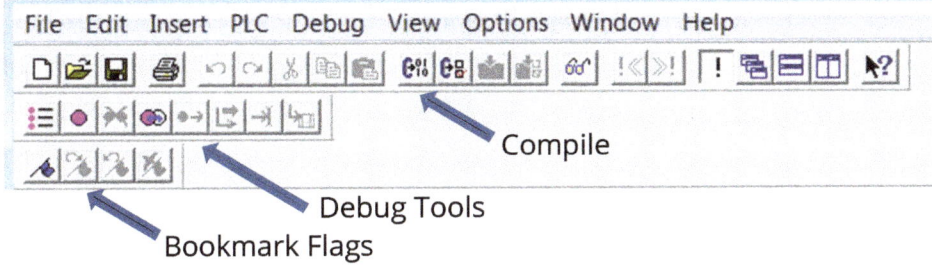

The SCL Editor Menu is similar to the standard editor, but it has several additional features. Bookmarks can be placed in the code so that the programmer can easily move from section to section. Debug tools are available to insert breakpoints in the program and run one section at a time.

After the SCL program has been written, it must be saved and compiled. The icon next to the Compile icon allows selected blocks to be compiled rather than compiling everything. If errors are found, there is a dialog that helps determine what they are.

After the block is compiled, it will appear in the Blocks folder of the Simatic Manager and is downloaded like any other block. If the block is double-clicked from within the manager, it opens in the SCL editor.

S7 Multi-language Example: Node Fault

The following example shows how the different Siemens languages work together. OB86 is a block that runs if there is a network problem or a failure in the PLC rack. The code in this example determines which rack or node is faulted in a Profibus or Profinet network.

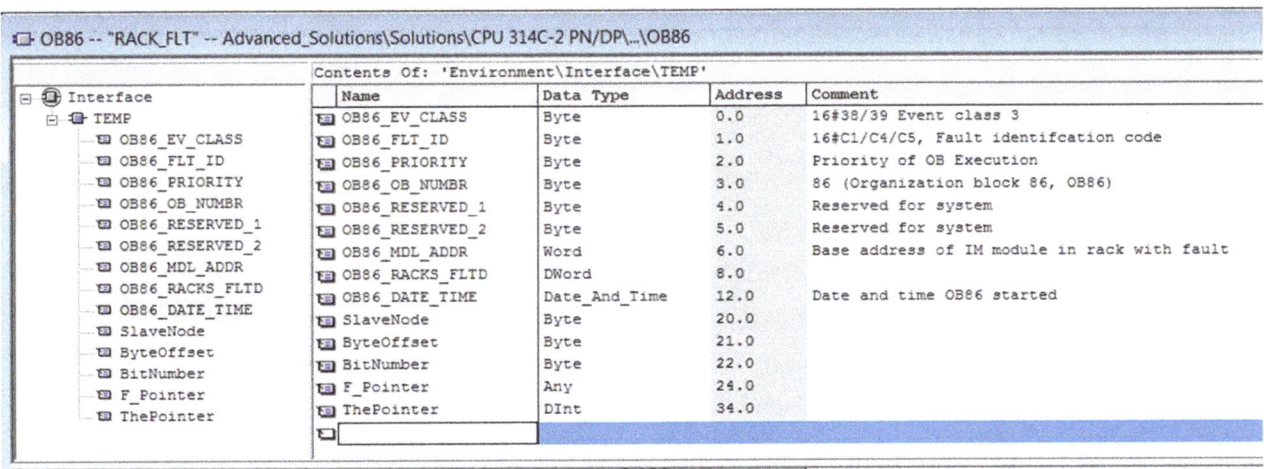

The TEMP variables in OB86 contain information about the reason the block was called. The variables that begin with "OB86_" are present in the block automatically, while the other variables were placed there by the programmer.

The first part of the code determines whether the OB86_EV_CLASS byte indicates that the node is exiting the faulted state (node return, B#16#38, Hex code 38) or entering the faulted state (Node fault, B#16#39, Hex code 39). The JC (Jump Conditional) instruction goes to the appropriate part of the code based on the number.

```
OB86 : "Loss Of Rack Fault"

Demonstration of a "Do it yourself" diagnostic program.
It assumes all possible PROFIBUS nodes are in use.

Network 1: Title:

This is demonstration code for use of pointers
It requires a DB named "DP_Faults" which contains an array of 126 BOOL
This code does not differentiate between DP networks, if you have more than one
Or a PN network if you have a PN / DP processor with IO on both.

The code below determines is it an incoming or outgoing fault.

        L       #OB86_EV_CLASS          #OB86_EV_CLASS   -- 16#38/39 Event class 3
        L       W#16#38
        ==I
        JC      L001

// The code below is for an incoming event - a rack fault.

        NOP     0
        L       LB      11
        T       #SlaveNode              #SlaveNode
        ITD
        L       L#8
        MOD
        T       #BitNumber              #BitNumber
        L       LB      11
        L       8
        /I
        SLD     3

        T       #ByteOffset             #ByteOffset
        L       P#0.0
        +I
        L       #BitNumber              #BitNumber
        +I
```

The section of code above calculates a pointer for indirect addressing. There are several types of pointers used in Step 7; they must be calculated in STL. Results are placed into the programmer-generated variables #SlaveNode, #BitNumber, and #ByteOffset. P# is the indicator that the variable is to be used as a pointer; in this case, the pointer is in the form P#<byte>.<bit> (4 bytes).

Pointers can be in three different formats; this is the simplest of the three. The other two formats are:

P#<Area><byte>.<bit> (6 bytes) where area is the register I, Q or M.
P#<Area><byte>.<bit> <length> (10 bytes) for moving blocks of data.

The data block is then opened and the appropriate bit is set, indicating the node that is faulted.

```
     OPN    "DP_Faults"                // Open the DB used as fault record
     L      #ByteOffset
     L      #BitNumber
     +D                                // Pointer to byte / bit offset
     T      #ThePointer
     S      DBX [#ThePointer]          // Set the bit that represents the rack number

// ============================================================

// This is for an outgoing (clearing) fault

L001: NOP   0                          //Start code to clear fault bits
     L      #OB86_EV_CLASS             // So the fault is cleared so we will clear the bit
     L      W#16#39
     ==I                               //Incoming event
     JC     L002

     L      LB  11                     //Get the node number (Local data byte 11 )
     T      #SlaveNode
     ITD
     L      L#8
     MOD                               // Calculate the bit number in the byte
     T      #BitNumber
     L      LB  11                     //DP Rack number in local stack
     L      8                          // Calculate the byte offset in the data block
     /I
     SLD    3

     T      #ByteOffset
     L      P#0.0
     +I
     L      #BitNumber
     +I

     OPN    "DP_Faults"
     L      #ByteOffset
     L      #BitNumber
     +D                                // Pointer to byte / bit offset
     T      #ThePointer                // Fault clear reset the bit
     R      DBX [#ThePointer]

L002: NOP   0                          //We're Done
```

The second half of the code performs the same function, calculating a pointer to reset the faulted bit.

This code does not include error-checking and doesn't account for multiple networks.

The SCL code below creates FB86, which uses the type and class of node fault to determine whether it is a Profibus or Profinet type fault. This is then used to fill the contents of an HMI Fault message.

```
FUNCTION_BLOCK FB86

VERSION : '0.1'
// Sample SCL for DP/PN Diagnostics
    VAR_INPUT
        FaultID : Byte;
        ClassID : Byte;
        Rack_Z23 : DWord;
    END_VAR

    VAR
        DP_RacksFaulted : Array[0..125] of Bool;
        PN_RacksFaulted : Array[0..125] of Bool;
    END_VAR

    VAR_TEMP
        RackID : Int;
        SystemID : Int;
    END_VAR

BEGIN
    RackID := 0;
    SystemID :=0;

  //Check for DP Fault codes
  IF FaultID = W#16#C4 OR  FaultID = W#16#C5 OR  FaultID = W#16#C6 OR  FaultID = W#16#C7 OR  FaultID = W#16#C8 THEN
        RackID := DWORD_TO_INT(Rack_Z23 & W#16#ff);
        SystemID := DWORD_TO_INT((SHR(IN:=Rack_Z23, N:= 8)) &W#16#F);

    CASE BYTE_TO_INT(ClassID) OF
      56 :  DP_RacksFaulted[RackID]:= FALSE;

      57 :  DP_RacksFaulted[RackID]:= TRUE;

    END_CASE;

    END_IF;


  // Check for PN Fault codes
    IF FaultID = W#16#CA OR  FaultID = W#16#CB OR  FaultID = W#16#CC OR  FaultID = W#16#CD THEN
        RackID := DWORD_TO_INT(Rack_Z23 & W#16#3fff);
        SystemID := DWORD_TO_INT((SHR(IN:=Rack_Z23, N:= 8))& W#16#F);

    CASE BYTE_TO_INT(ClassID) OF
      56 :  PN_RacksFaulted[RackID]:= FALSE;

      57 :  PN_RacksFaulted[RackID]:= TRUE;

    END_CASE;

    END_IF;

END_FUNCTION_BLOCK
```

Compiling the SCL code generates FB86, which is then downloaded to the processor. The call to FB86 is then placed in network 2 after the pointer logic in network 1 of OB86.

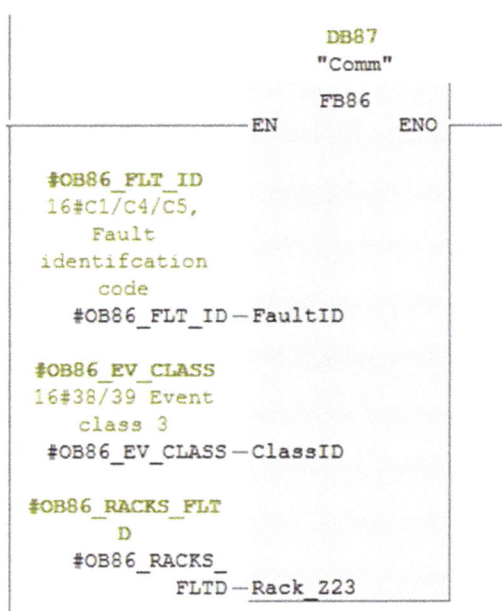

DB87 (Data Block 87) contains the FaultID, ClassID and Rack_Z23 input variables declared in the VAR_INPUT section of the SCL code, as well as the two arrays DP_RacksFaulted and PN_RacksFaulted. These are both arrays of BOOLs, which each contain one bit for every possible expansion rack in the system.

Between the two data blocks -- DB86 "DP_Faults" controlled by Network 1 and DB87 "Comms" controlled by FB86 -- the type of fault (PN/DP), node number and rack number of the fault can be determined and recorded. OB86_Date_Time can also be used from the OB86 TEMP variables to record the time at which the node faulted or returned.

S7 Graph (SFC)

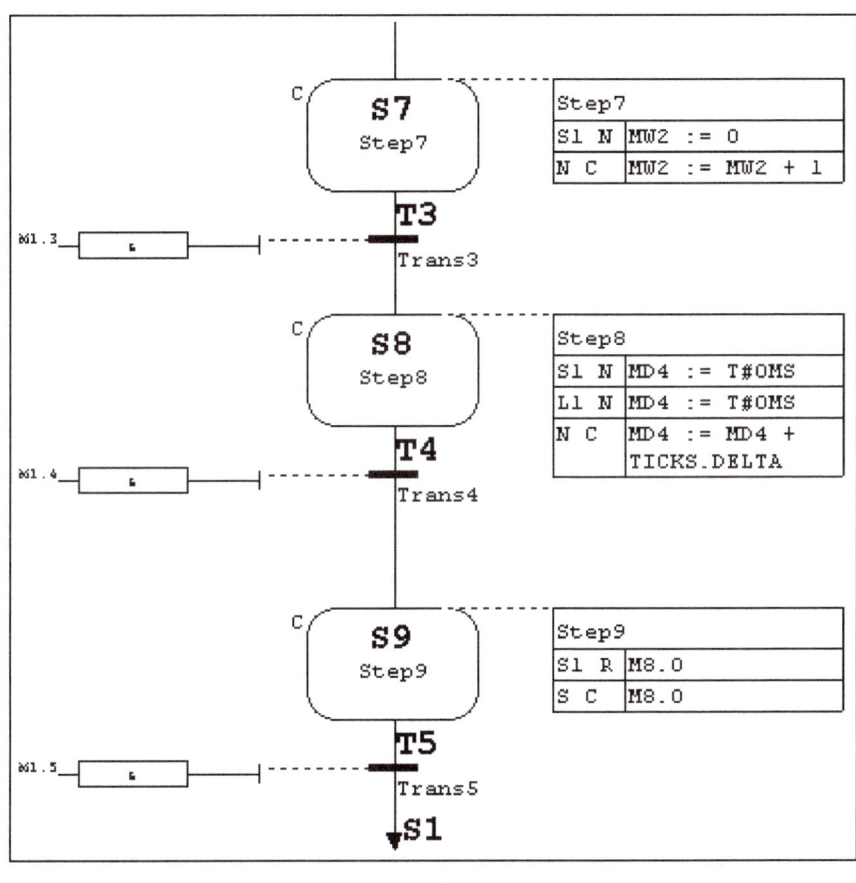

Graph, like SCL, is a compiled language that originates in the Sources folder in S7 Programs. It is Siemens' version of SFC, Sequential Function Charts, an IEC61131 PLC language. It consists of Steps (S#) and Transitions (T#), as described in the IEC61131 discussion in this book. There is an excellent example of a drilling operation in the Siemens "Help on S7-Graph: Programming Sequential Control Systems" article included with the programming software.

Siemens TIA Portal PLC Software

The programming software for the newer Siemens S7-TIA PLCs is called TIA Portal. This software can also be used to program S7-300 and 400 PLCs after 2007, but once the program has been converted, it will not be compatible with older Step 7 software.

As of November 2018, the most recent release of TIA Portal software was v15.

Siemens TIA Portal Hardware

PLC Platform	Range of usage
S7-1200	Small system Micro PLC, Modules can be added left and right of CPU.
S7-1500	Mid and high-end performance, rackless design compatible with S7-300 I/O modules.

There is a wide variety of network based I/O devices for Profibus and Profinet in the ET200 family.

S7-1200 Platform

The SIMATIC S7-1200 was the first product in the TIA (Totally Integrated Automation) family and replaced the obsolete S7-200. It mounts on standard DIN rail, and modules can be added with up to three communication modules (CM) and eight signal modules (SM). In addition, small signal boards can be attached to the front of the PLC for additional communications, I/O points or a battery module. Unlike the S7-300/400 and S7-1500 platforms, the S7-1200 does not support Statement List (STL) programming.

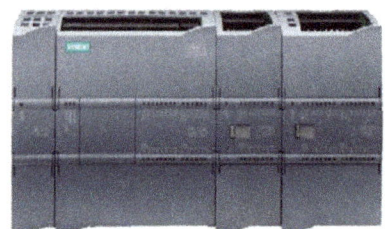

S7-1200 CPUs

CPU	Catalog Order number	Description
CPU 1211C	6ES7211-1AE40-0XB0	Entry CPU for Small Applications, 6 DC In, 4 DC Out, 2 0-10V Analog In
CPU 1212C	6ES7212-1AE40-0XB0	Compact CPU for Small Applications, 8 DC In, 6 DC Out, 2 0-10V Analog In
CPU 1214C	6ES7214-1AG40-0XB0	High Performance CPU for Small Applications, 14 DC In, 10 DC Out, 2 0-10V Analog In

| CPU 1215C | 6ES7215-1AG40-0XB0 | High Performance CPU for Small Applications, 14 DC In, 10 DC Out, 2 0-10V Analog In, 2 0-20mA Analog Out |
| CPU 1217C | 6ES7217-1AG40-0XB0 | High Performance CPU for Small Applications, 14 DC In, 10 DC Out, 2 0-10V Analog In, 2 0-20mA Analog Out, HSC Line Driver 1MHZ |

Part numbers listed above are for the "DC/DC/DC" versions. AC/DC/Relay and DC/DC/Relay versions also available.

SIPlus CPUs for extreme conditions and Failsafe CPUs for integrated safety are also available.

S7-1500 Platform

The SIMATIC S7-1500 is the fastest processor in the Siemens family. It has the same footprint as the S7-300 and is compatible with S7-300 I/O cards. A display on the front of the processor allows the diagnostics to be viewed for error messages. A built-in web server can be accessed in the processor to display status also. All processors also have Profinet ports.

S7-1500 CPUs

CPU	Catalog Order number	Description
CPU 1511C	6ES7511-1CK01-0AB0	Compact CPU, 175KB Program/1MB Data Memory
CPU 1512C	6ES7512-1CK01-0AB0	Compact CPU, 250KB Program/1MB Data Memory
CPU 1511F-1	6ES7511-1AK02-0AB0	Standard CPU, 150/225KB Program/1MB Data Memory
CPU 1513F-1	6ES7513-1AL02-0AB0	Standard CPU, 300/450KB Program/1.5MB Data Memory
CPU 1515F-2	6ES7515-1AM02-0AB0	Standard CPU, 500/750KB Program/3MB Data Memory
CPU 1516F-3	6ES7516-3AN01-0AB0	Standard CPU, 1/1.5MB Program/5MB Data Memory
CPU 1517F-3	6ES7517-3AP00-0AB0	Standard CPU, 2/3MB Program/8MB Data Memory
CPU 1518F-4	6ES7518-4AP00-0AB0	Standard CPU, 4/6MB Program/20MB Data Memory
CPU 1511TF-1	6ES7511-1UK01-0AB0	Technology CPU, 225/225KB Program/1MB Data Memory

CPU 1515TF-2	6ES7515-2UM01-0AB0	Technology CPU, 750/750KB Program/3MB Data Memory
CPU 1516TF-3	6ES7516-3TN00-0AB0	Technology CPU, 1.5/1.5MB Program/5MB Data Memory
CPU 1517TF-3	6ES7517-3UP00-0AB0	Technology CPU, 3/3MB Program/8MB Data Memory
CPU 1518F-4	6ES7518-4AX00-1AC0	MFP CPU, 4/6MB Program/20MB Data Memory, Exp. To 50/500MB, Programmable in C/C++

SIPlus CPUs for extreme conditions and Failsafe CPUs for integrated safety are also available.

Starting and Editing a Project with TIA Portal

TIA Portal is the newest software from Siemens and can be used to program all current S7 platforms.

Platform Differences

Many of the basic concepts of Siemens programming are the same as described in the Step 7 part of this book. There are some important differences and improvements that should be mentioned.

TIA addresses are all tag-based. There will be addresses behind the tags, but the address is much less important than it is in Step 7. All addresses MUST have a name.

TIA projects are in a file with the extension ap<xx>, where xx is the software version number. Since the version of software as of the writing of this manual is v15, the file saved under this version in this example is "DemoPRJ1.ap15".

Additional files are still kept in folders within the project folder and are archived and retrieved as with Step 7 files. Instead of the common .zip extension for archives (Siemens licensed the Step 7 archive utility from PKZip), files are now archived and retrieved to a file with the extension .zap<xx>, where, once again, the xx represents the software version. The archived version of the project in this example would then be "DemoPRJ1.zap15".

Projects can be converted from Step 7 to TIA and from earlier software versions to later versions. Intermediate steps must sometimes be taken; v12 converts to v13, then v13 SP1, and from there to v14 and v15.

One of the biggest differences in TIA Portal is that there is a single programming environment. Whereas Step 7 had the Manager, Hardware Configuration, Editor, Symbol Table and VAT Tables all in separate programs, all of these can be accessed from within the single program.

It is much easier to determine whether files have been downloaded or whether the files in the processor match the files saved on the computer. Colored icons appear next to blocks and folders while online.

The simulator, while still available, is more complicated than the Step7 simulator and therefore less user-friendly. It does, however, allow sequences of operations to be stepped through automatically, a powerful feature.

Creating a Project

The software can be opened by selecting it from the programs list, or it can be placed on the desktop. The icon shows the version of the software, as shown at left; multiple versions can be installed on the same computer.

When TIA Portal is opened, it defaults to the Portal View. The project can be named and saved from this screen, and a tutorial is available for new users.

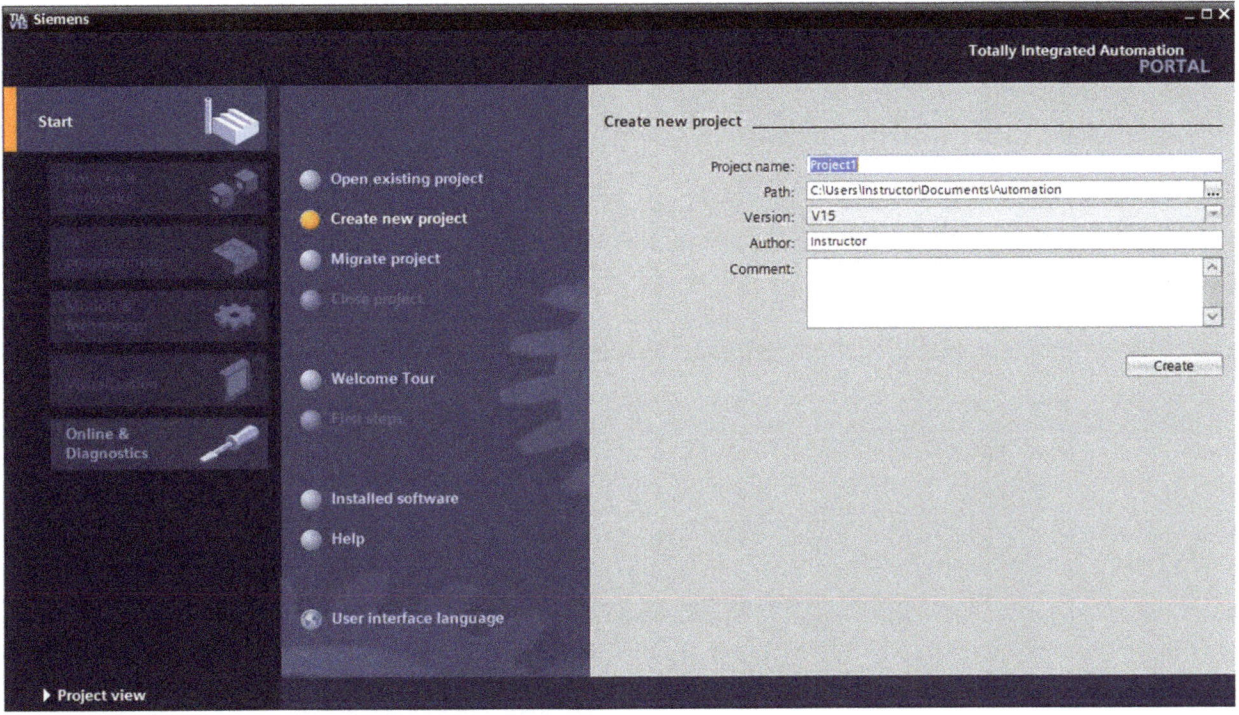

This is also the screen where a Step 7 project or WinCC HMI file can be converted to a TIA Portal file (Migrate project).

The Welcome Tour, a tutorial for new users, requires a connection to the internet to run.

The default path for saving files is in the Users\Username\Documents\Automation folder; if the programmer wishes to save the project to a different location, the software will remember the path. After pushing the "Create" button, the software will place the required files and folders in the selected location; this takes more time than Step 7.

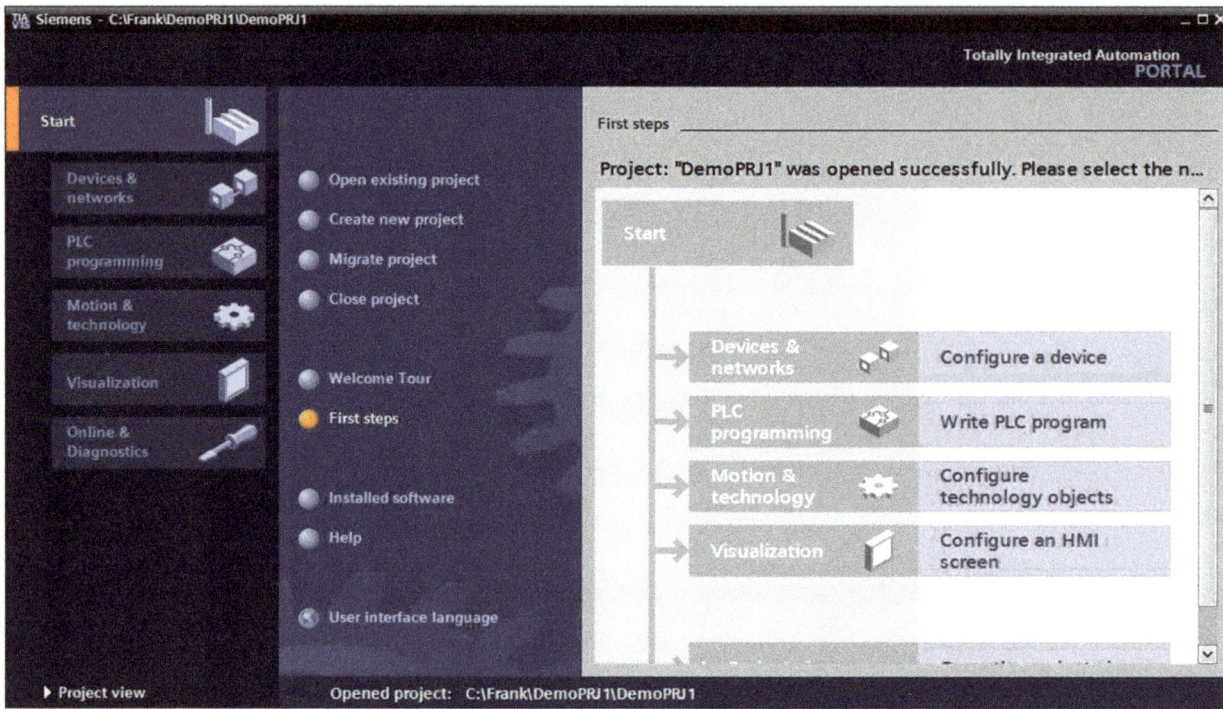

After creating the project, the programmer can continue configuration from the portal view or click on the label at the lower left of the screen, which changes the environment to the more familiar Project View.

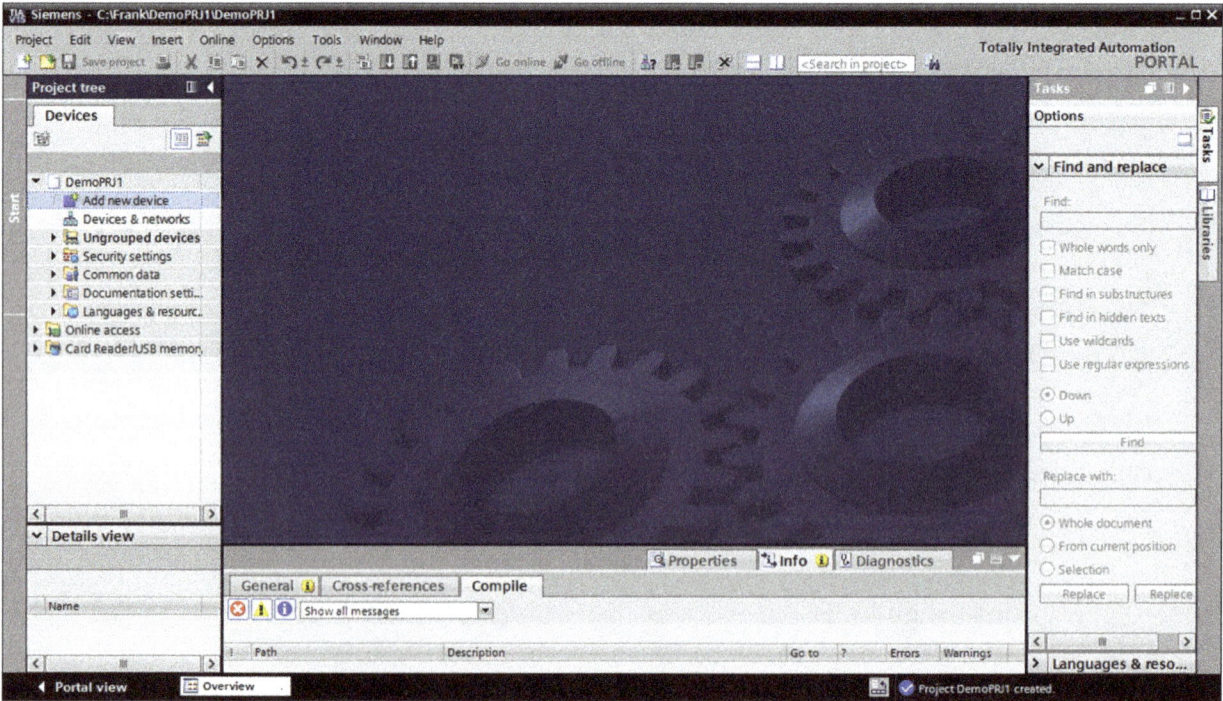

The Project View has a Project tree to the left of the screen, which contains folders for all of the devices, blocks and tables for the project. Until hardware is configured, many of the folders are not yet present.

Configuring Hardware

As with all PLCs, the first thing that needs to be done when creating a new project is to configure the hardware. Selecting "Add New Device" will open the hardware configuration. Double-clicking the Device Configuration icon under the PLC's name after the hardware has been configured will allow the hardware to be modified.

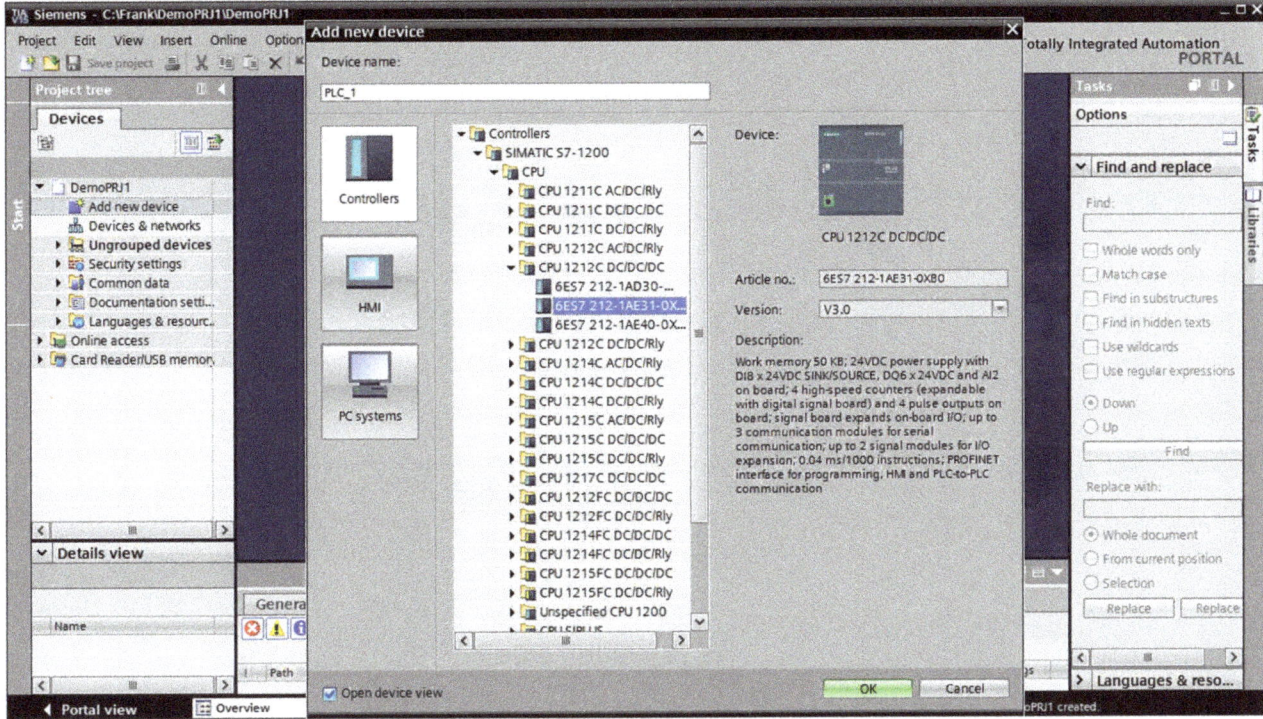

Selecting a processor for the project allows the processor to be named and the firmware selected. Detailed information about the different processors is listed below the device picture in the description.

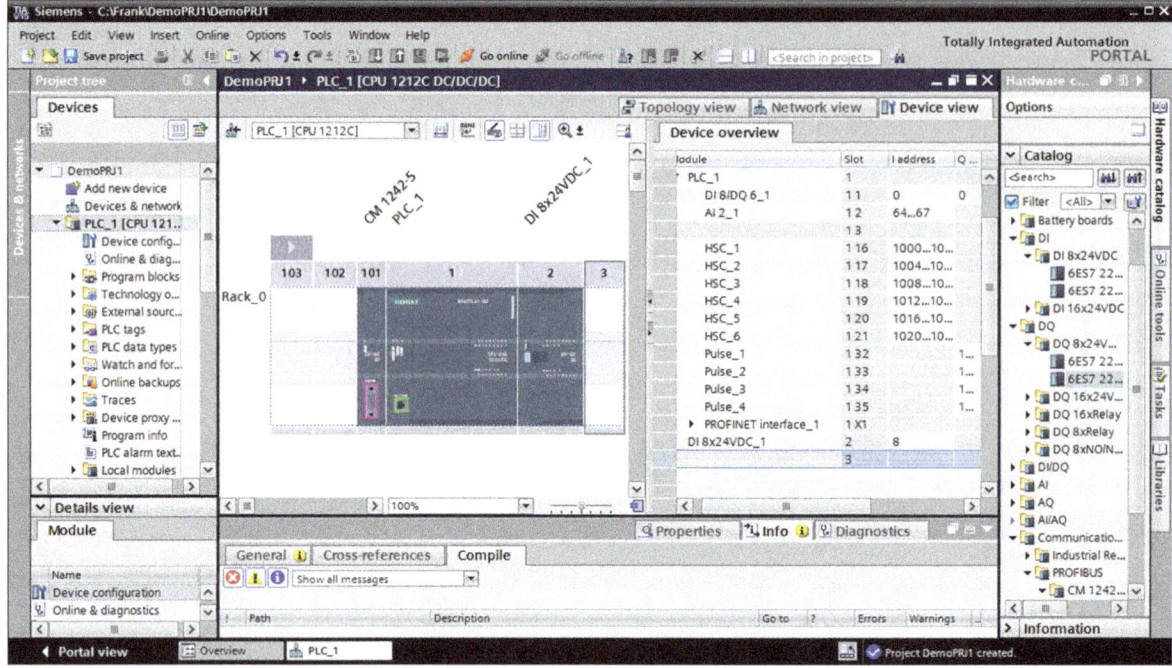

The Hardware Configuration screen has three tabs: Device View, Network View, and Topology View. The Device View allows cards to be selected for the different processors from the folders on the right. Selecting different parts of the hardware, such as the card and ports, brings up detailed information about the component in the General tab at the bottom of the window.

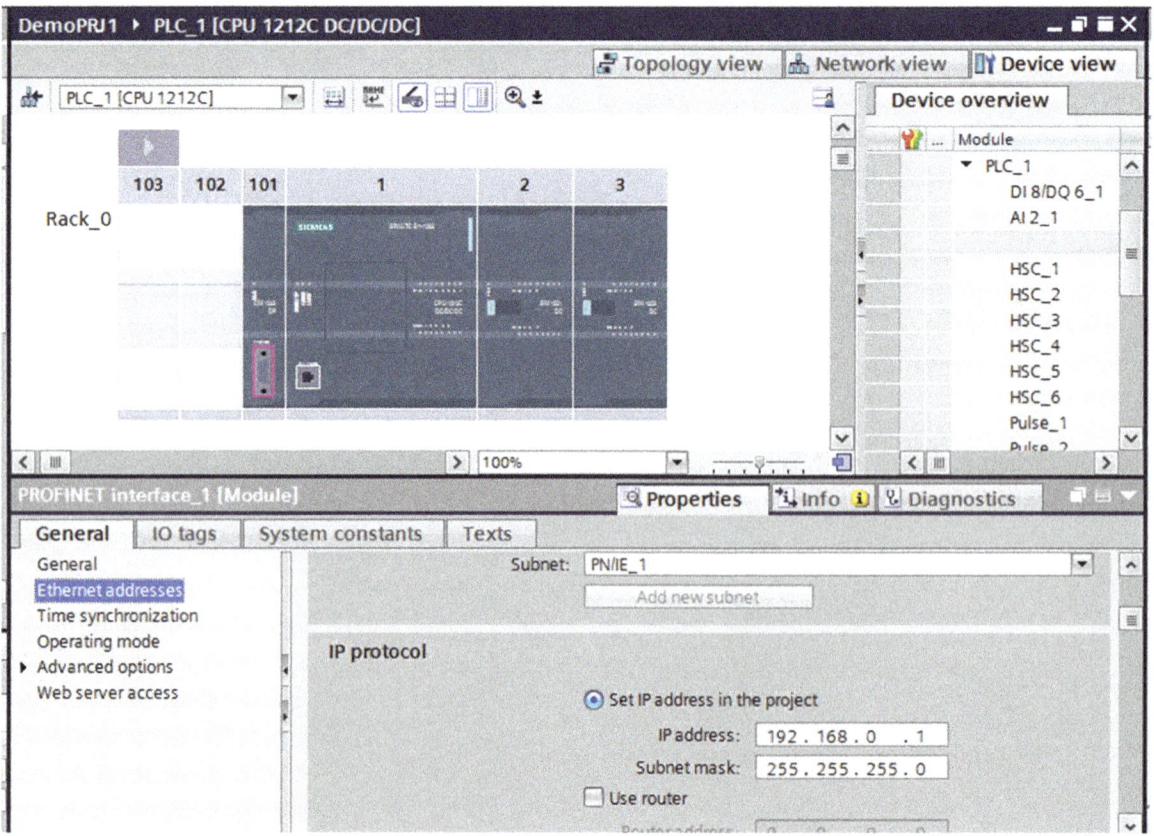

The Ethernet address and network configuration can be found under the General tab.

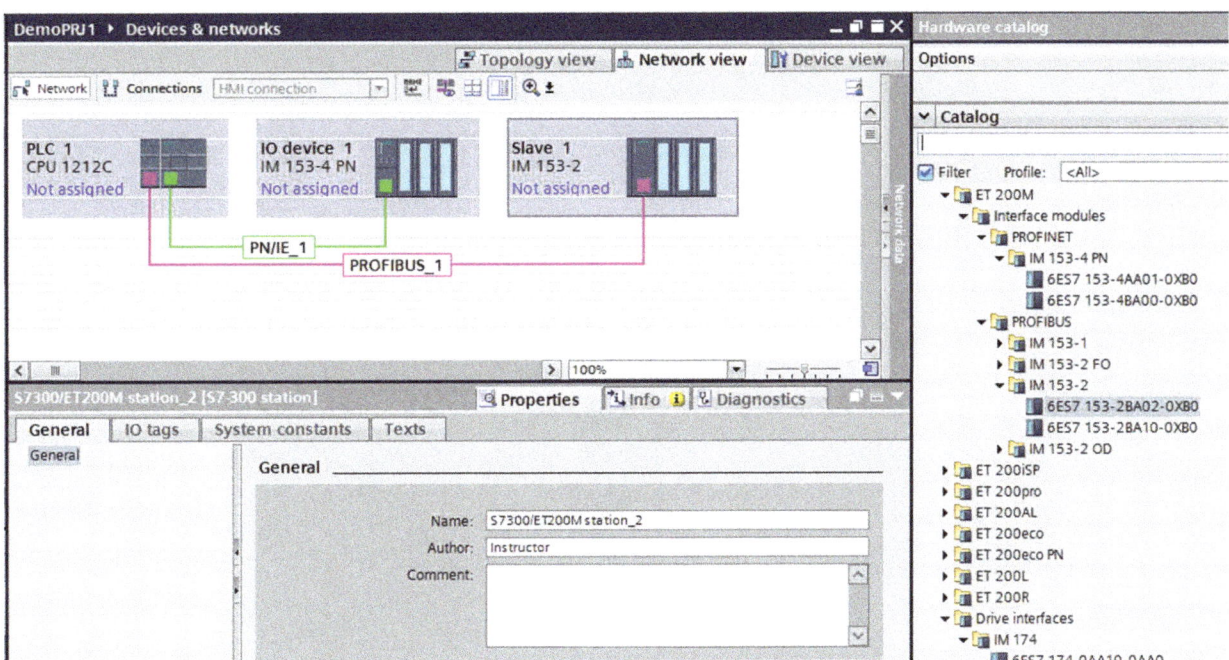

Selecting "Add New Subnet" on the port configuration allows components to be added on the Network View screen shown above. Selecting the device and returning to the device view allows remote devices to be further configured.

The Network View shows all of the logical subnets of the project, while the Topology view shows passive components, such as switches, media converters and cables.

After configuring the hardware, it can be downloaded to the processor after going online.

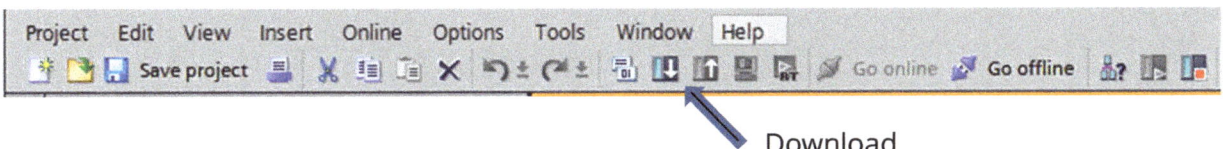

Download

Writing the Program

Blocks and Structures in TIA Portal are as described in the Step 7 programming section of this book, with a few important differences. Organization blocks have the same general function as in Step 7, but they are selected by function from menus rather than by number. The Main block is OB1 as before, and it runs continuously. Other continuous blocks can be added; they will automatically be numbered above OB123.

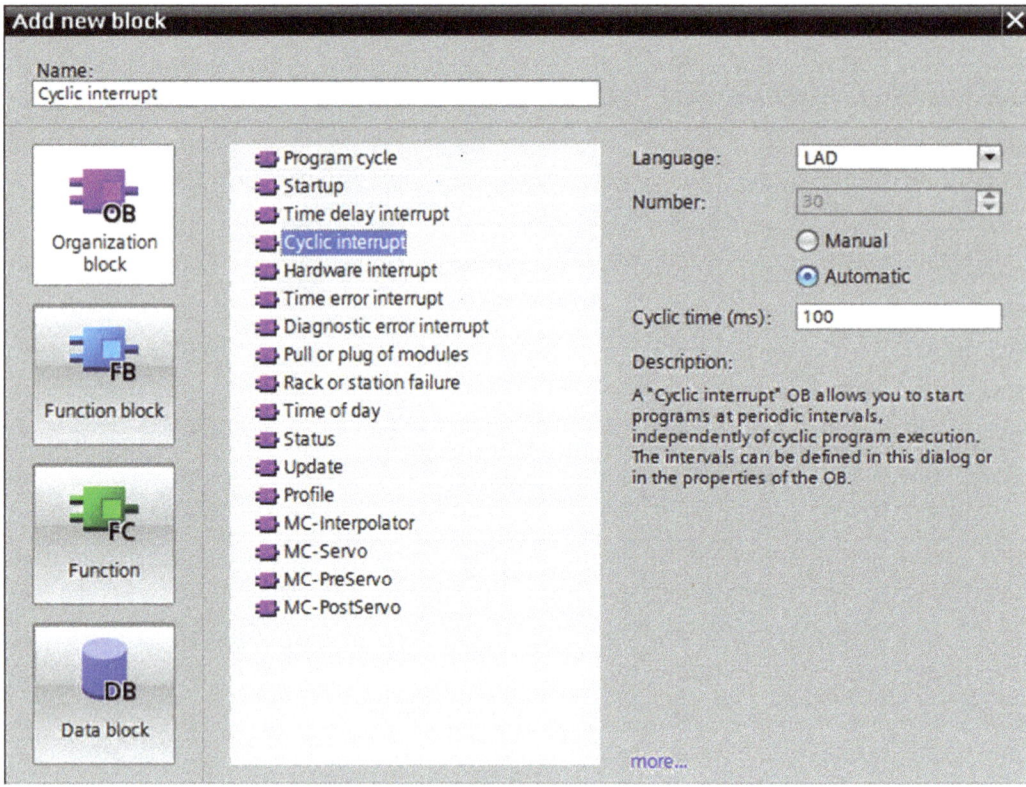

Cyclic OBs are in the 30 range as with Step 7; however, they do not have to be configured in hardware. The time period is set from the block configuration screen as shown above.

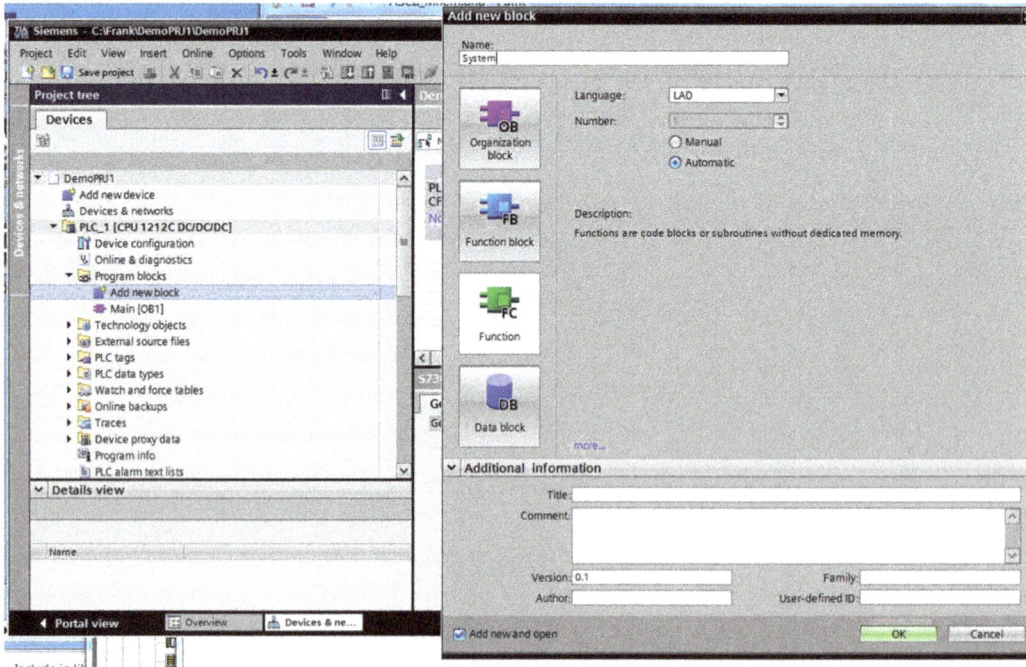

Functions and Function Blocks (FCs and FBs) are added in much the same way. The block can be numbered automatically or the programmer can choose the number.

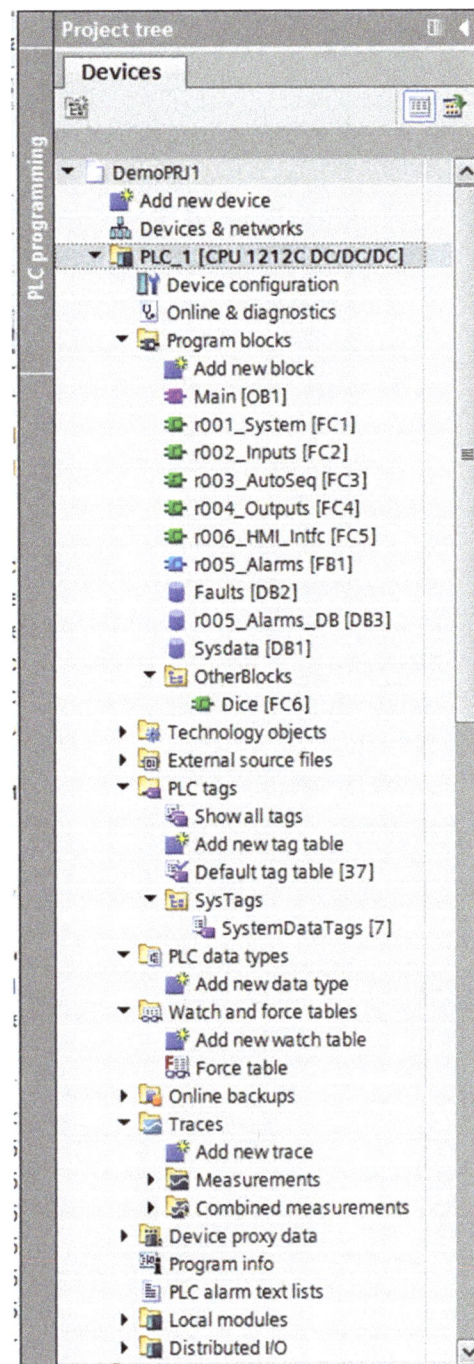

After adding blocks and tags to the project, the Project Tree shows all of the objects in the project. Groups can be added to the folders as shown; the groups "OtherBlocks" and "SysTags" have been added to the folders as shown. These groups are only for organizational purposes, i.e. for the programmer. The group in the PLC tags folder do not make the tags local, and the block group does not separate the blocks programatically from the other blocks. This is simply a technique to make blocks and tags easier to find.

Note also that blocks are listed by type and then in alphabetical order in the tag folders. For this reason, the programmer may want to place a character or number in front of the tag name to list blocks in the order in which they are called.

Other folders in the Project tree contain objects similar to those found in the Simatic Manager of Step 7:

Watch and Force Tables - these are similar to the VAT tables and force tables of Step 7. They exist only in the programming device and are not downloaded to the processor.

PLC Data Types – These are the same as the data type in the Simatic Manager and are user defined (UDTs). They are used to define tags and are not in the processor.

External Source Files – Similar to the sources folder in Step 7, these files can be imported from other projects and may be SCL, DB and .udt files.

Other folders introduce new objects that were not part of the Step 7 software:

Technology Objects – these include motion control, PID and barcode reader interfaces.

Traces – A trace is a graphical tool that trends tags. It is downloaded to the processor. This allows it to run even when the programming device (computer) is not connected; it is retentive and can be saved and uploaded as a measurement.

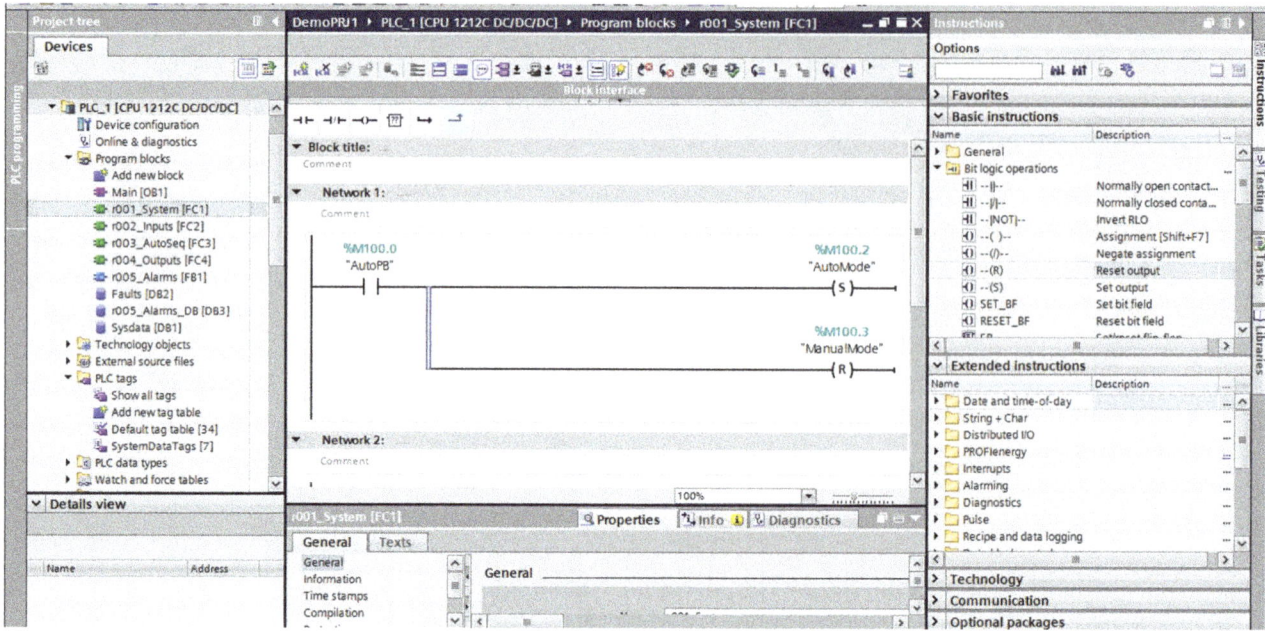

After creating blocks, double-clicking the icon in the blocks folder will open it for editing. A simple menu of contacts, a coil and a configurable "box" instruction appears at the top of the editor, and the full set of instructions is located to the right of the edit window.

As instructions are entered, addresses can be typed in as direct addresses for marker memory, as shown, or a tag name can be entered. If the name does not exist yet, right-clicking on it and selecting "Define Tag" will place it in the tag table.

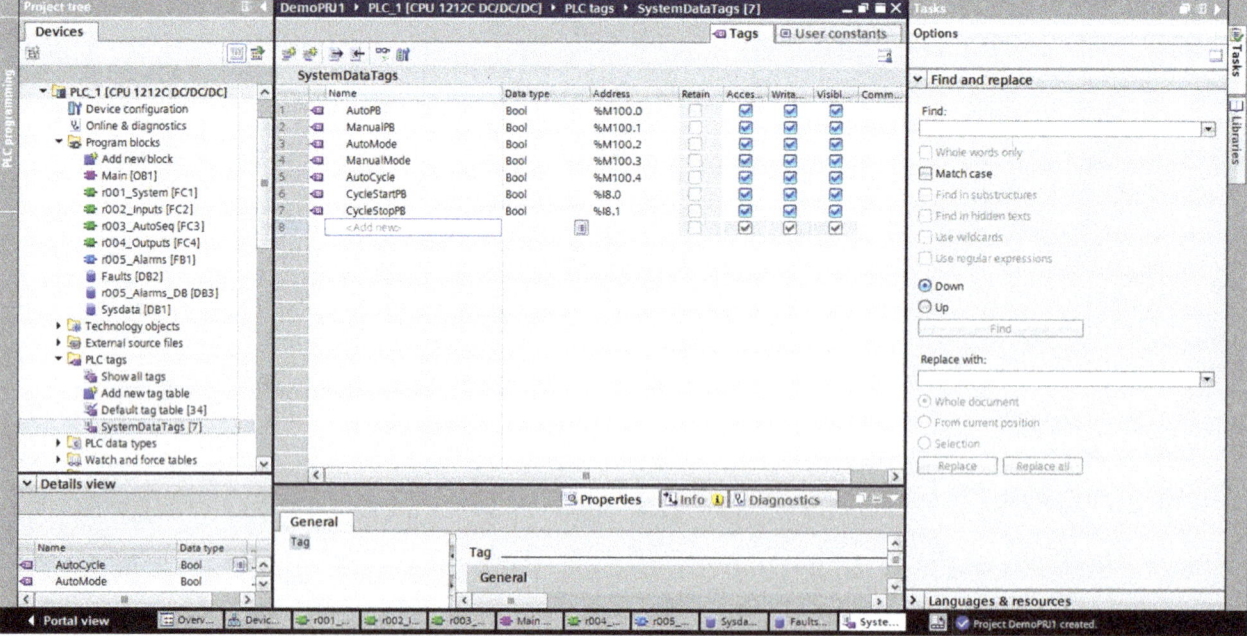

As in Step 7 software, tags can be typed into the tag table directly before selecting them or added as they are created from the editor. All addresses must have tags assigned, however, unlike symbols.

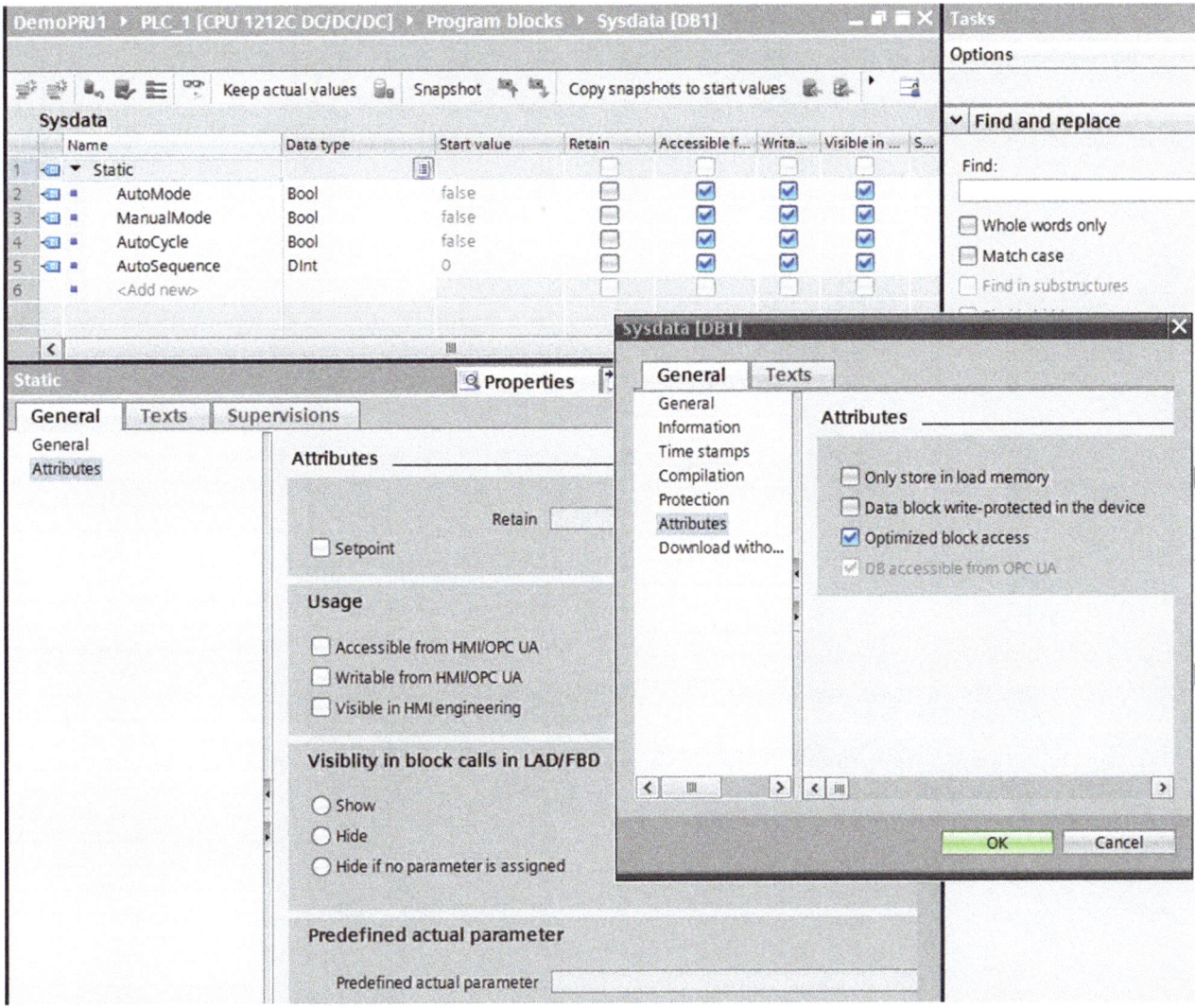

Data Blocks are treated differently in TIA than in Step 7. Elements can be individually assigned accessibility from external devices or be assigned as a setpoint. DBs are also set by default as "Optimized". This makes them faster to access and smaller in size, but they cannot be directly accessed by address if created as an optimized block. This means that if pointers are to be used in STL for indirect addressing, DBs must be created without optimized block access. The window with the attributes selection for optimization is accessed by right-clicking the block in the folder and selecting "Properties".

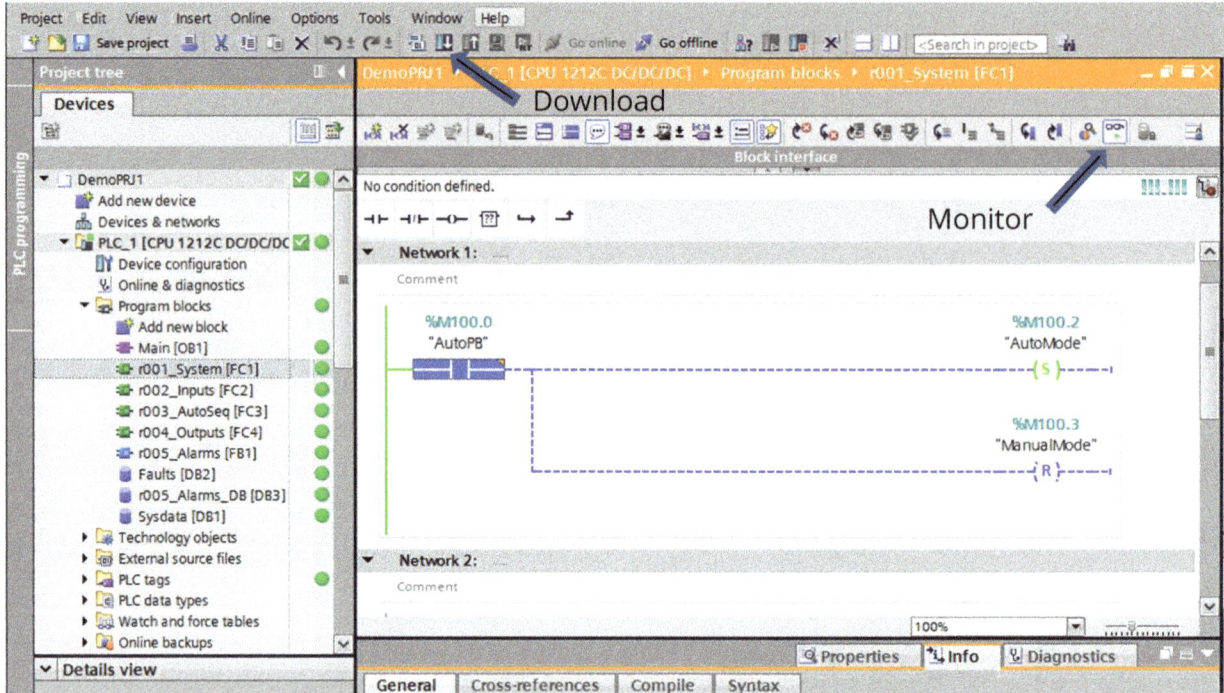

After downloading the blocks, selecting the icon with the glasses allows the block to be monitored. The green dots in the Project tree show that all of the blocks in the folders match the blocks in the processor.

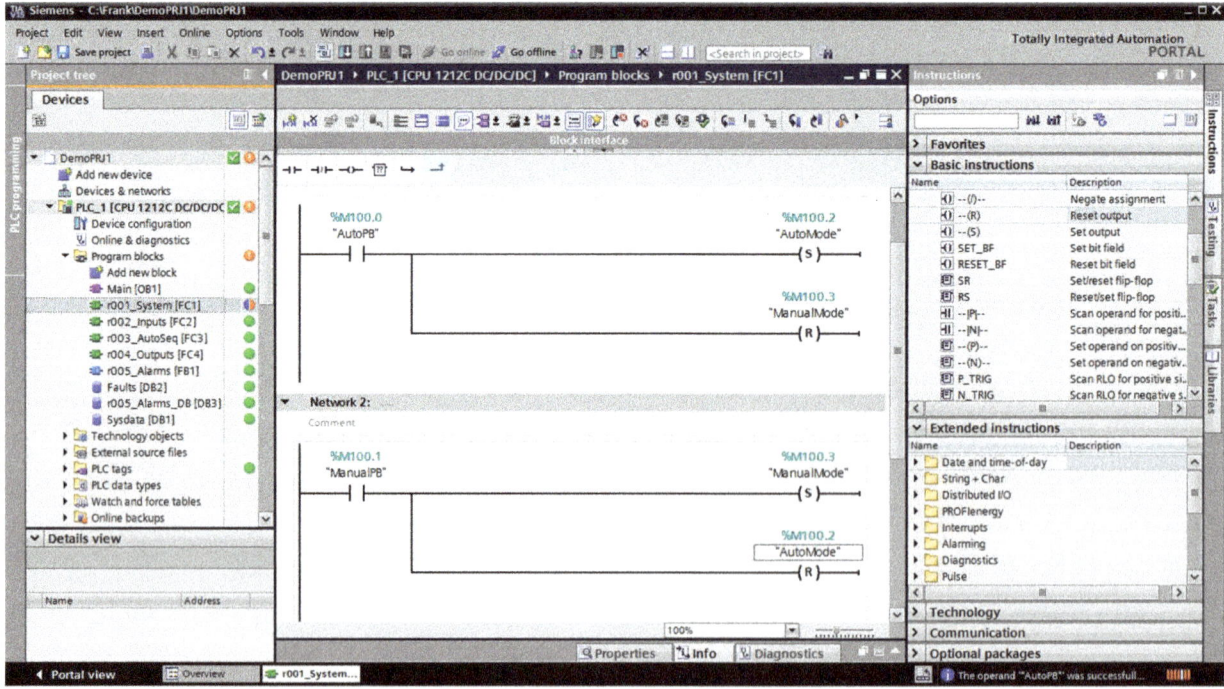

If the block is edited, the orange color in the header will disappear and a blue and orange dot appears in the project tree next to the function, indicating that the block no longer matches the one in the PLC.

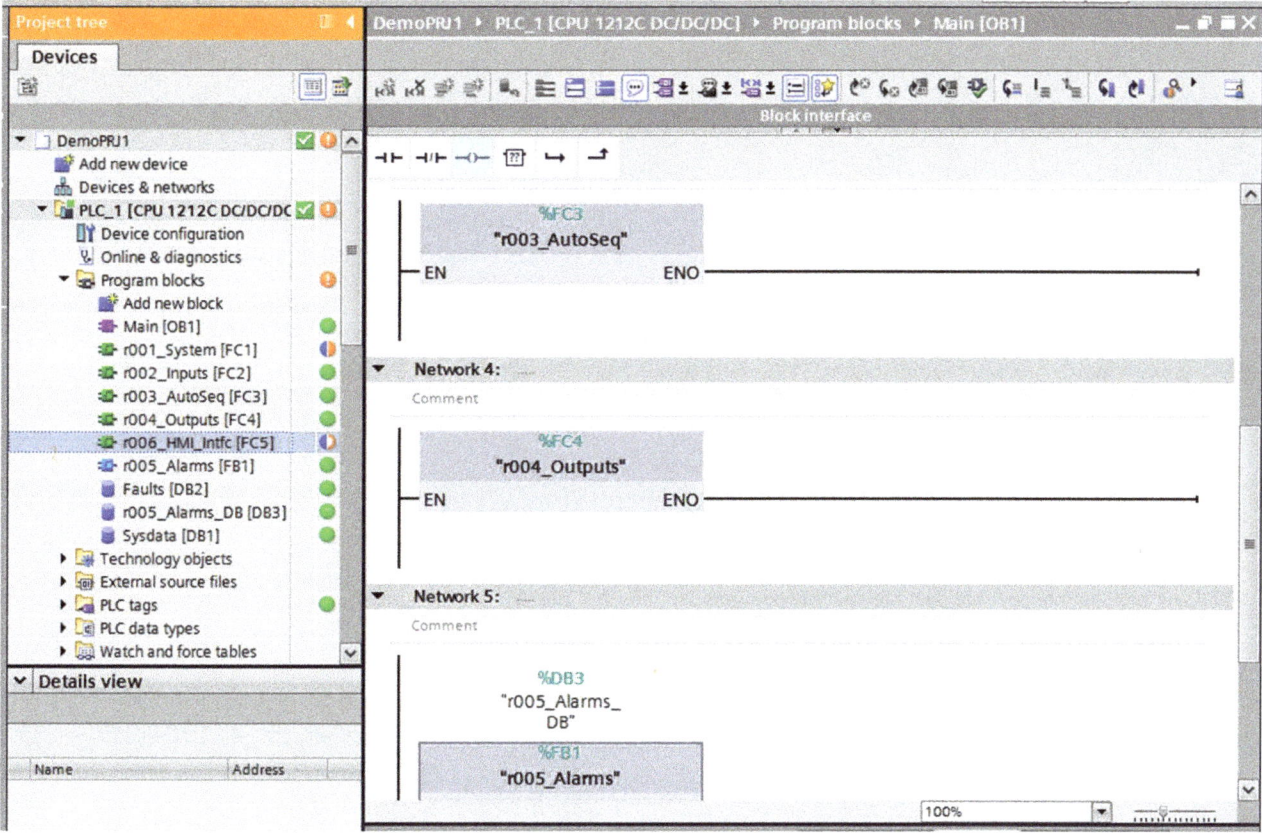

The icon for a new block that has not been downloaded at all is also blue and orange but has a white center in the orange field. Downloading in TIA Portal automatically sends all of the blocks to the processor, unlike Simatic Manager in Step 7.

Overall TIA Portal is an improvement over Step 7, but as with anything newer and with more features, it can be more complex.

Communications- Siemens Set PG-PC Interface

The setup for Siemens communications drivers can be found under Options in the Simatic Manager or in the block editor.

1. Select "Set PG/PC Interface" to open the dialog.

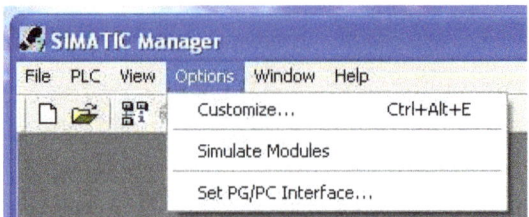

2. There will be a number of drivers available for selection. For Ethernet, select the TCP/IP driver.

If Auto is selected, you will not be able to communicate through a bridge or router. If you are using a router, select the TCP/IP driver that ends with .1.

3. Clicking "Properties" allows modifications to the properties of the driver and to the network card (Network properties)

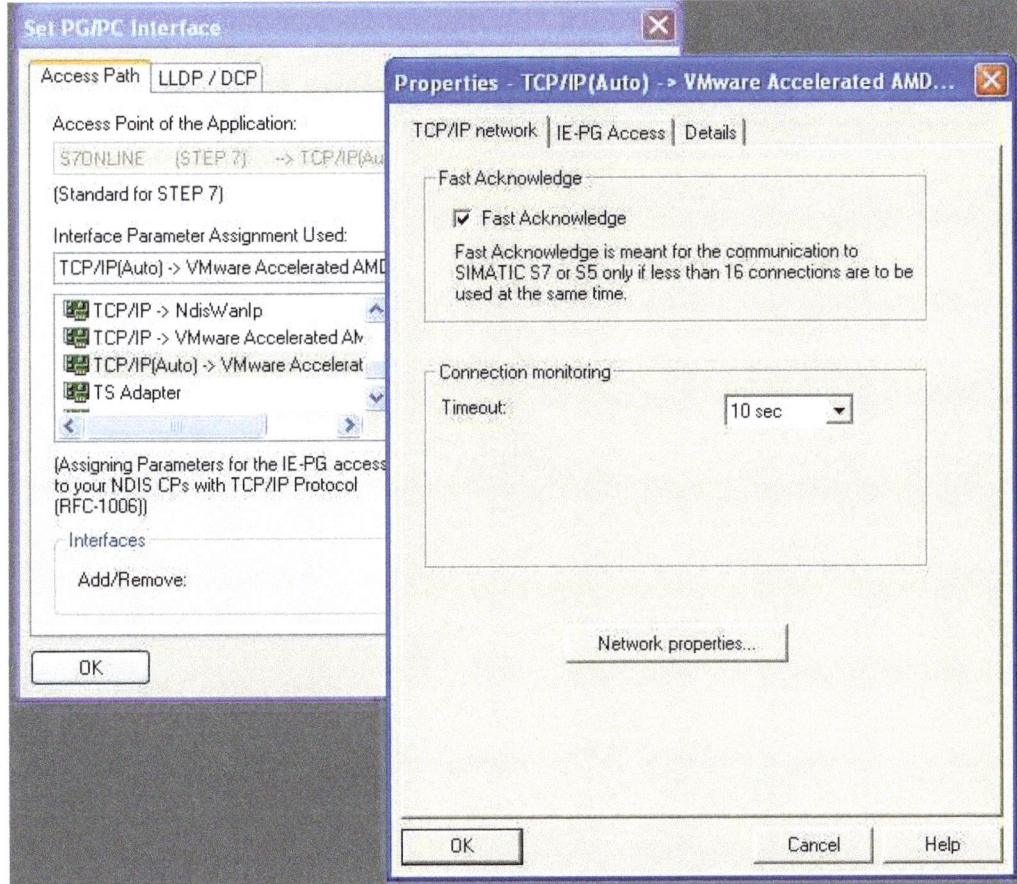

To see the processors available to connect to, select "Display Accessible Nodes" under the PLC tab. Siemens finds devices by MAC address rather than IP, so even devices outside of the Ethernet mask (domain) can be seen.

Other drivers that can be selected from the Set PG/PC Interface dialog include MPI (Serial RS232, 9 pin port) and Profibus (RS485). Both require a special adapter.

Communications- Siemens TIA Portal

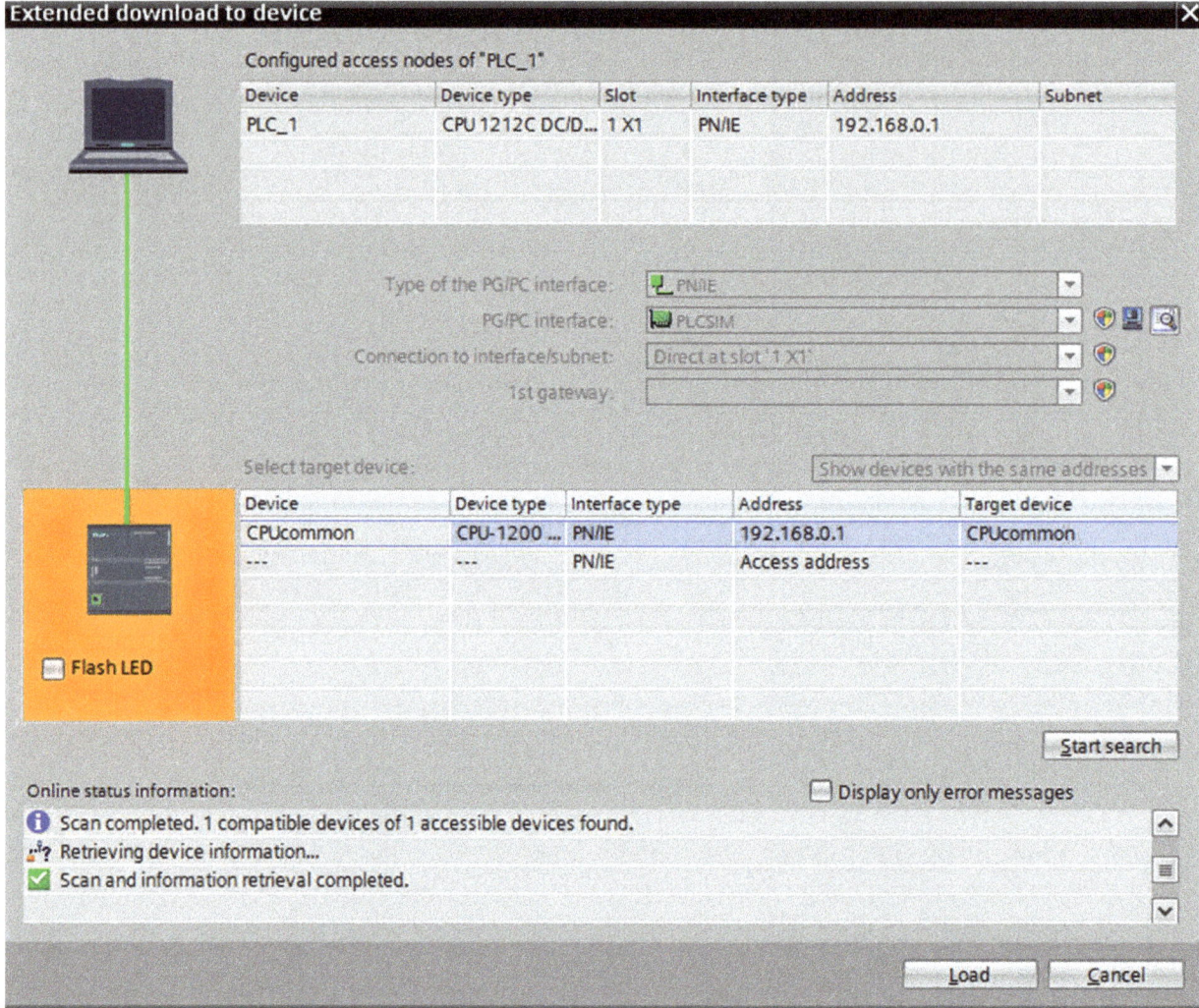

Selecting the download icon in TIA Portal opens a dialog where different types of drivers can be selected from a list titled "Type of the PG/PC Interface". The PLC name appears in the configured access nodes list, and the target device can be found by clicking "Start search". If multiple processors are present on the network, the LEDs can be flashed to help identify the target.

After the correct driver has been set, unless hardware is downloaded, the processor will not be stopped.

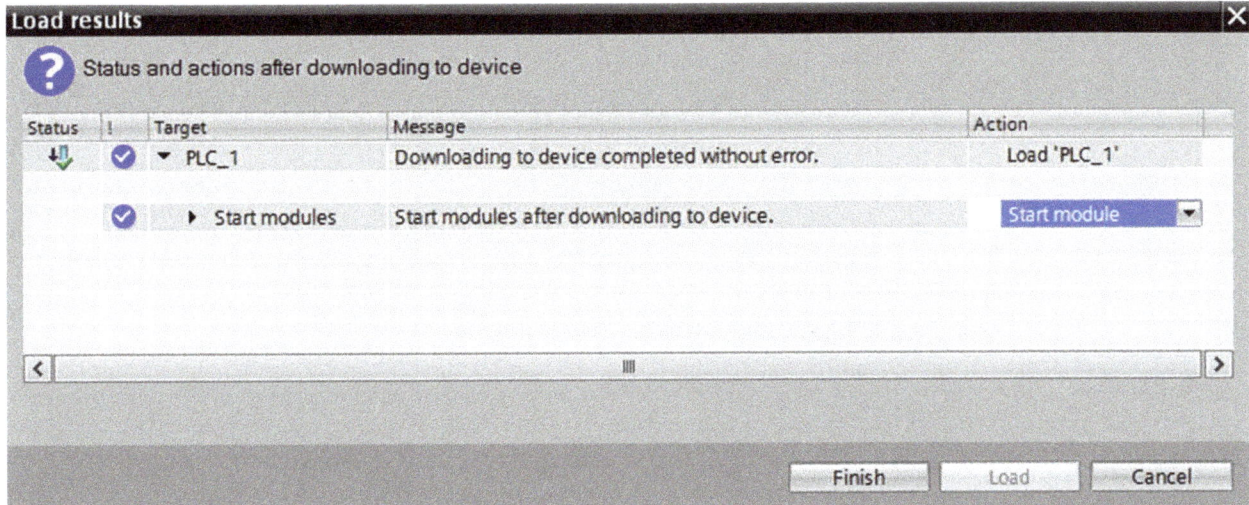

If hardware was downloaded, another screen will appear prompting the user to restart the modules. As with Step 7, downloading software does not require stopping the processor.

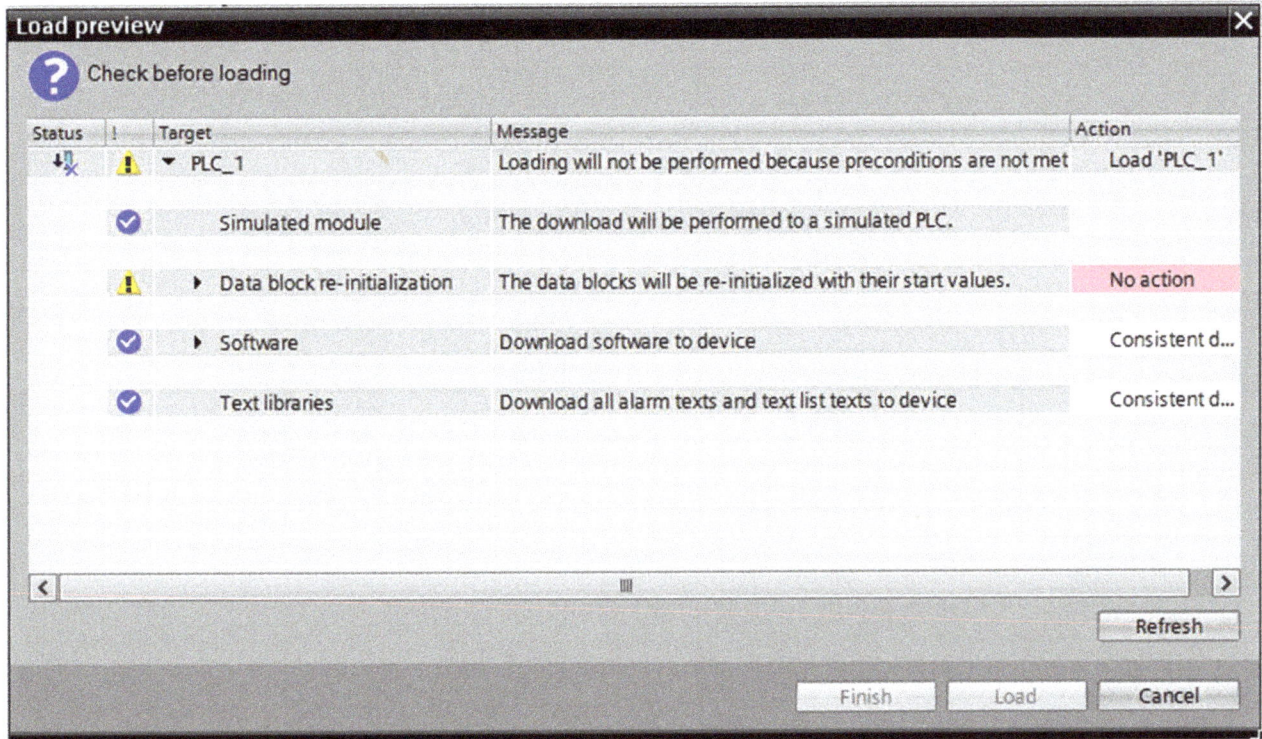

If data blocks are changed, they must be re-initialized, which potentially changes the contents of the data blocks. Since this can disrupt the proper operation of the program, the user is prompted to re-initialize the block (indicated by the red field that reads "no action").

Major PLC Platforms

Following is a list of some of the major PLC manufacturers, website and their country of origin.

ABB – Switzerland	http://new.abb.com/plc
Allen-Bradley – USA	http://ab.rockwellautomation.com/Programmable-Controllers
Automation Direct-USA	https://www.automationdirect.com/
B&R – Germany	http://www.br-automation.com/en-us/products/control-systems/
Beckhoff – Germany	http://www.beckhoff.com/
Bosch – Germany	http://www.boschrexroth.com/
GE Fanuc – USA	http://www.geautomation.com/products/programmable-automation-controllers
Hitachi – Japan	http://www.hitachi-ies.co.jp/english/products/plc/index.htm
Idec – Japan	http://us.idec.com/Home.aspx
Keyence – Japan	http://www.keyence.com/products/controls/plc-package/index.jsp
Koyo – Japan	http://www.koyoele.co.jp/english/product/plc/
Mitsubishi – Japan	http://www.mitsubishielectric.com/fa/products/cnt/plc/
Omron – Japan	https://industrial.omron.us/en/products/programmable-logic-controllers
Panasonic – Japan	https://na.industrial.panasonic.com/products/industrial-automation/factory-automation-devices/programmable-controllers
Siemens – Germany	http://w3.siemens.com/mcms/programmable-logic-controller/en/Pages/Default.aspx
Toshiba – Japan	http://www.toshiba.com/tic/industrial-systems/plcs

Others (smaller brands or may be obsolete): Cutler-Hammer, Eagle-Signal (Eptak), Giddings & Lewis, Philips, Square D, Texas Instruments, Triconex, Velocio, Vipa, Westinghouse

ASCII Tables

Table 1 - Basic ASCII

Dec	Hx	Oct	Char		Dec	Hx	Oct	Html	Chr	Dec	Hx	Oct	Html	Chr	Dec	Hx	Oct	Html	Chr
0	0	000	NUL	(null)	32	20	040	&#32;	Space	64	40	100	&#64;	@	96	60	140	&#96;	`
1	1	001	SOH	(start of heading)	33	21	041	&#33;	!	65	41	101	&#65;	A	97	61	141	&#97;	a
2	2	002	STX	(start of text)	34	22	042	"	"	66	42	102	&#66;	B	98	62	142	&#98;	b
3	3	003	ETX	(end of text)	35	23	043	&#35;	#	67	43	103	&#67;	C	99	63	143	&#99;	c
4	4	004	EOT	(end of transmission)	36	24	044	&#36;	$	68	44	104	&#68;	D	100	64	144	&#100;	d
5	5	005	ENQ	(enquiry)	37	25	045	&#37;	%	69	45	105	&#69;	E	101	65	145	&#101;	e
6	6	006	ACK	(acknowledge)	38	26	046	&	&	70	46	106	&#70;	F	102	66	146	&#102;	f
7	7	007	BEL	(bell)	39	27	047	'	'	71	47	107	&#71;	G	103	67	147	&#103;	g
8	8	010	BS	(backspace)	40	28	050	&#40;	(	72	48	110	&#72;	H	104	68	150	&#104;	h
9	9	011	TAB	(horizontal tab)	41	29	051	&#41;	)	73	49	111	&#73;	I	105	69	151	&#105;	i
10	A	012	LF	(NL line feed, new line)	42	2A	052	&#42;	*	74	4A	112	&#74;	J	106	6A	152	&#106;	j
11	B	013	VT	(vertical tab)	43	2B	053	&#43;	+	75	4B	113	&#75;	K	107	6B	153	&#107;	k
12	C	014	FF	(NP form feed, new page)	44	2C	054	&#44;	,	76	4C	114	&#76;	L	108	6C	154	&#108;	l
13	D	015	CR	(carriage return)	45	2D	055	&#45;	-	77	4D	115	&#77;	M	109	6D	155	&#109;	m
14	E	016	SO	(shift out)	46	2E	056	&#46;	.	78	4E	116	&#78;	N	110	6E	156	&#110;	n
15	F	017	SI	(shift in)	47	2F	057	&#47;	/	79	4F	117	&#79;	O	111	6F	157	&#111;	o
16	10	020	DLE	(data link escape)	48	30	060	&#48;	0	80	50	120	&#80;	P	112	70	160	&#112;	p
17	11	021	DC1	(device control 1)	49	31	061	&#49;	1	81	51	121	&#81;	Q	113	71	161	&#113;	q
18	12	022	DC2	(device control 2)	50	32	062	&#50;	2	82	52	122	&#82;	R	114	72	162	&#114;	r
19	13	023	DC3	(device control 3)	51	33	063	&#51;	3	83	53	123	&#83;	S	115	73	163	&#115;	s
20	14	024	DC4	(device control 4)	52	34	064	&#52;	4	84	54	124	&#84;	T	116	74	164	&#116;	t
21	15	025	NAK	(negative acknowledge)	53	35	065	&#53;	5	85	55	125	&#85;	U	117	75	165	&#117;	u
22	16	026	SYN	(synchronous idle)	54	36	066	&#54;	6	86	56	126	&#86;	V	118	76	166	&#118;	v
23	17	027	ETB	(end of trans. block)	55	37	067	&#55;	7	87	57	127	&#87;	W	119	77	167	&#119;	w
24	18	030	CAN	(cancel)	56	38	070	&#56;	8	88	58	130	&#88;	X	120	78	170	&#120;	x
25	19	031	EM	(end of medium)	57	39	071	&#57;	9	89	59	131	&#89;	Y	121	79	171	&#121;	y
26	1A	032	SUB	(substitute)	58	3A	072	&#58;	:	90	5A	132	&#90;	Z	122	7A	172	&#122;	z
27	1B	033	ESC	(escape)	59	3B	073	&#59;	;	91	5B	133	&#91;	[	123	7B	173	&#123;	{
28	1C	034	FS	(file separator)	60	3C	074	&#60;	<	92	5C	134	&#92;	\	124	7C	174	&#124;		
29	1D	035	GS	(group separator)	61	3D	075	&#61;	=	93	5D	135	&#93;	]	125	7D	175	&#125;	}
30	1E	036	RS	(record separator)	62	3E	076	&#62;	>	94	5E	136	&#94;	^	126	7E	176	&#126;	~
31	1F	037	US	(unit separator)	63	3F	077	&#63;	?	95	5F	137	&#95;	_	127	7F	177	&#127;	DEL

Source: www.LookupTables.com

Table 2 - Extended ASCII

Dec	Chr	Dec	Chr	Dec	Chr	Dec	Chr	Dec	Chr	Dec	Chr	Dec	Chr	Dec	Chr
128	Ç	144	É	160	á	176	░	192	└	208	╨	224	α	240	≡
129	ü	145	æ	161	í	177	▒	193	┴	209	╤	225	ß	241	±
130	é	146	Æ	162	ó	178	▓	194	┬	210	╥	226	Γ	242	≥
131	â	147	ô	163	ú	179	│	195	├	211	╙	227	π	243	≤
132	ä	148	ö	164	ñ	180	┤	196	─	212	╘	228	Σ	244	⌠
133	à	149	ò	165	Ñ	181	╡	197	┼	213	╒	229	σ	245	⌡
134	å	150	û	166	ª	182	╢	198	╞	214	╓	230	µ	246	÷
135	ç	151	ù	167	º	183	╖	199	╟	215	╫	231	τ	247	≈
136	ê	152	ÿ	168	¿	184	╕	200	╚	216	╪	232	Φ	248	°
137	ë	153	Ö	169	⌐	185	╣	201	╔	217	┘	233	Θ	249	·
138	è	154	Ü	170	¬	186	║	202	╩	218	┌	234	Ω	250	·
139	ï	155	¢	171	½	187	╗	203	╦	219	█	235	δ	251	√
140	î	156	£	172	¼	188	╝	204	╠	220	▄	236	∞	252	ⁿ
141	ì	157	¥	173	¡	189	╜	205	═	221	▌	237	φ	253	²
142	Ä	158	₧	174	«	190	╛	206	╬	222	▐	238	ε	254	■
143	Å	159	ƒ	175	»	191	┐	207	╧	223	▀	239	∩	255	

Source: www.LookupTables.com

PLC Hardware and Programming Exercise Solutions

Exercise 1 p. 48

Q5 –   The Control level.

Exercise 2 p. 57

Q1 –   Decimal: 27,831   Hexadecimal: 6CB7   Octal: 66,267
This number cannot be converted to BCD, two of the binary groups are higher than 1001.

Q2 –   Signed Integer: -27,255

Q3 –   BCD: 0100_0001_0111 = 417   Binary: 1 1010 0001   Hexadecimal: 1A1

Q4 –   Binary: 10 1010 1001 1110   Decimal: 10,910

Q5 –   There are 4 Bytes in a Double Integer

Q6 –   Some brands of PLC call a Tag a Symbol, but symbols usually are a shortcut to a numerical data address register.

Exercise 3 p. 64

Q1 –   Decimal, Octal and Hexadecimal.

Q2 –   The Main Routine

Q3 –   The hardware

Q4 –   Modules, communication ports, I/O networks, addresses

Exercise 4 p. 74

Q1 –   Instruction List (IL), Ladder (LAD), Function Block Diagram (FBD), Structured Text (ST), Sequential Function Charts (SFC)

Q2 –   1. Read physical inputs to Input Image Table. 2. Solve Logic 3. Write Output Image Table to physical outputs. 4. "Housekeeping" tasks.

Q3 –   Yes, online editing.

Exercise 5 p. 80

Q1 -

Rung 1:
- Auto Pushbutton `Auto_Pushbutton <Digital_InputCard1_Pt.8>` ─┤ ├─
  - parallel: Automatic Mode `Auto_Mode <MemoryBit[7].1>` ─┤ ├─
- Machine Fault `Fault <MemoryBit[7].11>` ─┤/├─
- Manual Pushbutton `Manual_Pushbutton <Digital_InputCard1_Pt.9>` ─┤/├─
- Automatic Mode `Auto_Mode <MemoryBit[7].1>` ─( )─

Rung 2:
- Manual Pushbutton `Manual_Pushbutton <Digital_InputCard1_Pt.9>` ─┤ ├─
  - parallel: Manual or Maintenance Mode `Manual_Mode <MemoryBit[7].10>` ─┤ ├─
  - parallel: Machine Fault `Fault <MemoryBit[7].11>` ─┤ ├─
- Auto Pushbutton `Auto_Pushbutton <Digital_InputCard1_Pt.8>` ─┤/├─
- Manual or Maintenance Mode `Manual_Mode <MemoryBit[7].10>` ─( )─

Q2 -

Rung 1:
- Start Motor 1 Pushbutton `Start_Pushbutton <Digital_InputCard1_Pt.10>` ─┤ ├─
  - parallel: Motor 1 `Motor1 <Digital_OutputCard1_Pt.6>` ─┤ ├─
- Stop Motor 1 Pushbutton `Stop_Pushbutton <Digital_InputCard1_Pt.11>` ─┤/├─
- Motor 1 Module Faulted `Motor1_Fault <MemoryBit[7].12>` ─┤/├─
- Motor 1 `Motor1 <Digital_OutputCard1_Pt.6>` ─( )─

Rung 2:
- Motor 1 Overload Tripped `Motor1_Ovld <Digital_InputCard1_Pt.12>` ─┤ ├─
  - parallel: Motor 1 Guard Door Opened `Motor1_GdDoor <Digital_InputCard1_Pt.13>` ─┤ ├─
- Motor 1 Module Faulted `Motor1_Fault <MemoryBit[7].12>` ─(L)─

Rung 3:
- Motor 1 Fault Reset Pushbutton `Motor1_FltRstPB <Digital_InputCard1_Pt.14>` ─┤ ├─
- Motor 1 Overload Tripped `Motor1_Ovld <Digital_InputCard1_Pt.12>` ─┤/├─
- Motor 1 Guard Door Opened `Motor1_GdDoor <Digital_InputCard1_Pt.13>` ─┤/├─
- Motor 1 Module Faulted `Motor1_Fault <MemoryBit[7].12>` ─(U)─

Rung 4:
- Motor 1 Module Faulted `Motor1_Fault <MemoryBit[7].12>` ─┤ ├─
- Red Motor Fault Light `RedLight <Digital_OutputCard1_Pt.7>` ─( )─

Rung 5:
- Motor 1 `Motor1 <Digital_OutputCard1_Pt.6>` ─┤ ├─
- Green Motor Running Light `GreenLight <Digital_OutputCard1_Pt.8>` ─( )─

Exercise 6 p. 85

Q1 –

Q2 –

Q3 –

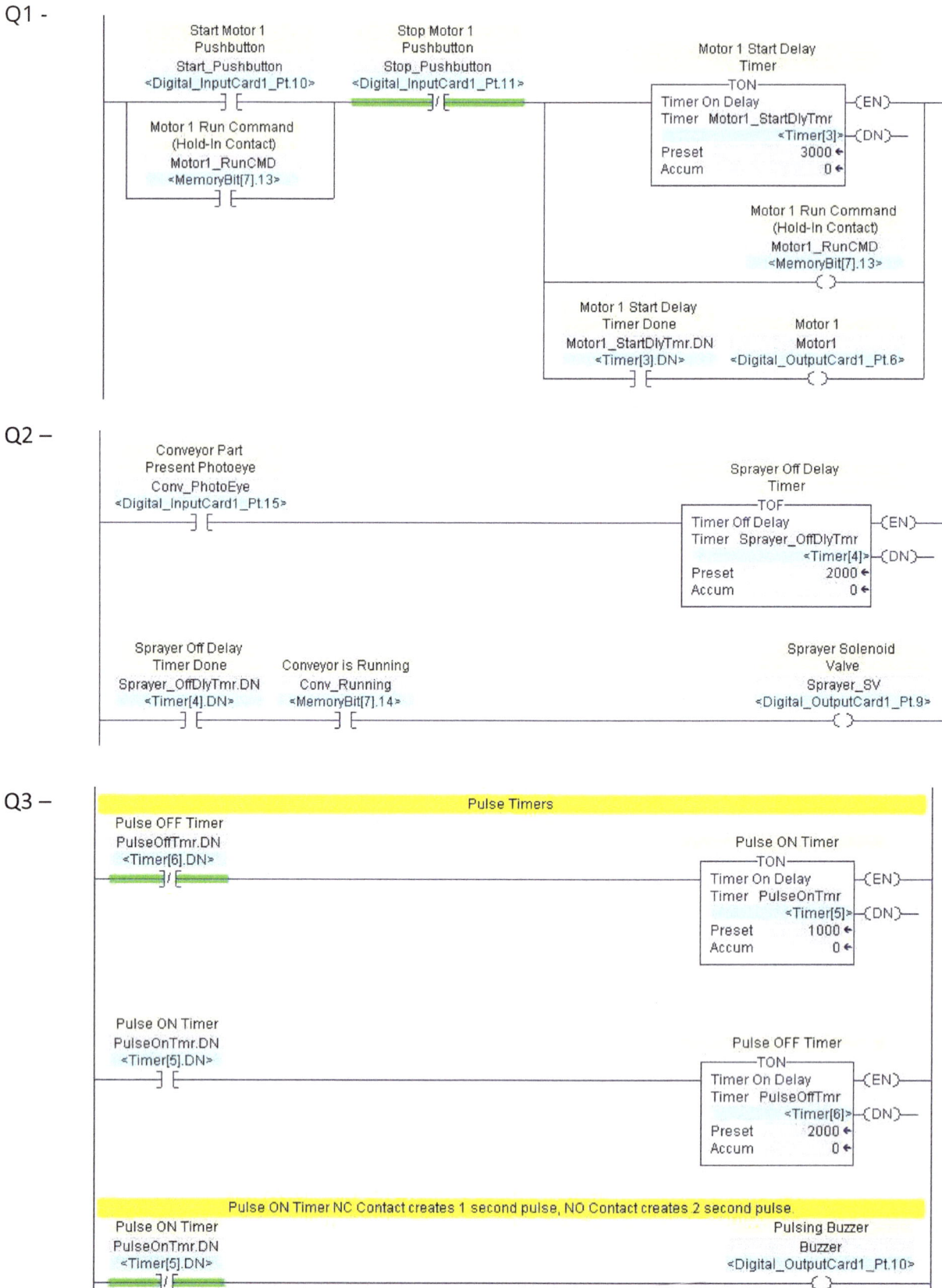

Q4 – See pages 314 and 315 of this book for an example of Auto Cycle Start and Stop logic.

Exercise 7 p. 90

Q1 -

Q2 -

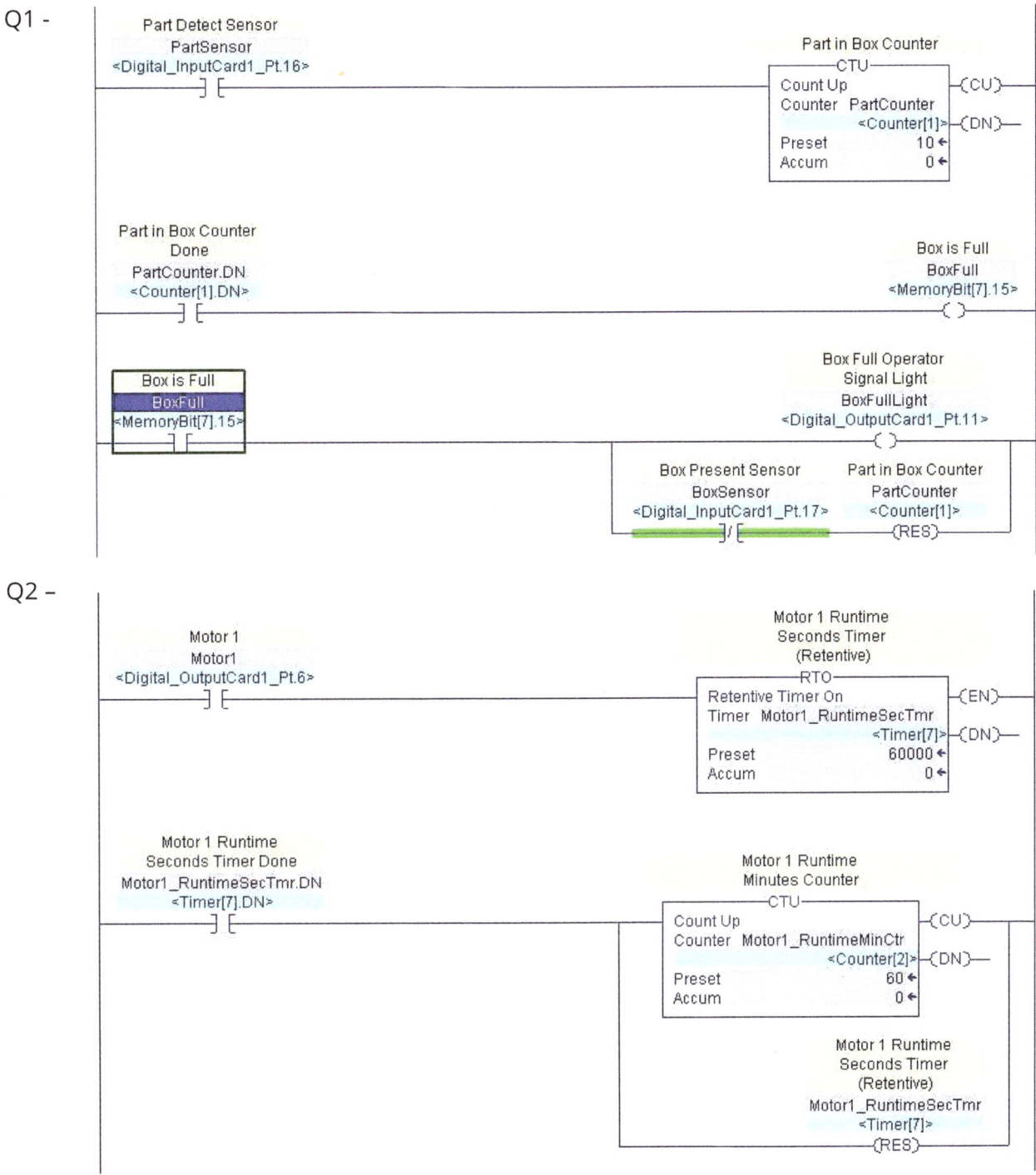

Exercise 7 p. 90 (Q2) continued

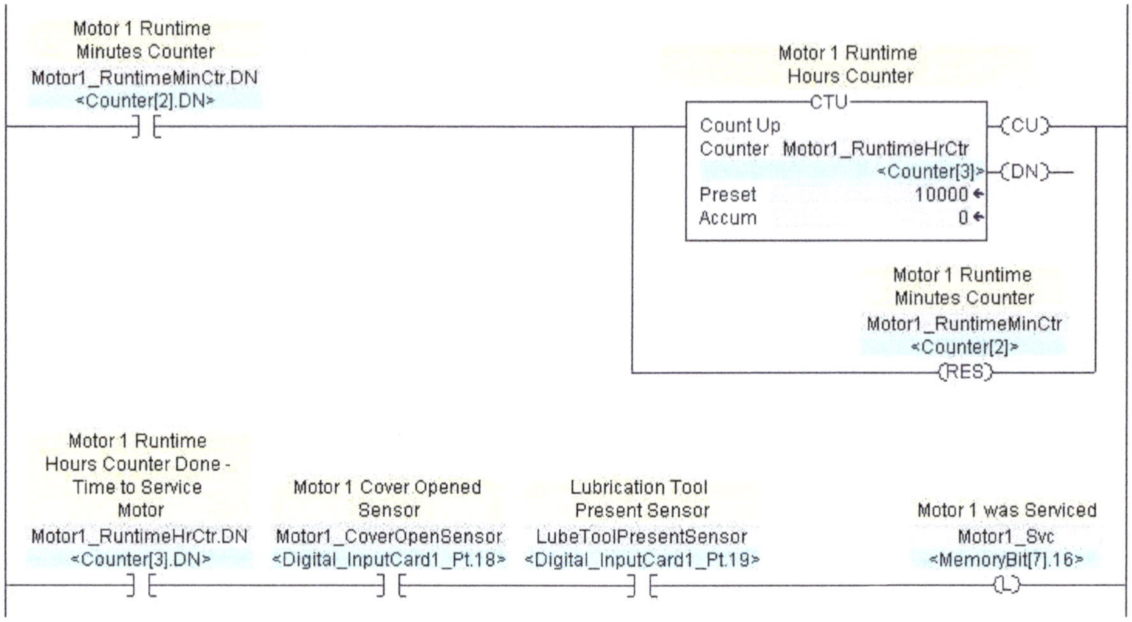

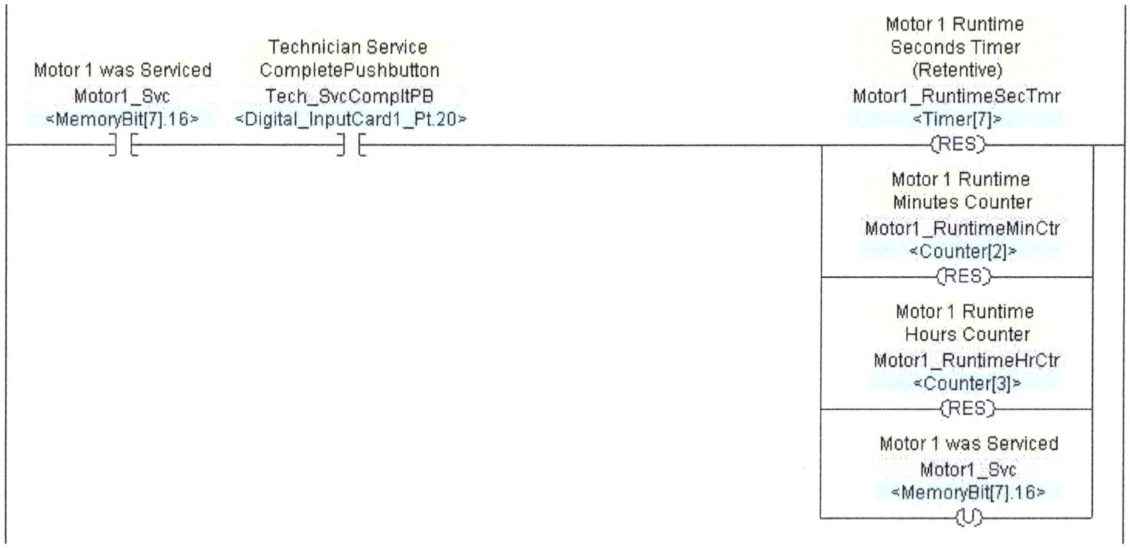

[Ladder logic diagram showing Motor 1 Runtime Minutes Counter (Counter[2].DN) contact feeding into Motor 1 Runtime Hours Counter CTU (Motor1_RuntimeHrCtr, Counter[3], Preset 10000, Accum 0) with CU and DN outputs, and a RES on Motor1_RuntimeMinCtr (Counter[2]).

Second rung: Motor1_RuntimeHrCtr.DN (Counter[3].DN), Motor1_CoverOpenSensor (Digital_InputCard1_Pt.18), LubeToolPresentSensor (Digital_InputCard1_Pt.19) contacts in series energizing Motor1_Svc (MemoryBit[7].16) output latch.

Third rung: Motor1_Svc (MemoryBit[7].16) and Tech_SvcCompltPB (Digital_InputCard1_Pt.20) contacts in series, outputs RES on Motor1_RuntimeSecTmr (Timer[7]), RES on Motor1_RuntimeMinCtr (Counter[2]), RES on Motor1_RuntimeHrCtr (Counter[3]), and unlatch Motor1_Svc (MemoryBit[7].16).]

Exercise 8 p. 97

Q1 -

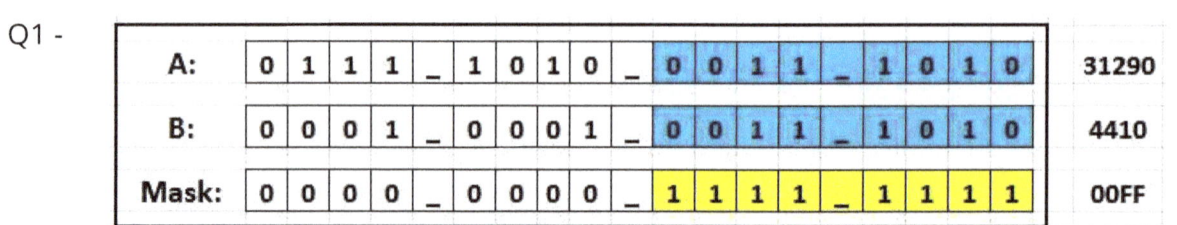

Yes, they are equal through the mask.

Exercise 8 p. 97 Continued

Q2 –

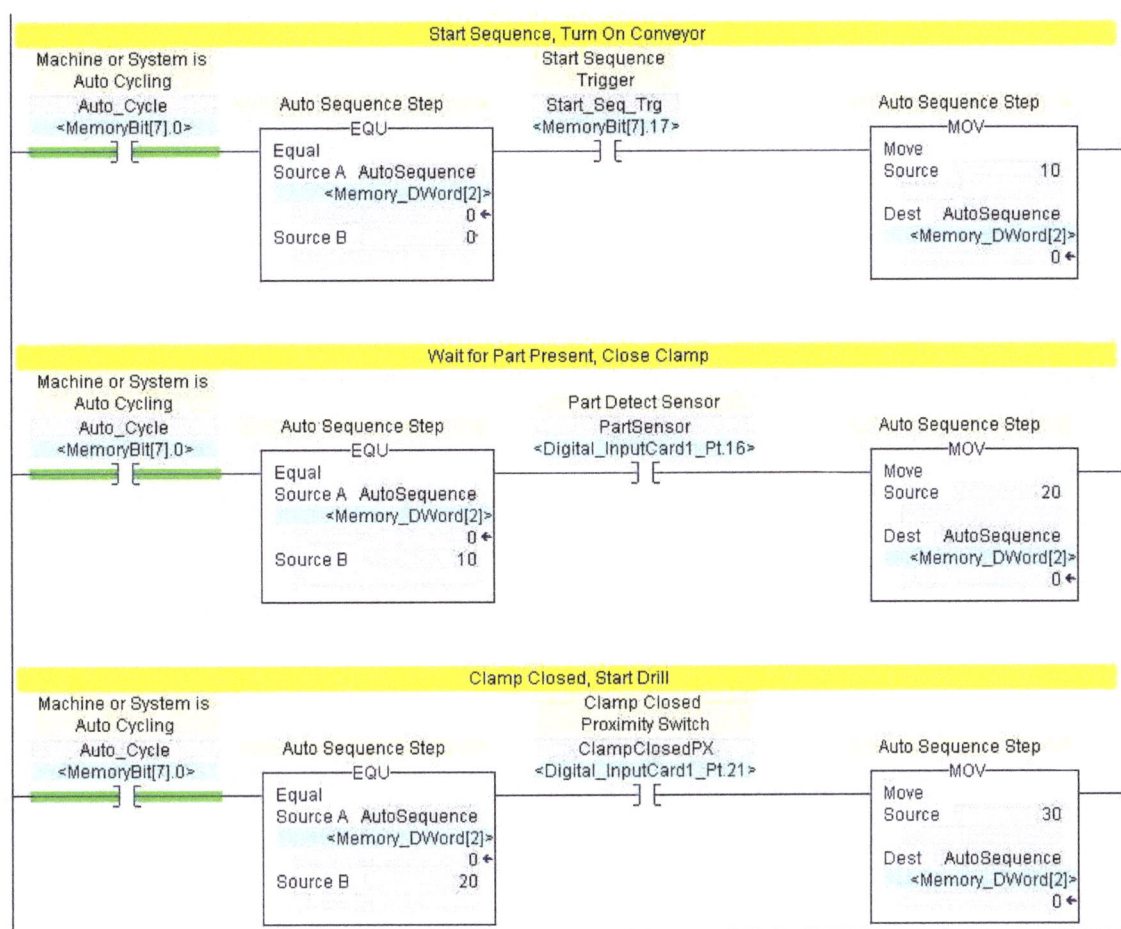

Auto sequences increment by 10 so that an extra step (i.e. Step 15) can be inserted if necessary.

Exercise 9 p. 102

Q1 -   1750/31,760 = 0.0551 (Scaling Factor)

Test: 31,760 * 0.0551 = 1749.976. Close enough! So, <Sensor Value> * 0.0551 = RPM.

Percent = RPM/Max RPM * 100%

(Continued next page)

Exercise 9 p. 102 Q1 Continued

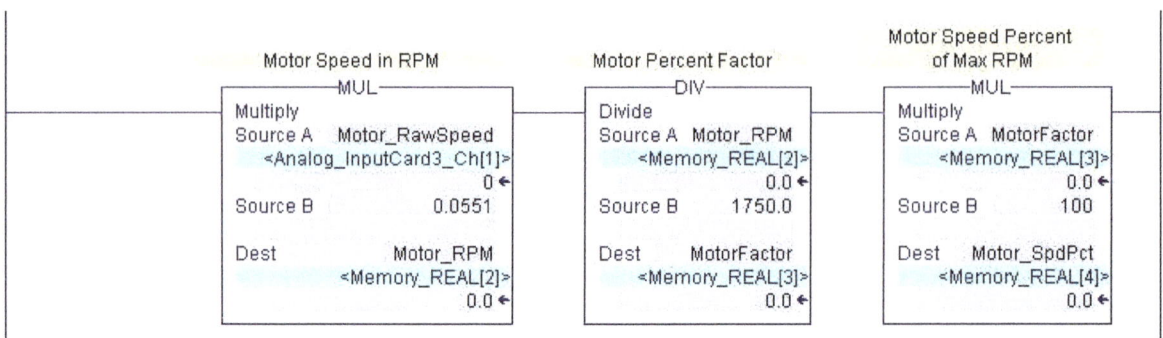

Q2 –

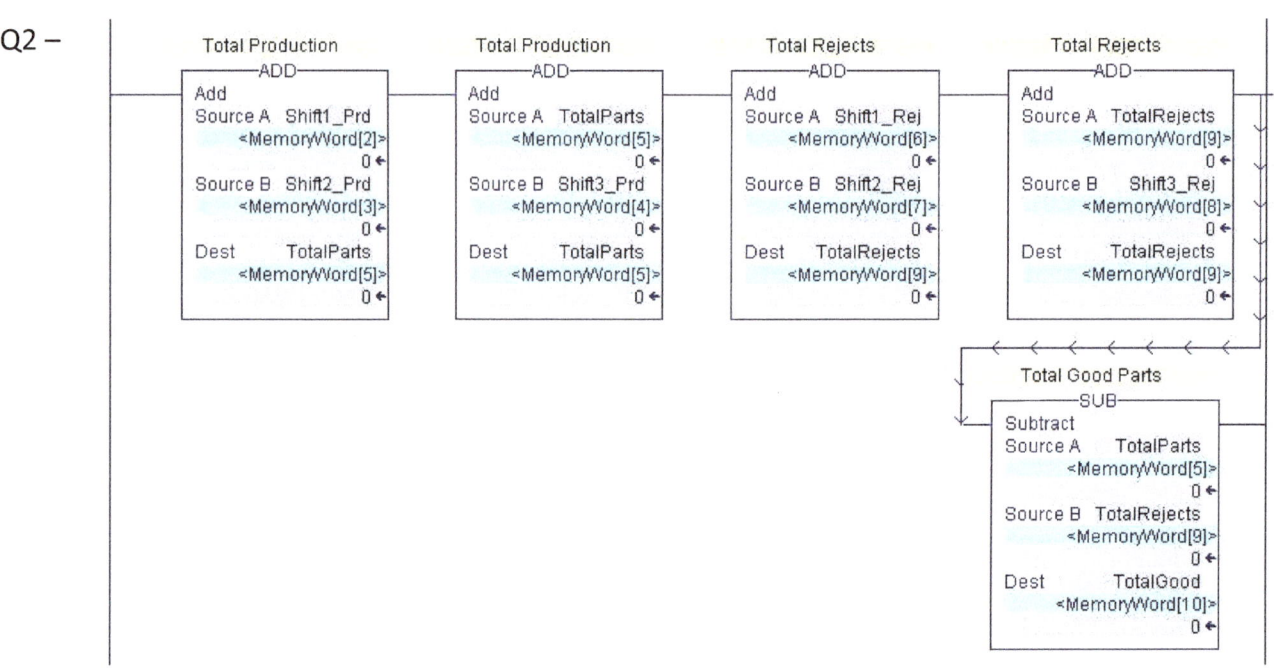

Exercise 10 p. 105

Q1 –   M = (Y2-Y1)/(X2-X1) = (6000-0)/(24780-96) = 0.24307

B = Y1 – (M * X1) = 0 – (0.24307 * 96) = -23.33472

Liters = Gallons * 3.78541

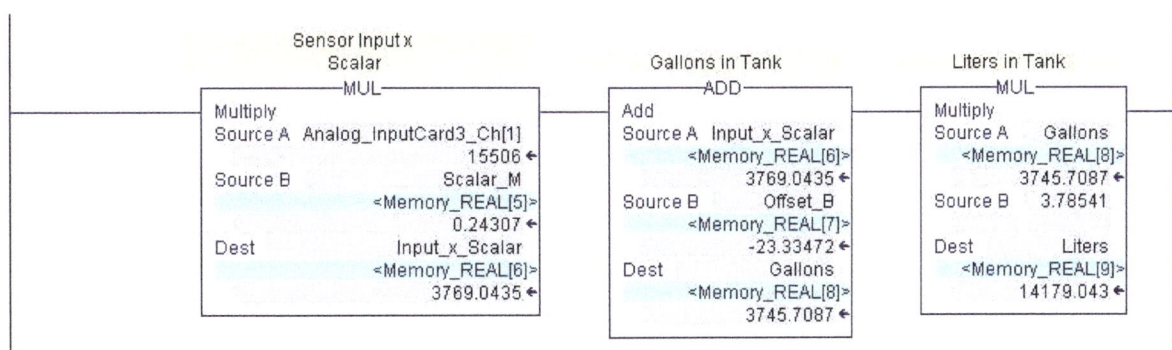

Exercise 11 p. 113

Q1 –   Trigonometric functions are usually used in motion control and positioning applications; they calculate geometrical coordinates.

Q2 -   47 _G_ 6F _o_ 6F _o_ 64 _d_ 20 ___ 4A _J_ 6F _o_ 62 _b_ 21 _!_
Good Job!

Q3 -   Yes, Jump instructions can jump backward. If so, a method to exit the loop is necessary, such as incrementing a counter.

Q4 -   FIFO: First-In First -Out   LIFO: Last-In First-Out

Q5 -   A Sequencer monitors and controls repeatable operations

Exercise 12 p. 120

Q1 -   False

Q2 -   True

Q3 -   Q98.5 "PP04" will be off.

Q4 -   Q98.5 will be on.

Q5 -   Look for coils and operations affecting digital and analog outputs.

Q6 -   Physical inputs and contacts with no coil.

The Art of Programming Lab Solution

There is no single solution for a program; however, this code satisfies the requirements for the Lab Exercise.

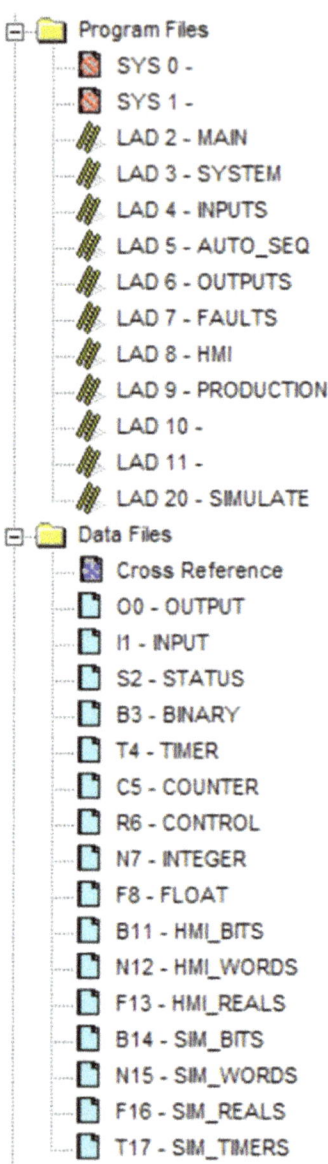

The Color Sorter lab was programmed on an Allen-Bradley MicroLogix 1400 processor. The software used was RSLogix500 Micro, which is a full-function package used for the MicroLogix controllers only.

Routines and files are shown to the left. To enable the programmer to do all programming online rather than having to download, the routines shown are already in the processor, as well as all of the data files listed. The names for Ladder 3-9 were named after the start of programming.

The HMI files (B11, N12, F13) and the Simulation files (B14, N15, F16, and T17) are separated from the standard data files so that the programmer does not accidentally use addresses assigned to simulations or the HMI in their program.

The code in this example was run on a Fischertechnik brand "Color Sorting with Conveyor" and operates correctly. As with all programs, it can be refined and improved; however, it was important to go through the same steps that a student would use in completing the exercise. It took the author about 2 days (16 hours) to complete the exercise.

Ladder 2 – Main Routine

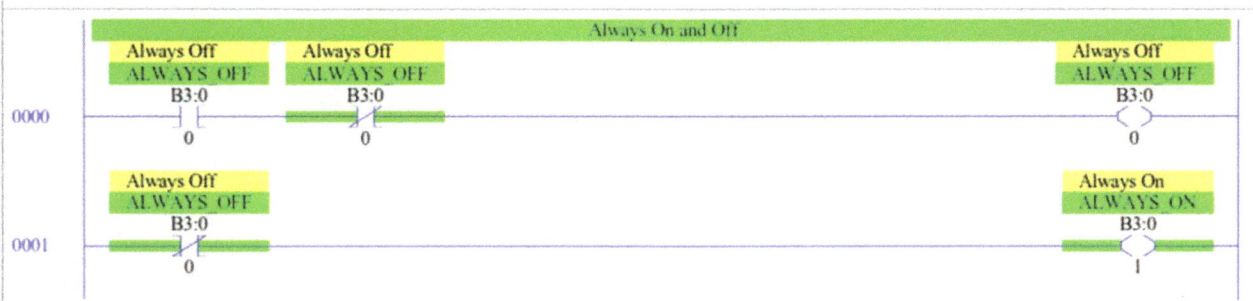

Ladder 3 - System

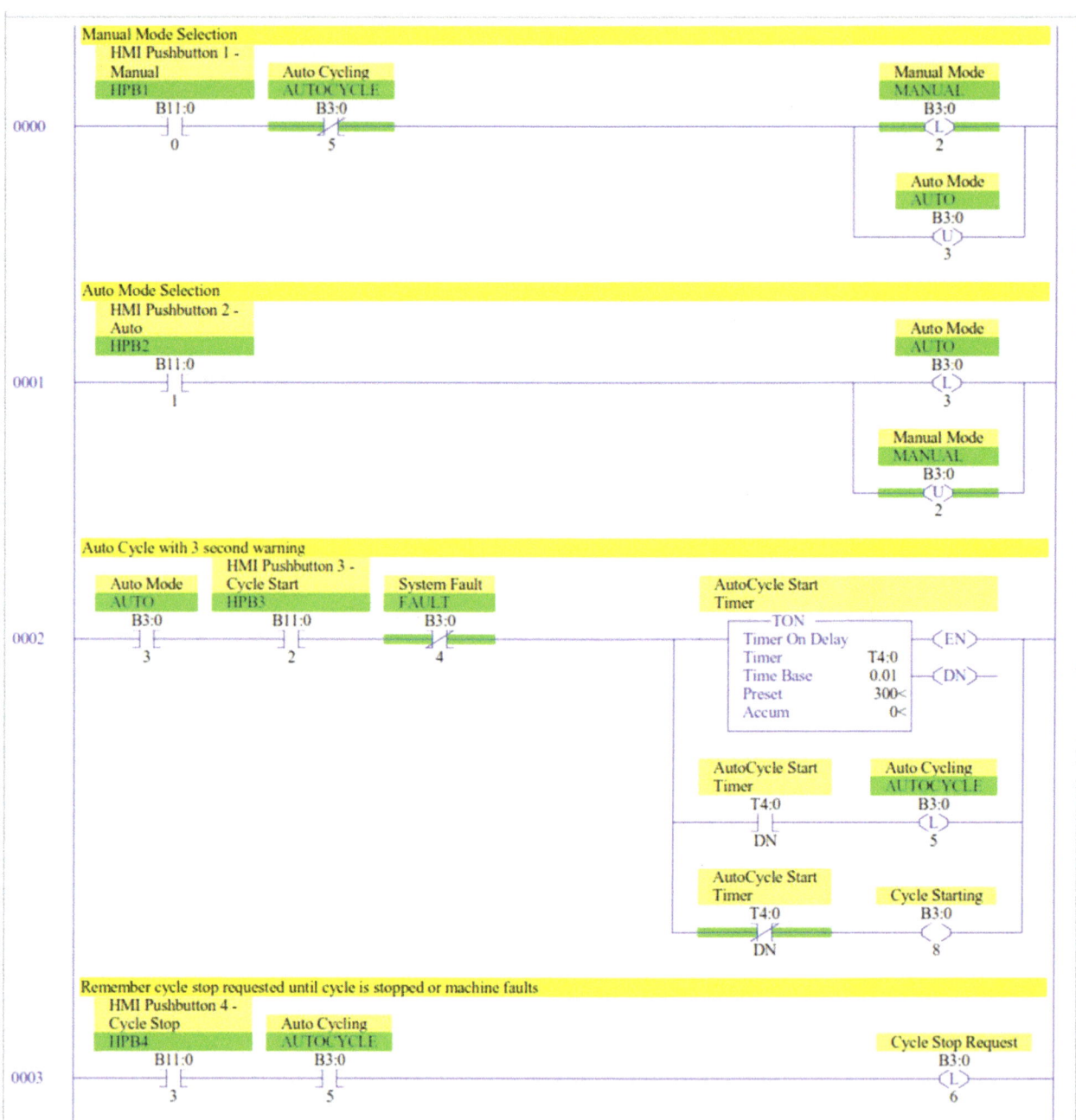

Ladder 3 System continued

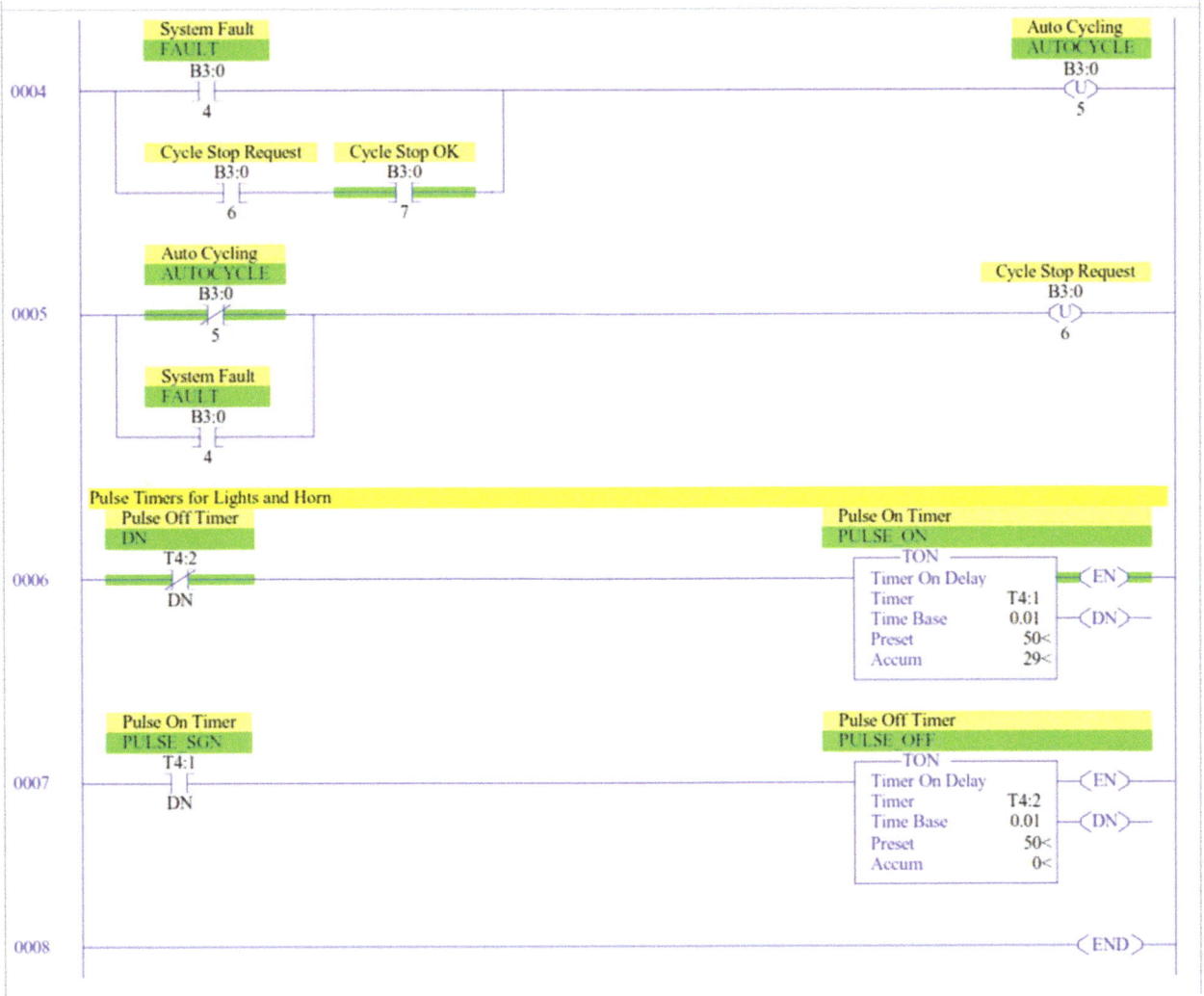

Ladder 4 - Inputs

Note that the photoeye status bits are reversed; parts are present when the eye is off (blocked).

Ladder 5 – Auto Sequence

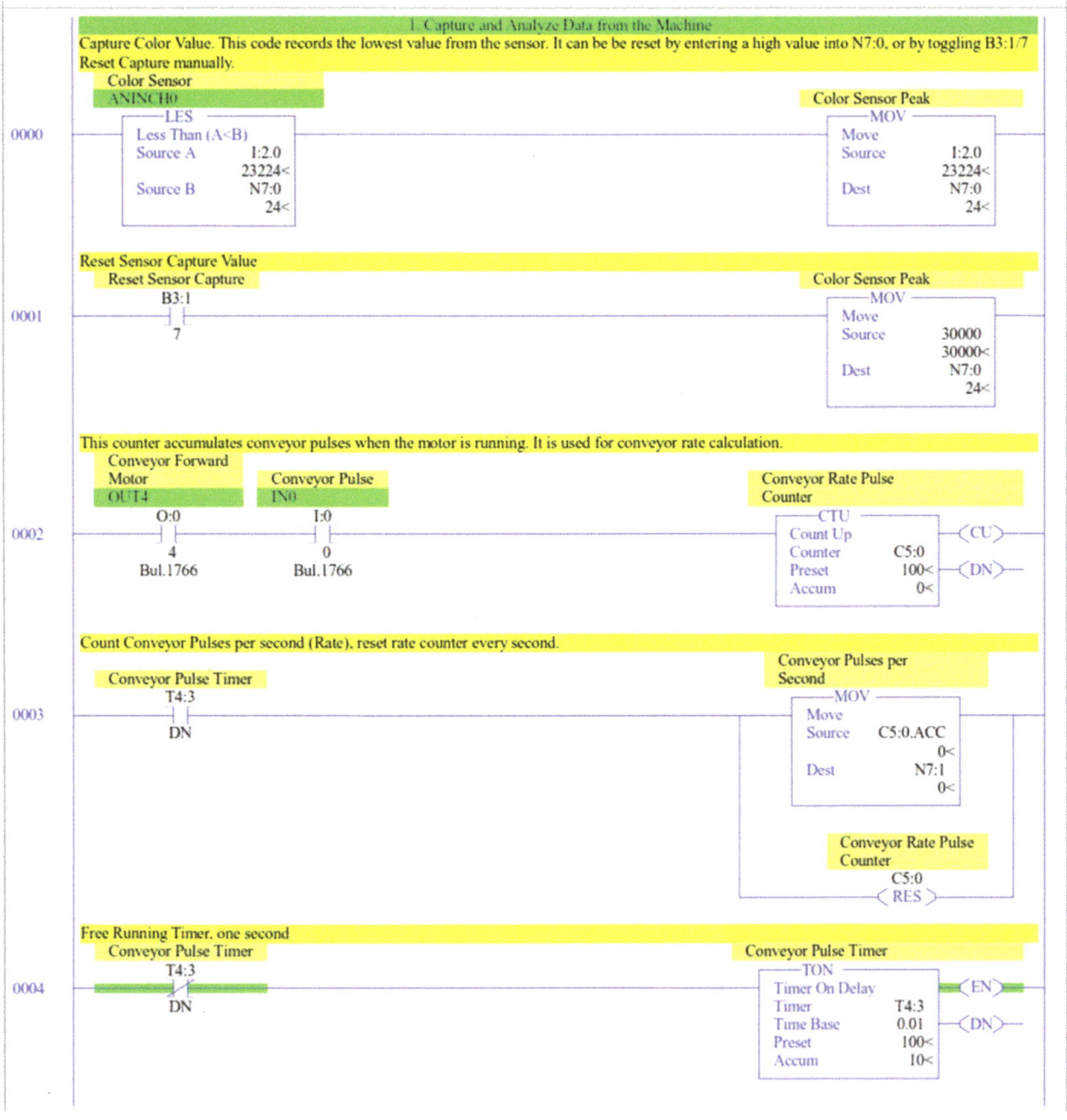

Ladder 5 Auto Sequence Continued

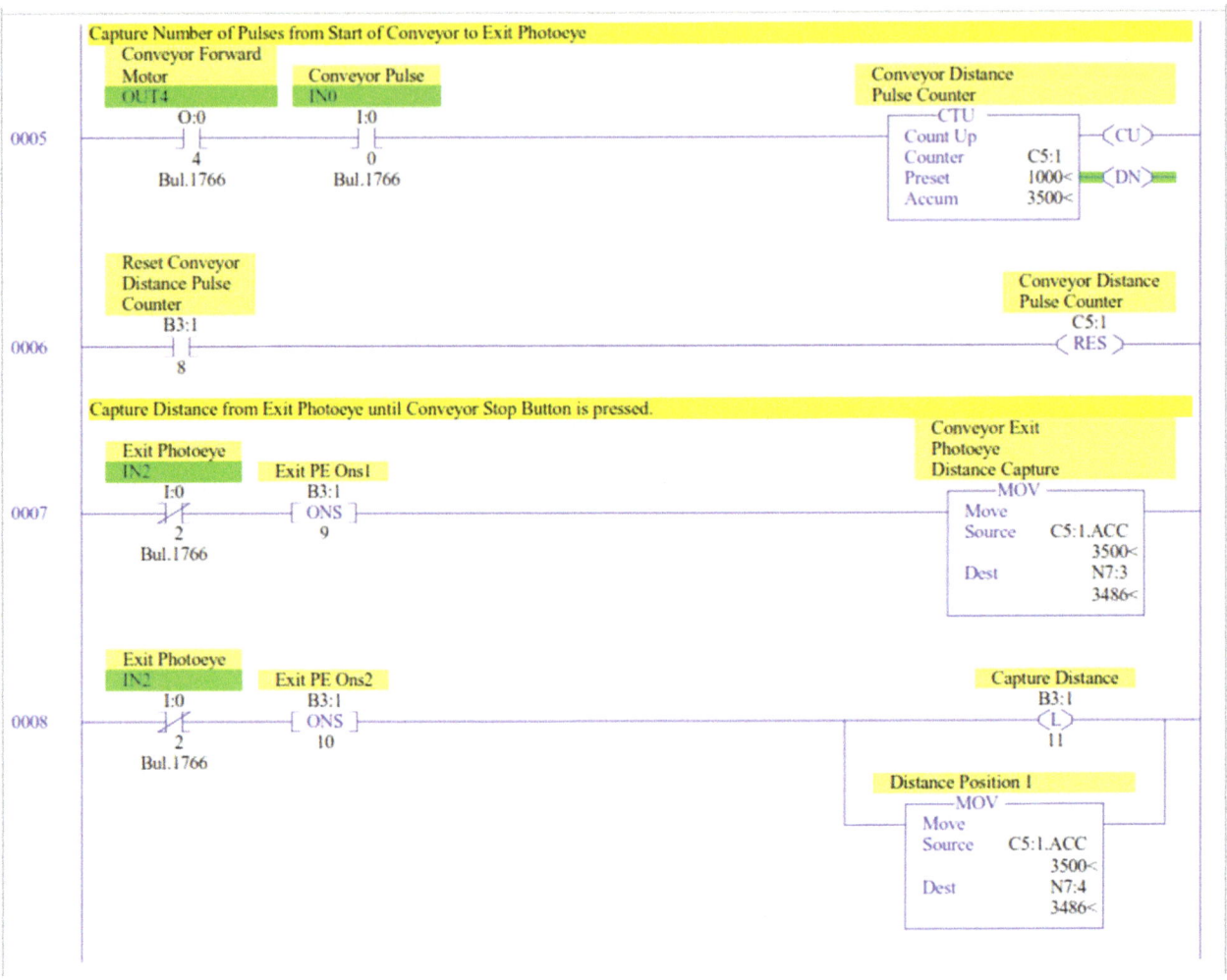

Ladder 5 Auto Sequence Continued

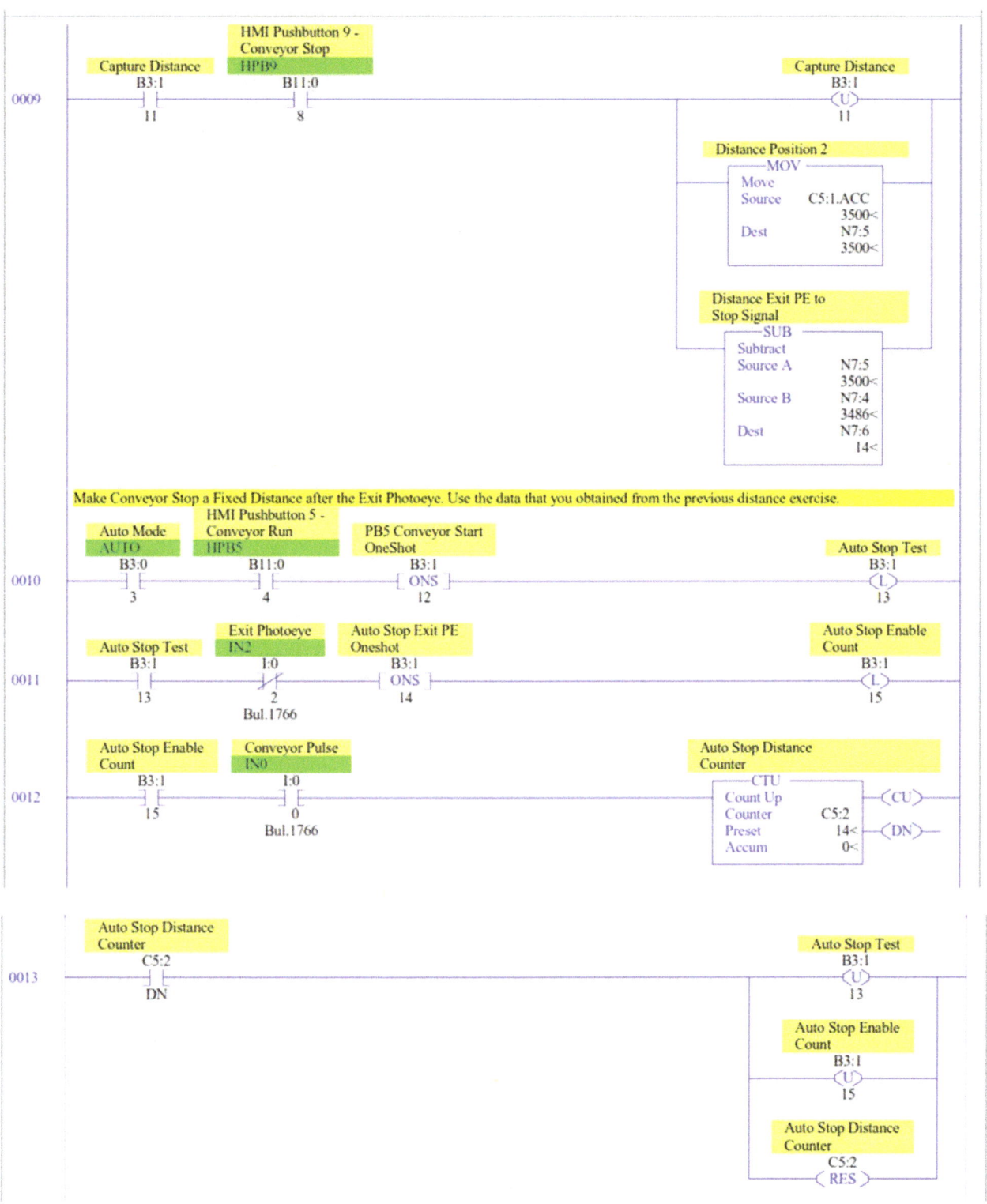

Ladder 5 Auto Sequence Continued

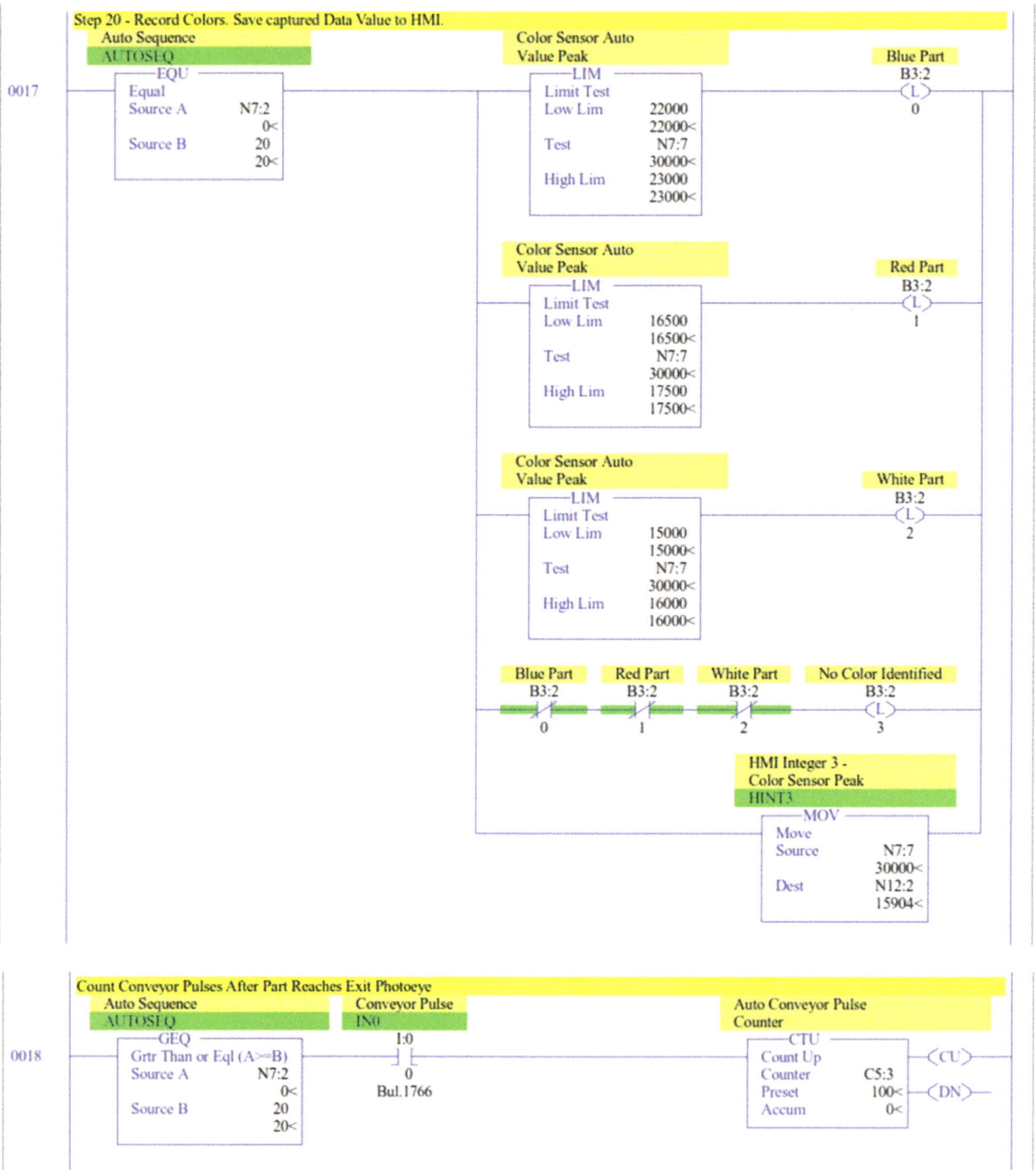

Ladder 5 Auto Sequence Continued

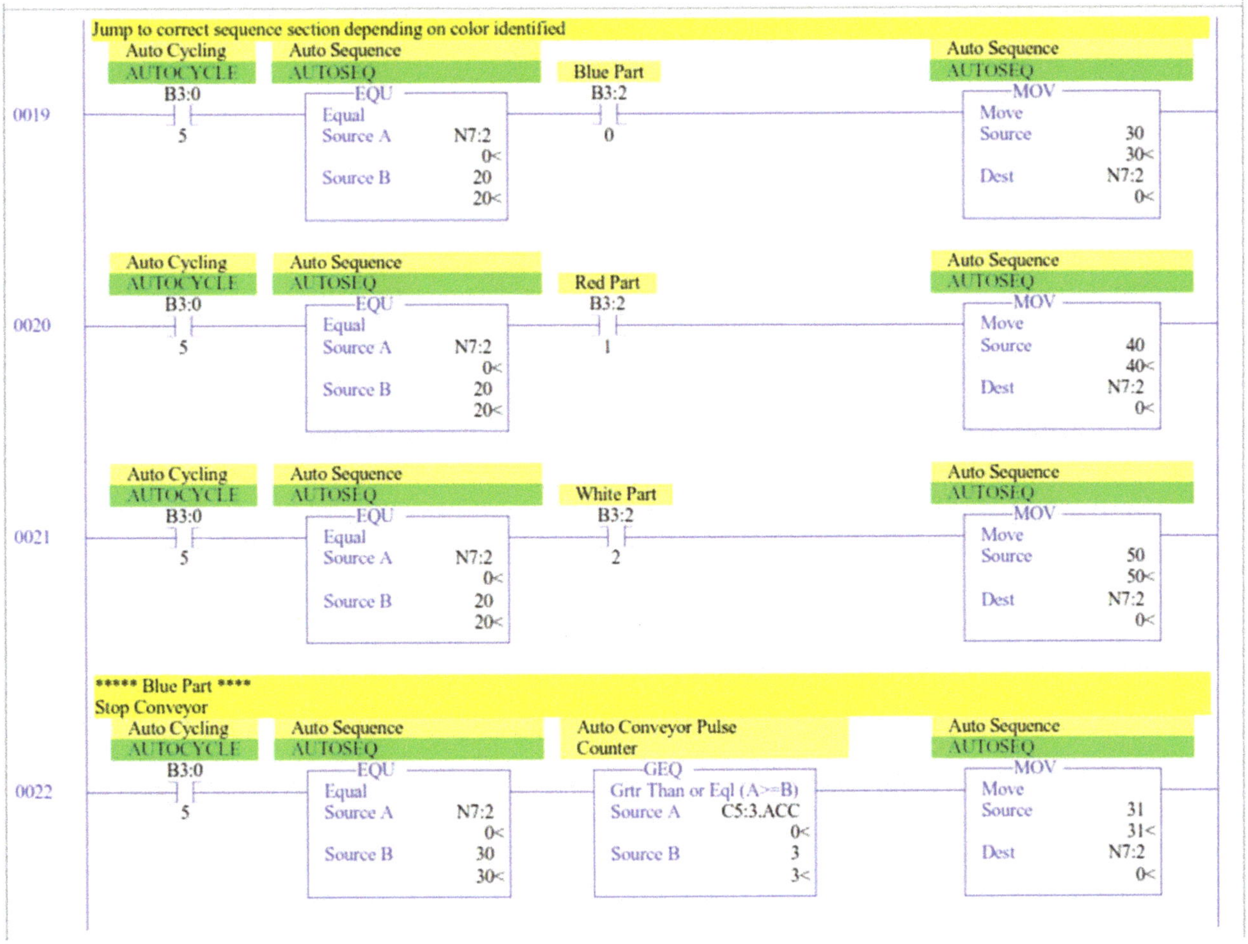

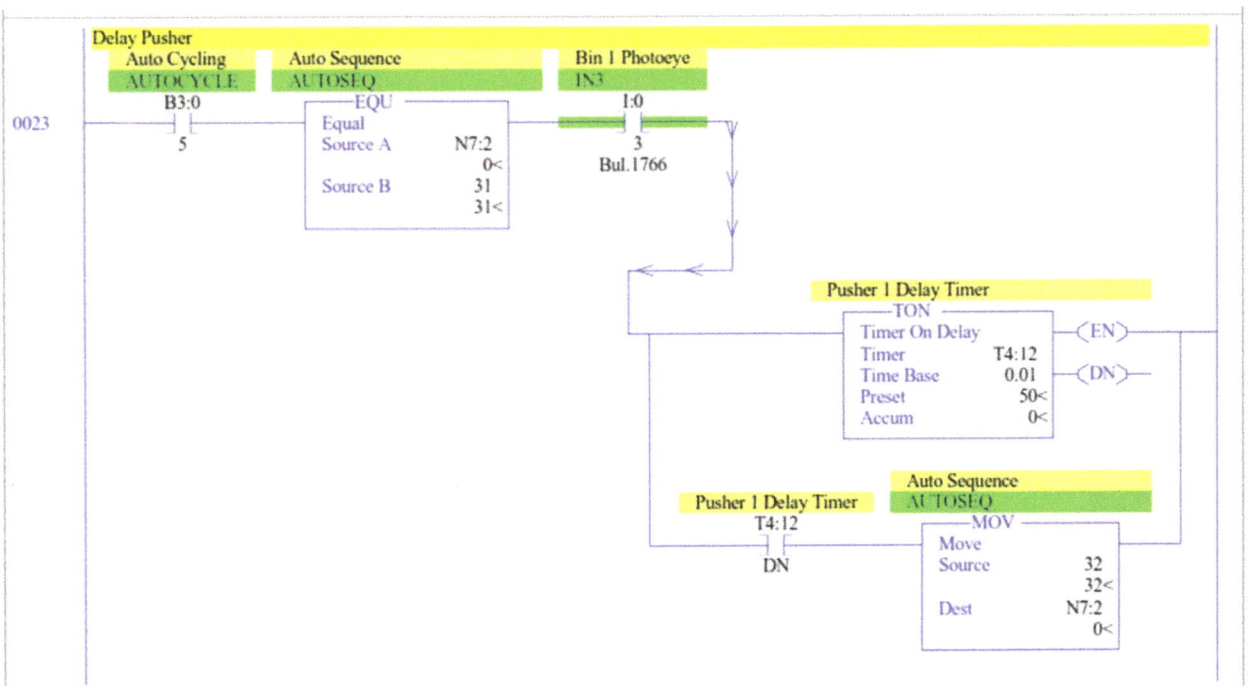

Ladder 5 Auto Sequence Continued

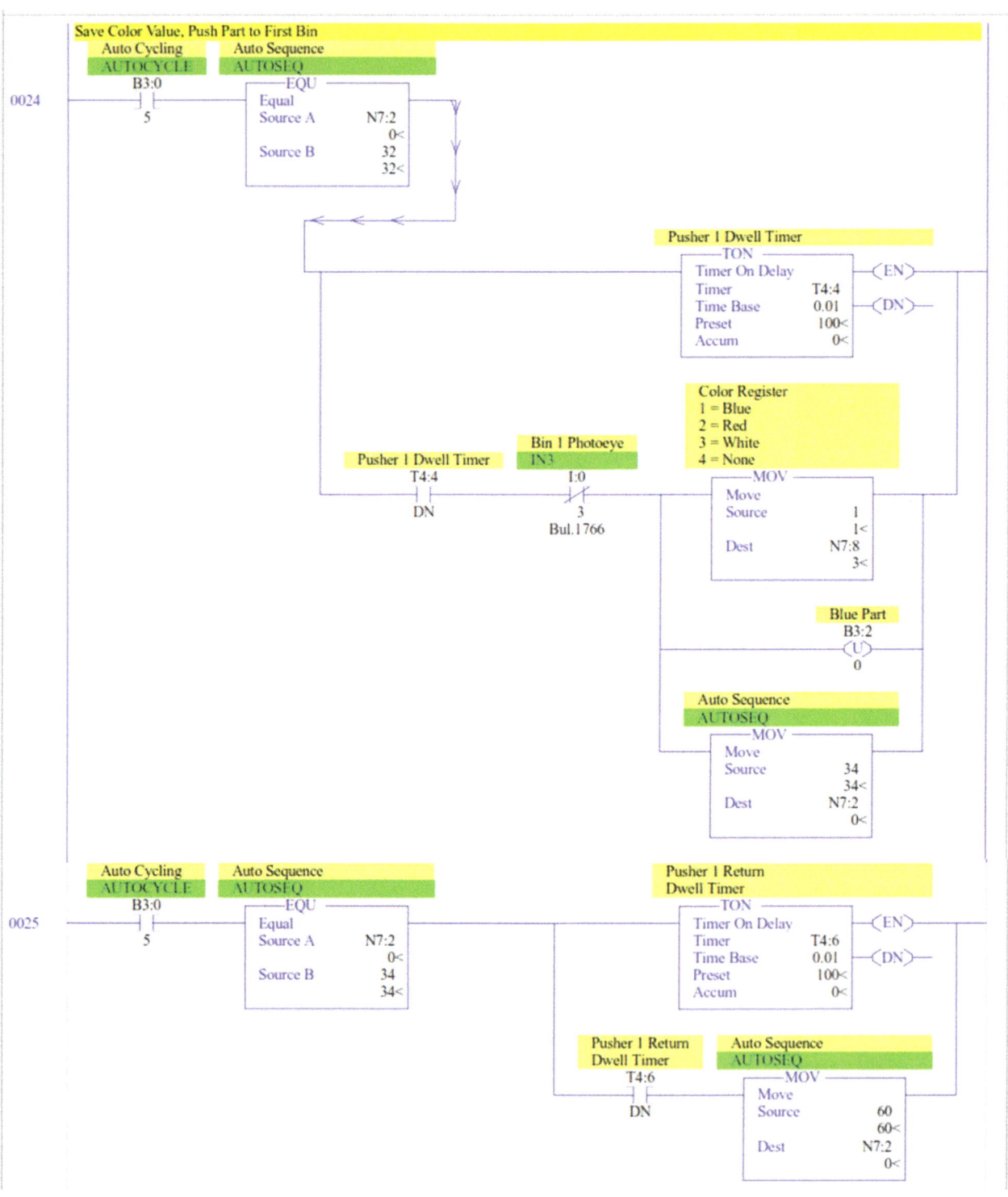

Ladder 5 Auto Sequence Continued

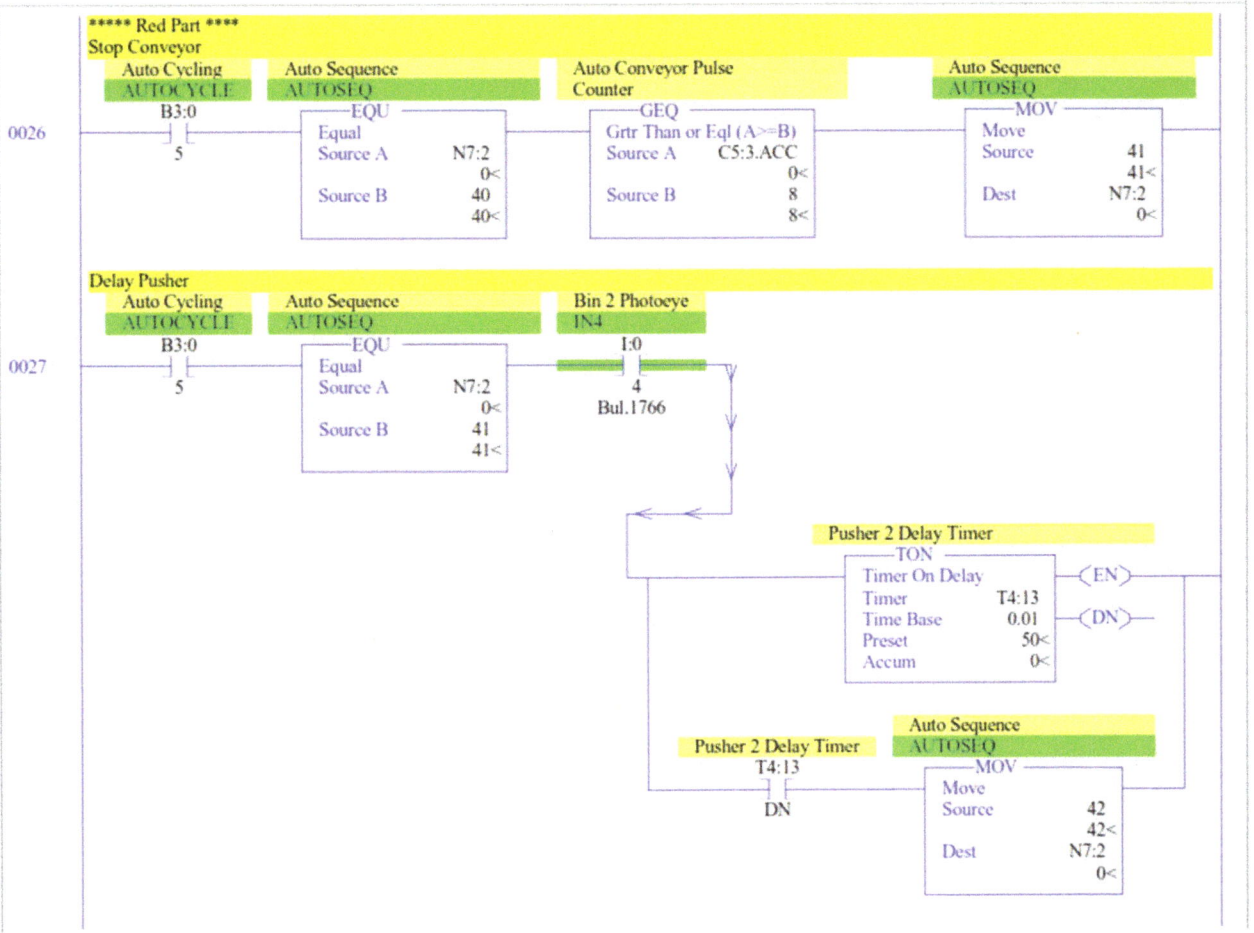

Ladder 5 Auto Sequence Continued

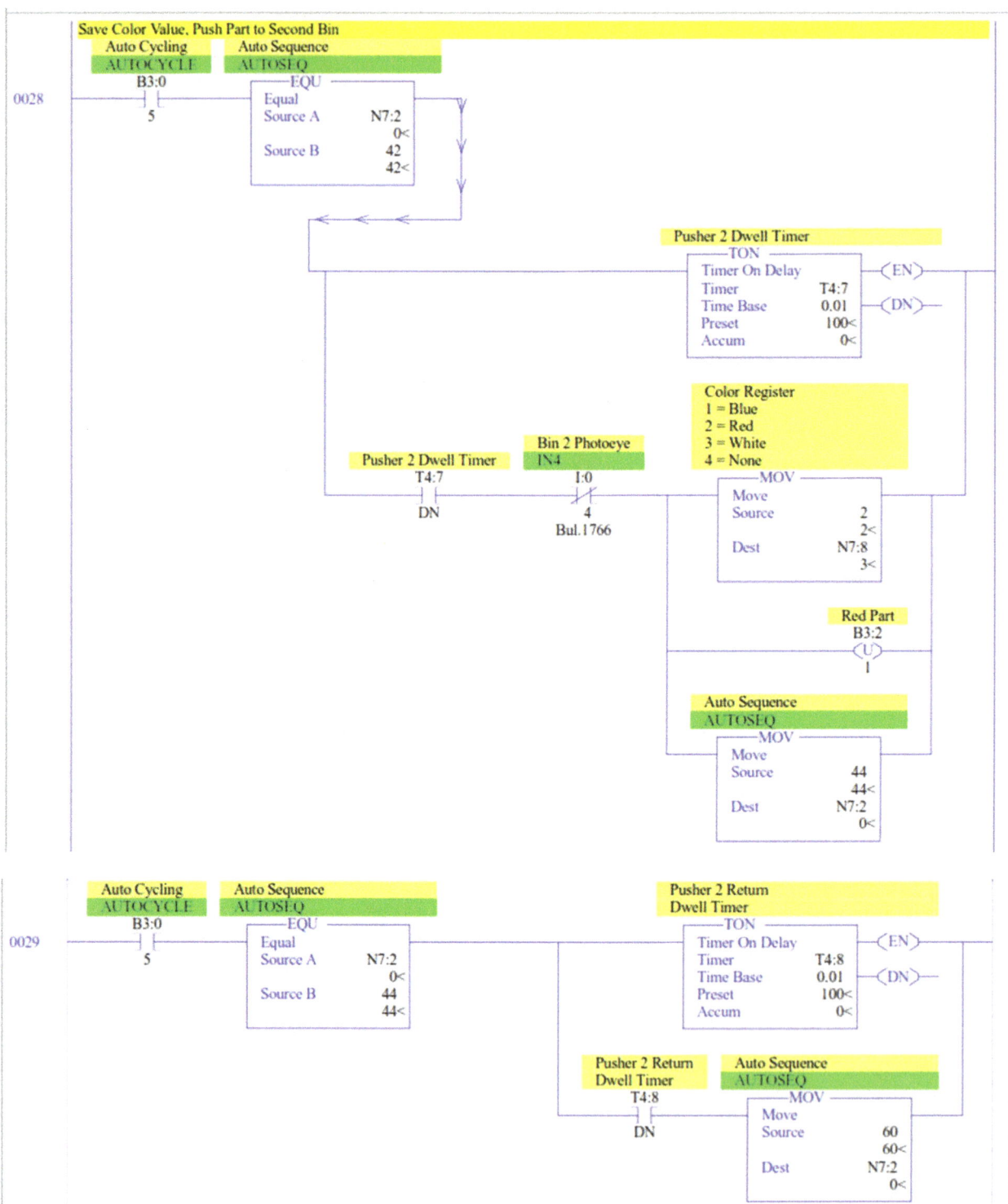

Ladder 5 Auto Sequence Continued

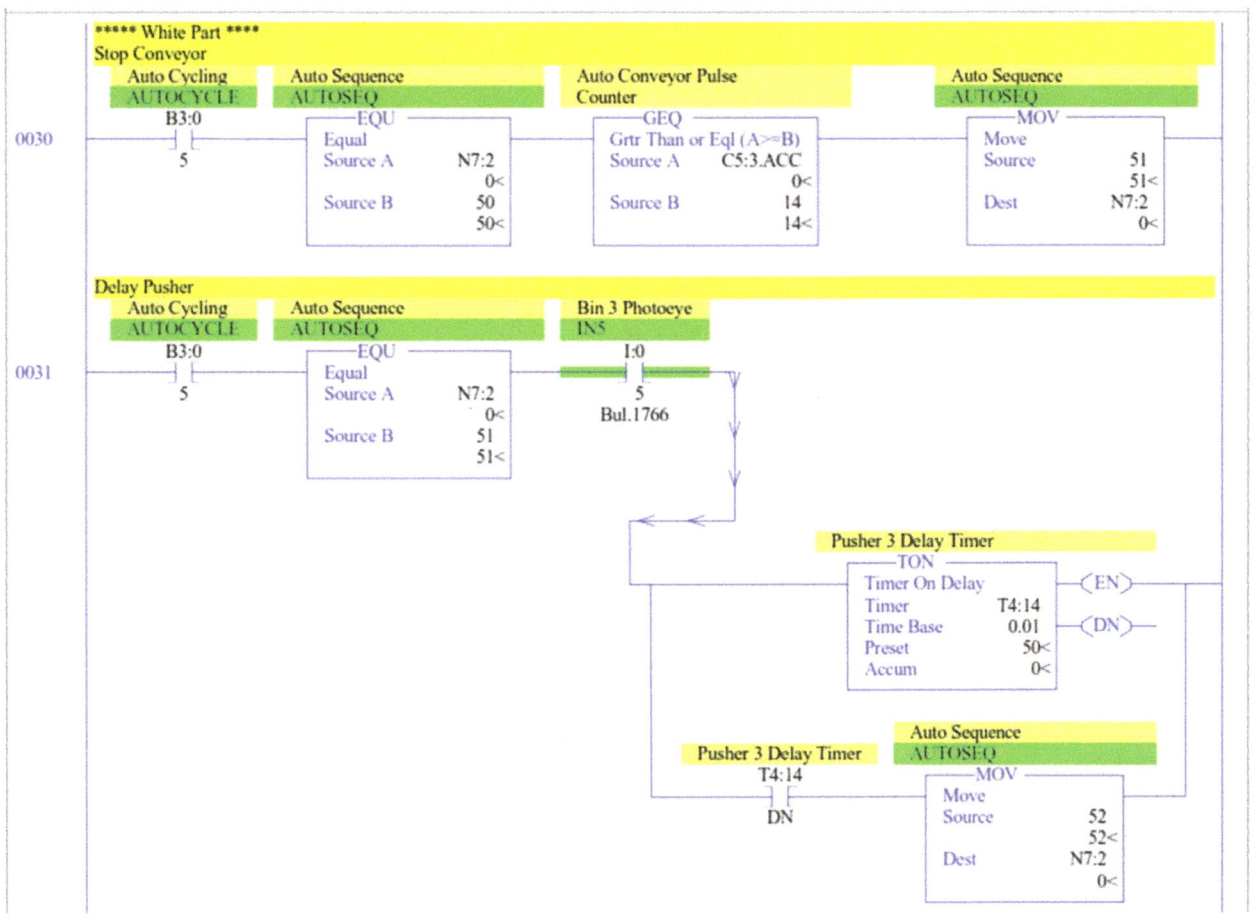

Ladder 5 Auto Sequence Continued

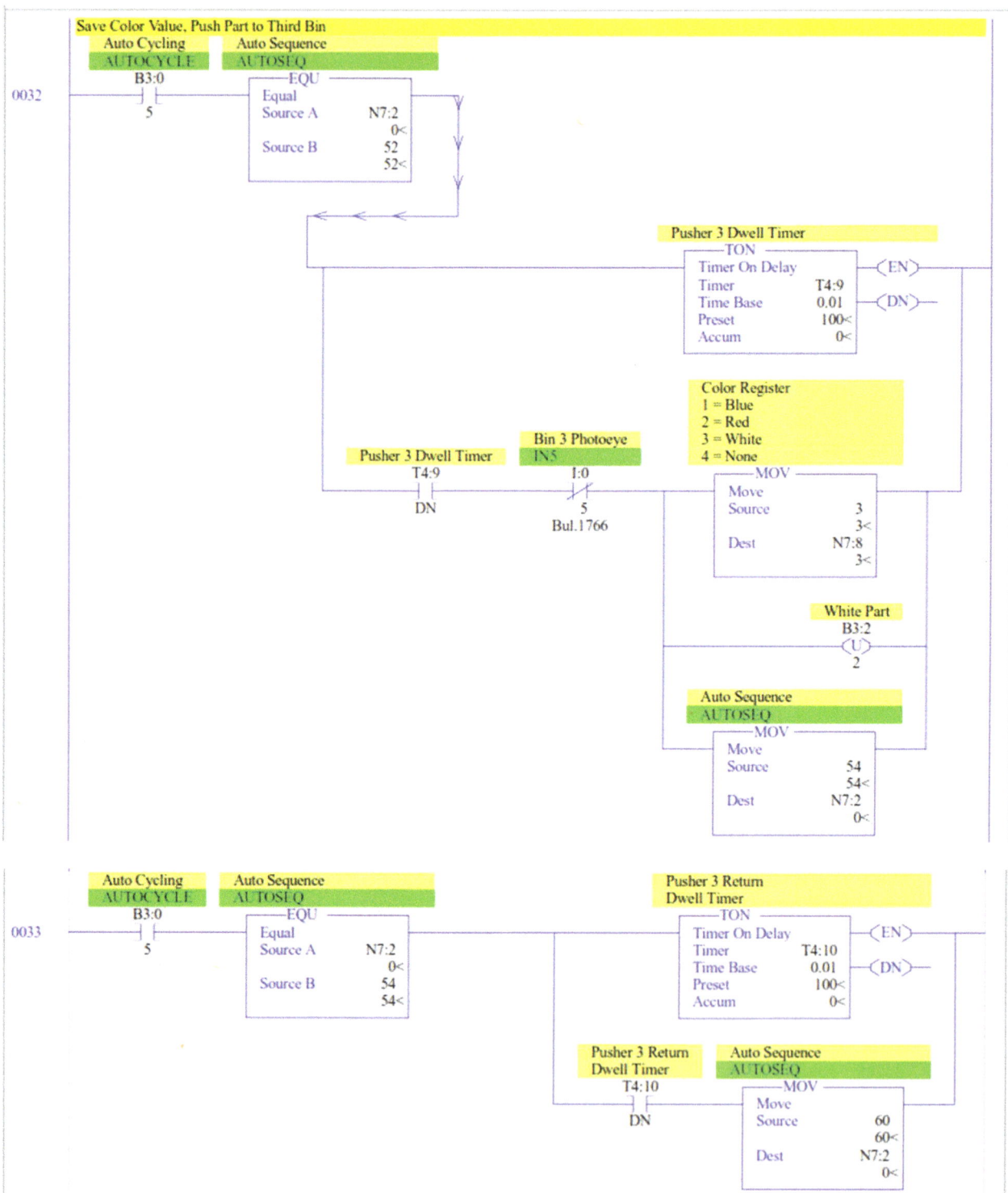

Ladder 5 Auto Sequence Continued

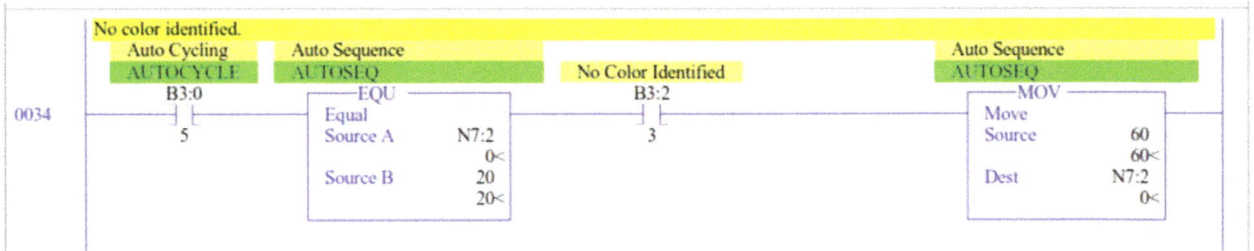

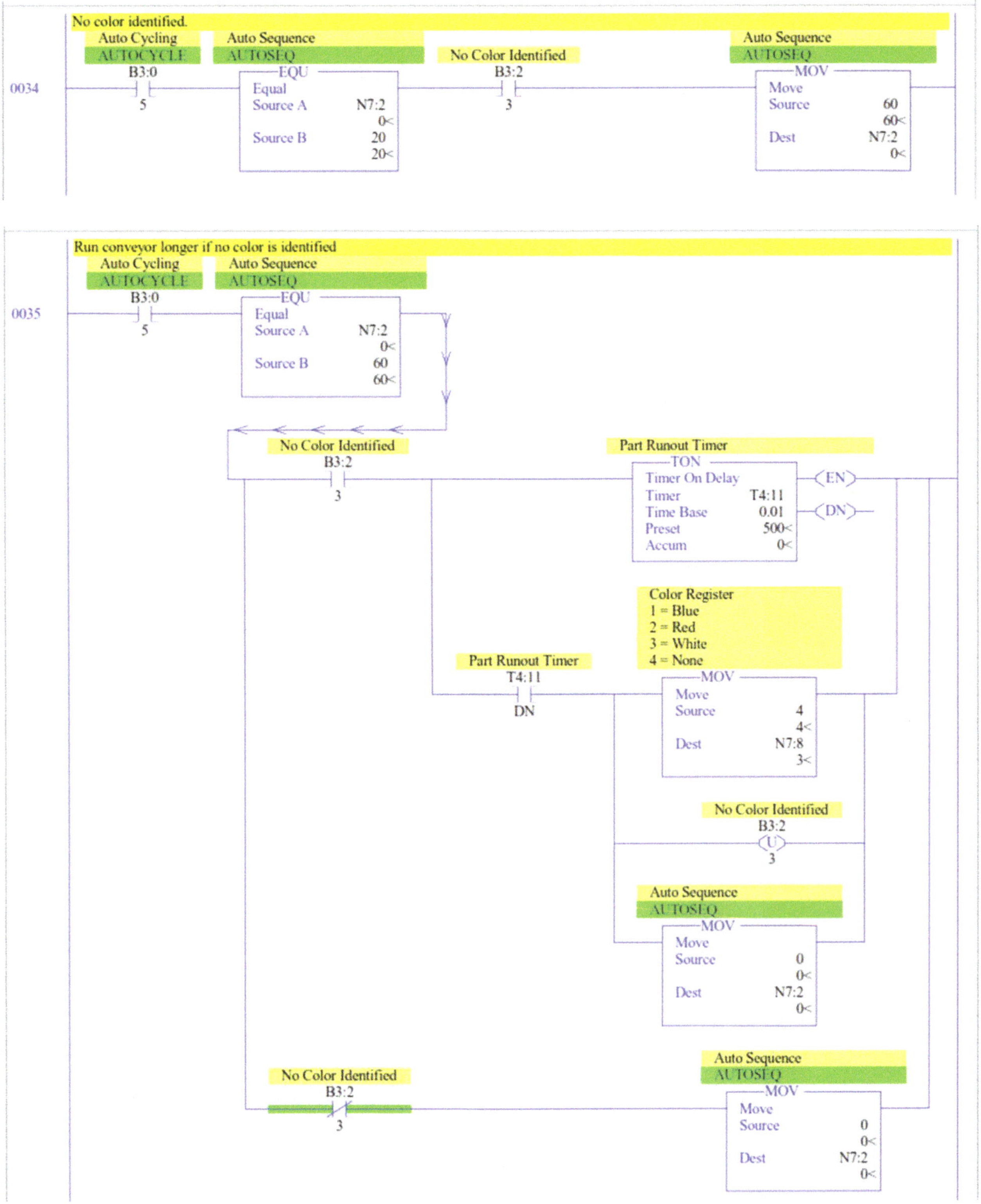

Ladder 5 Auto Sequence Continued

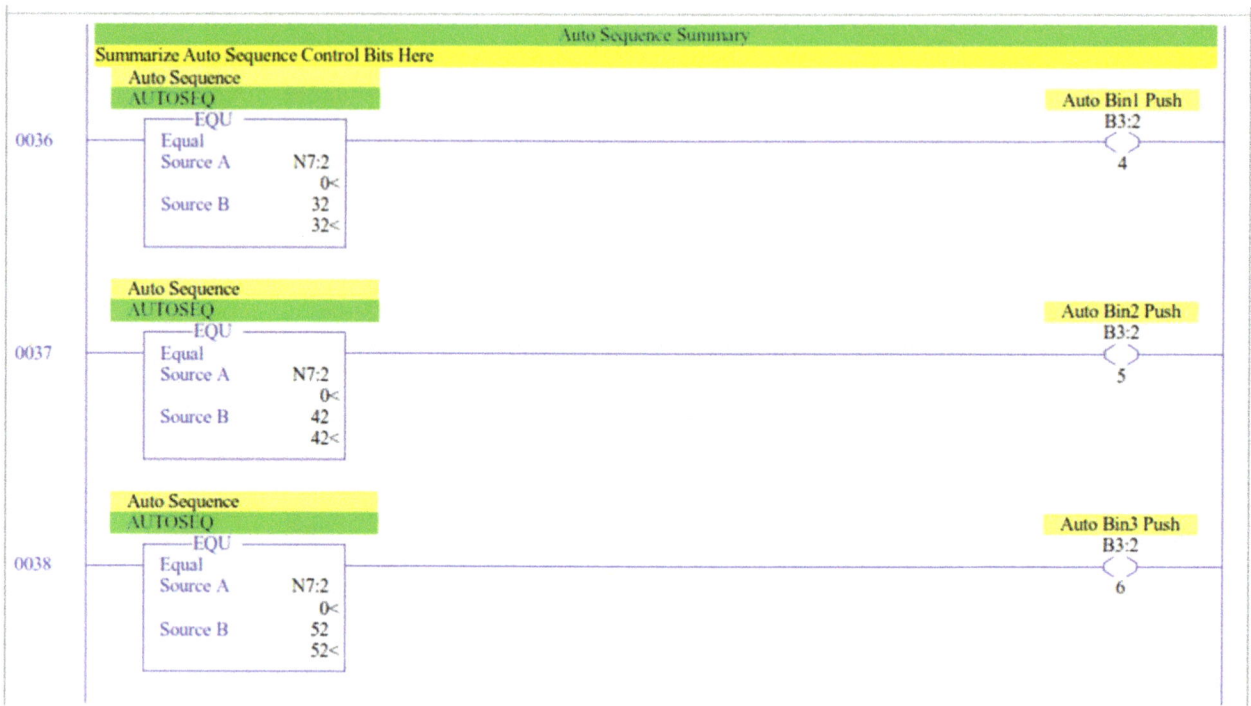

Ladder 5 Auto Sequence Continued

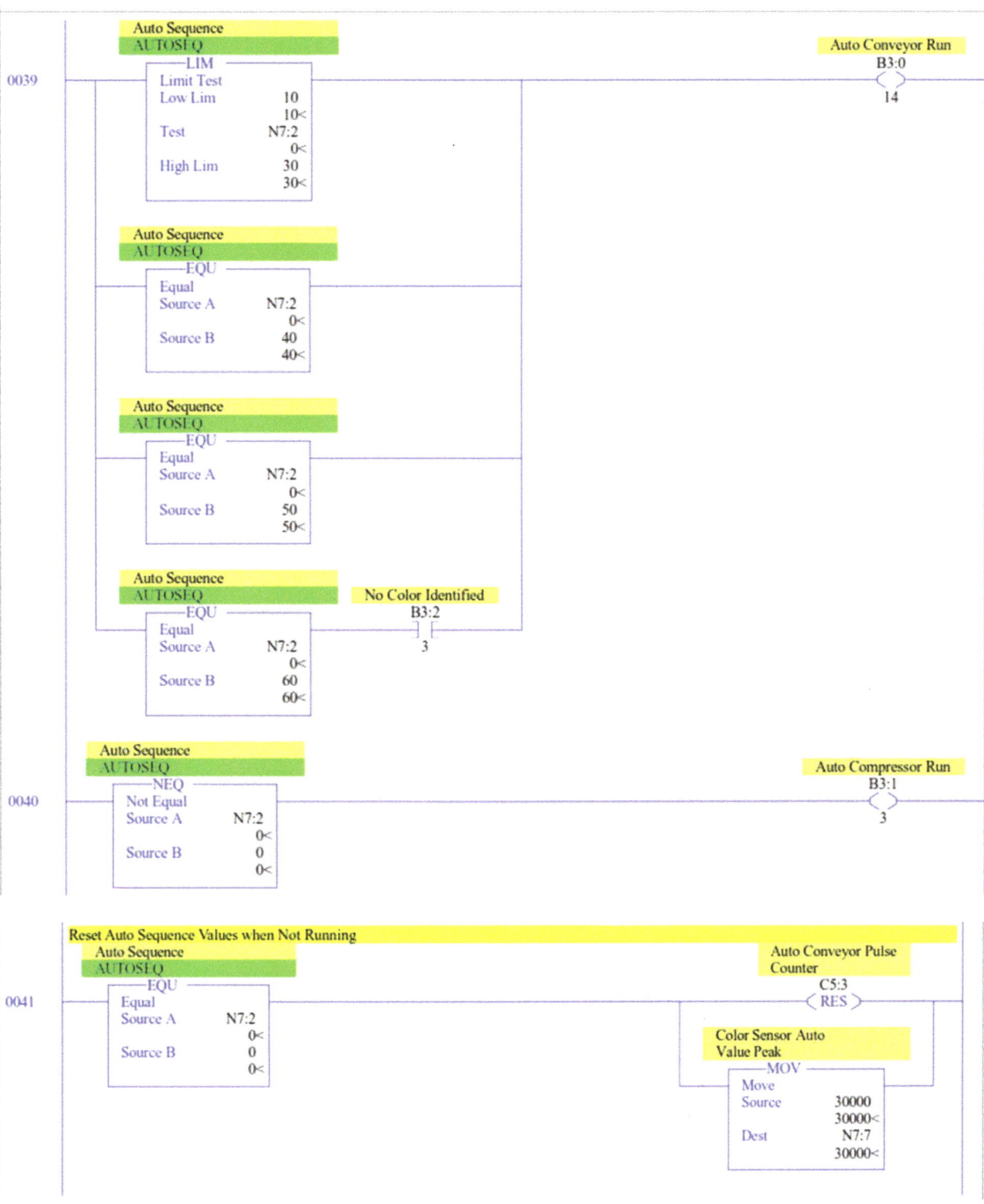

Ladder 5 Auto Sequence Continued

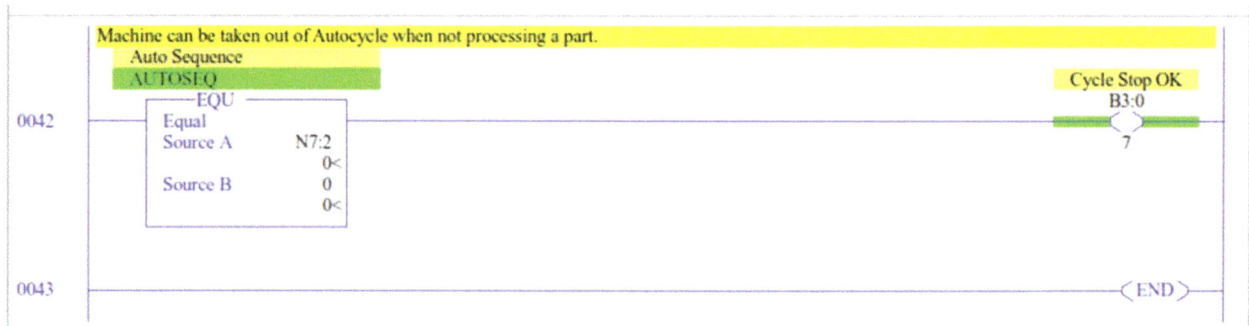

Ladder 6 – Outputs

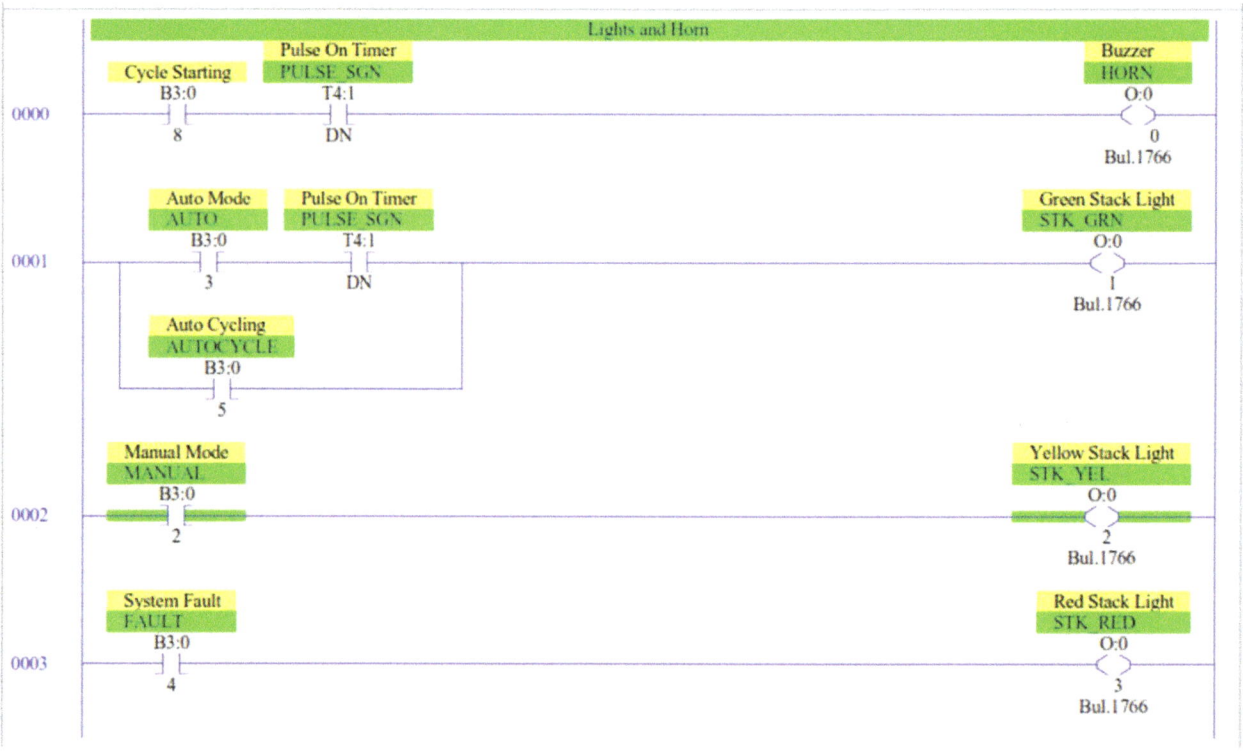

Ladder 6 Outputs Continued

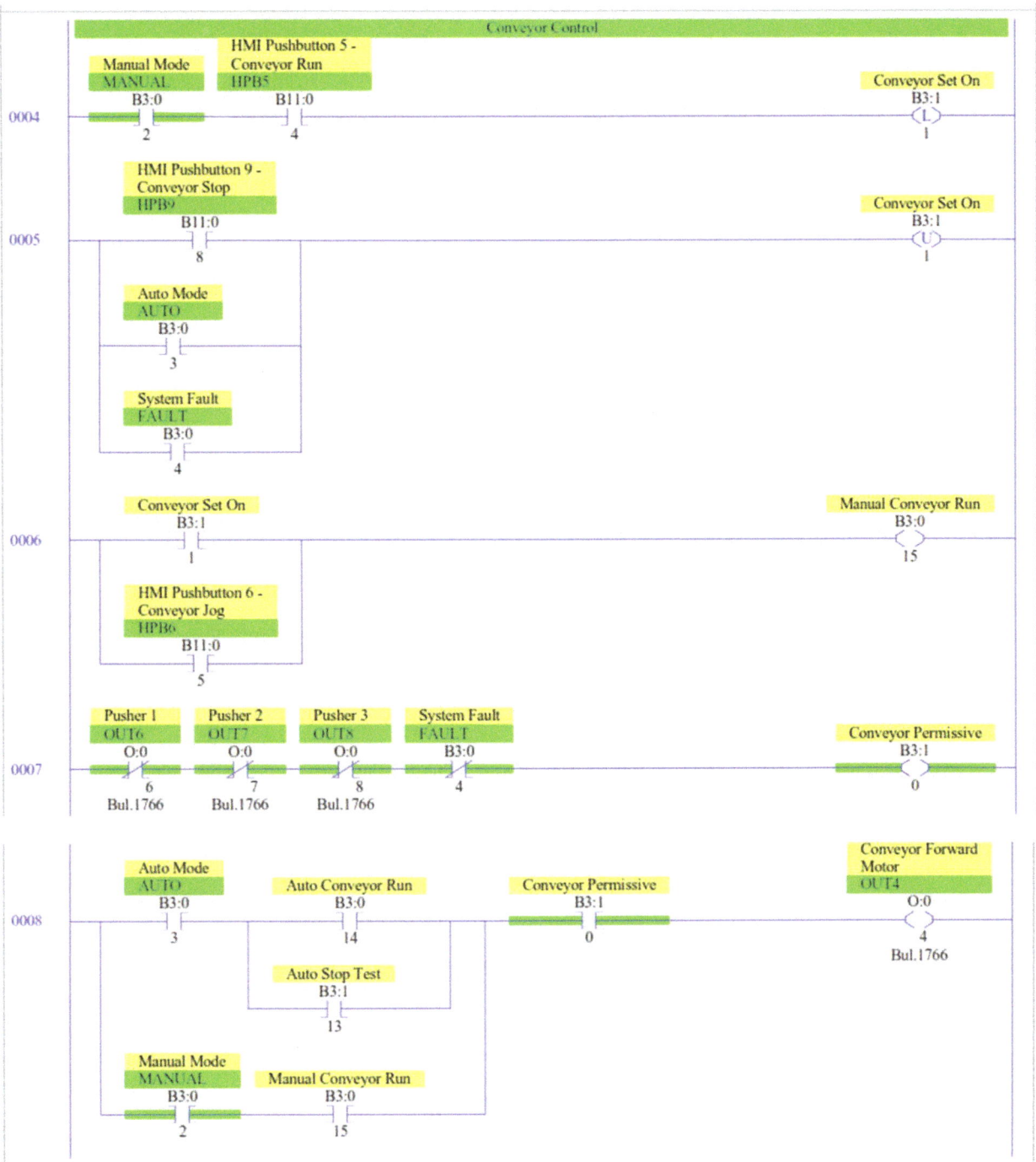

Ladder 6 Outputs Continued

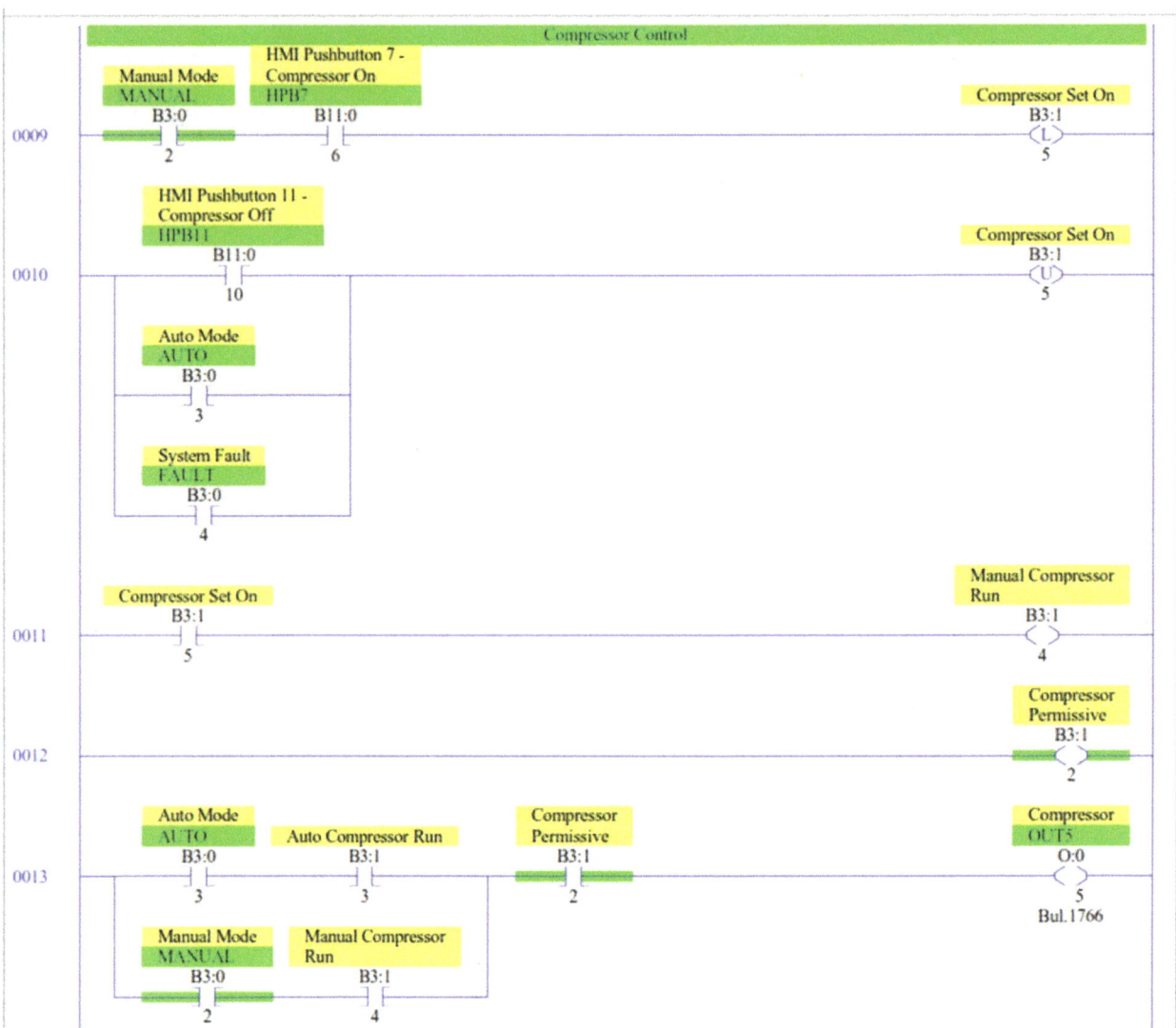

Ladder 6 Outputs Continued

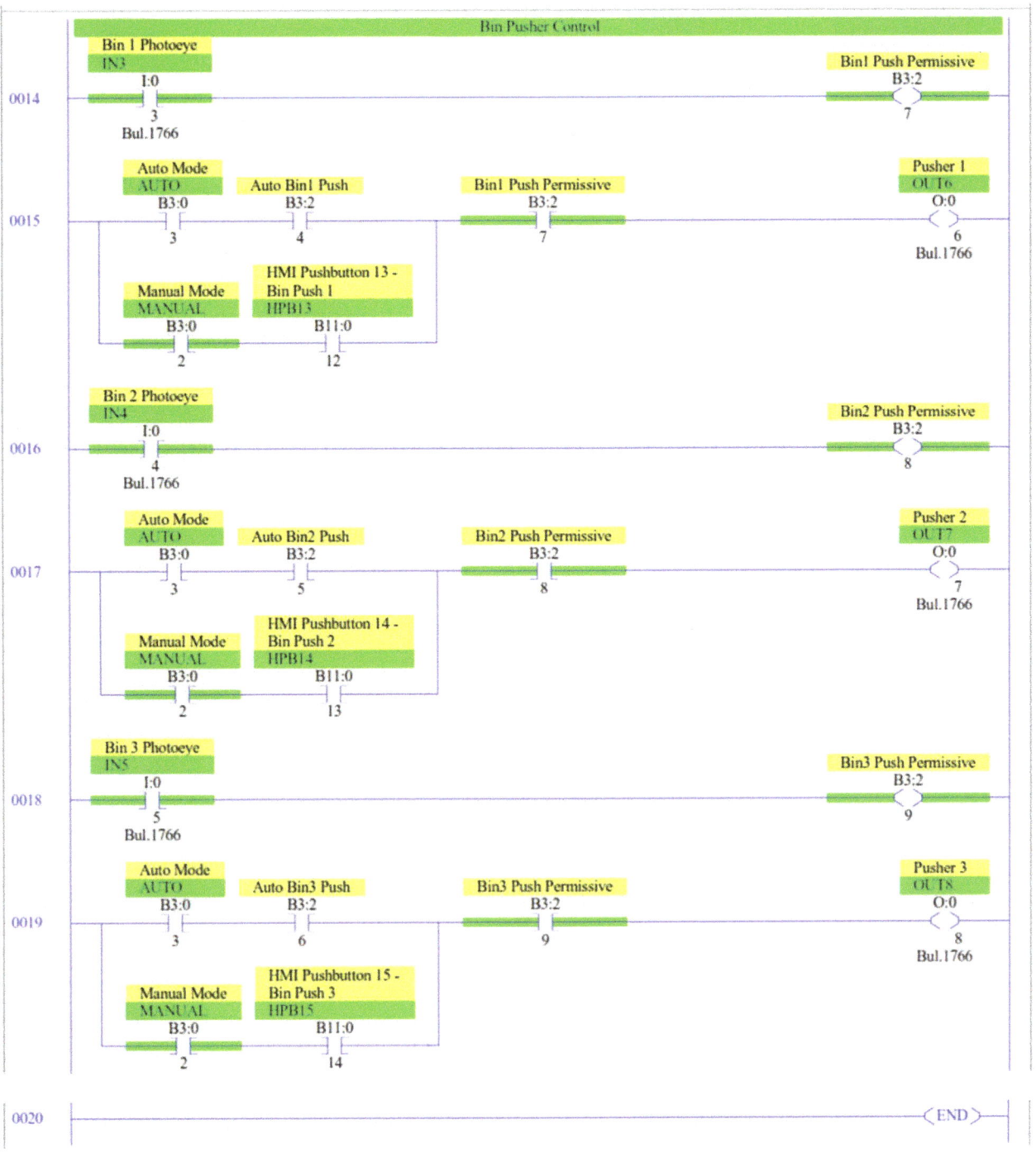

Ladder 7 – Faults

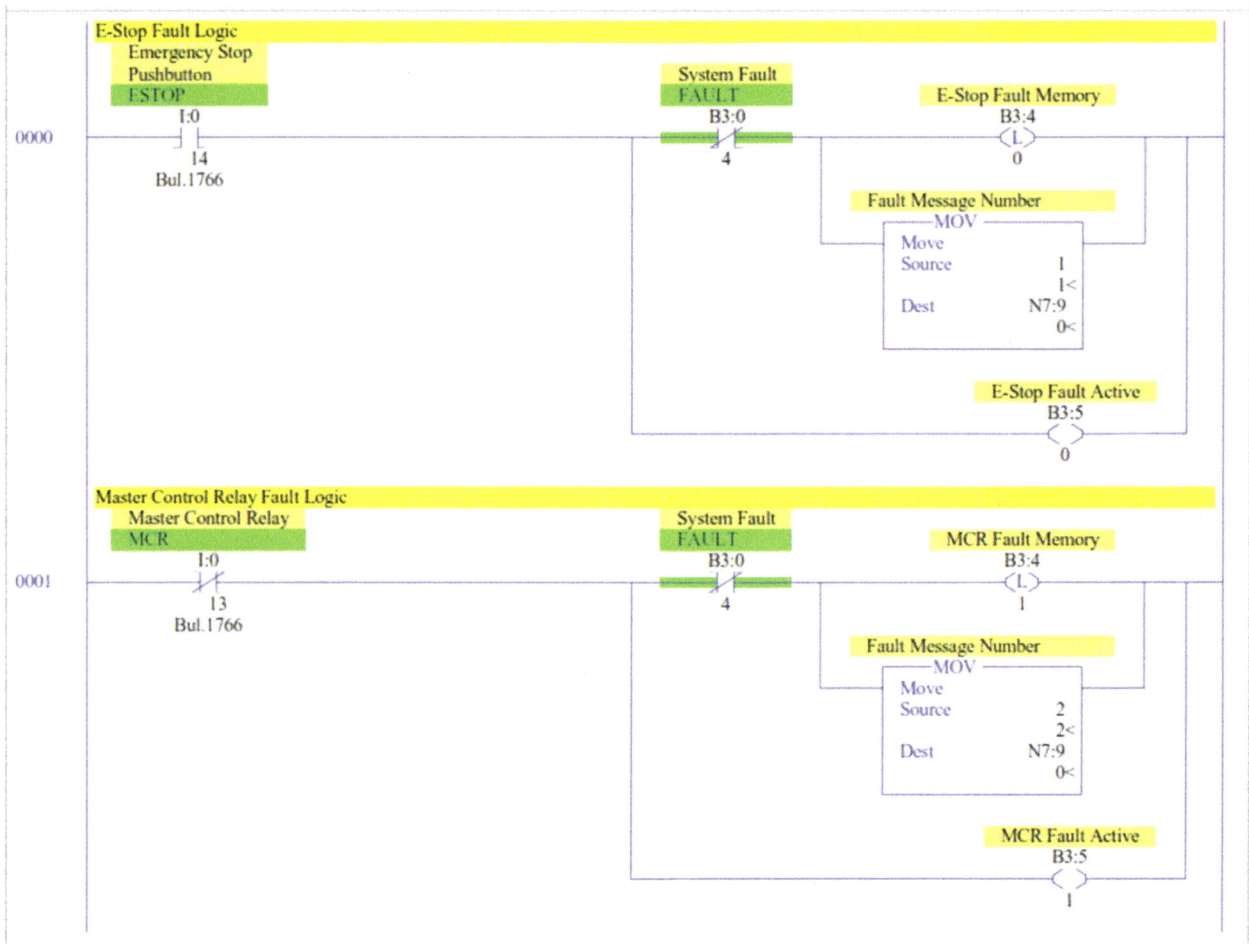

Ladder 7 Faults Continued

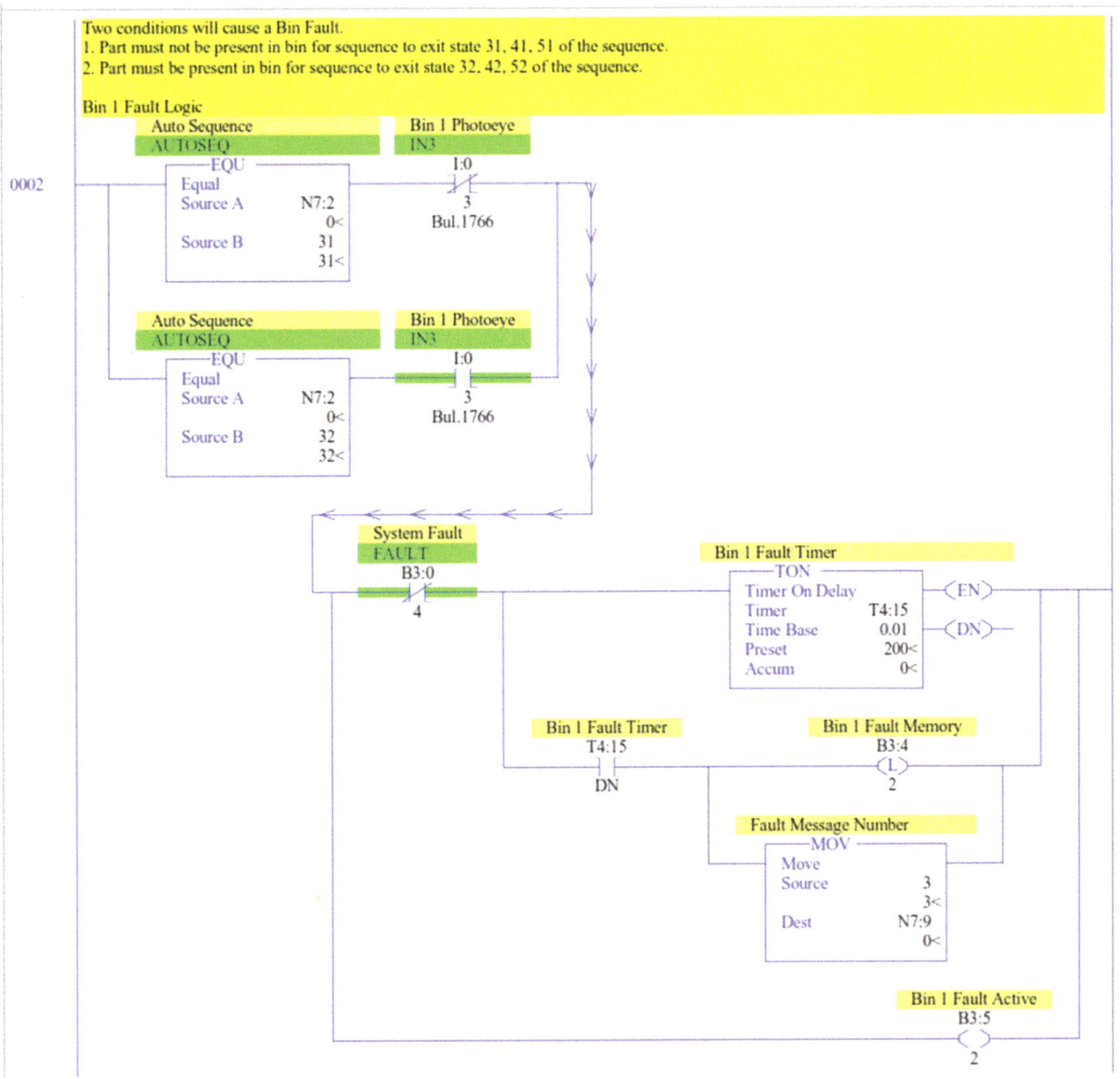

Ladder 7 Faults Continued

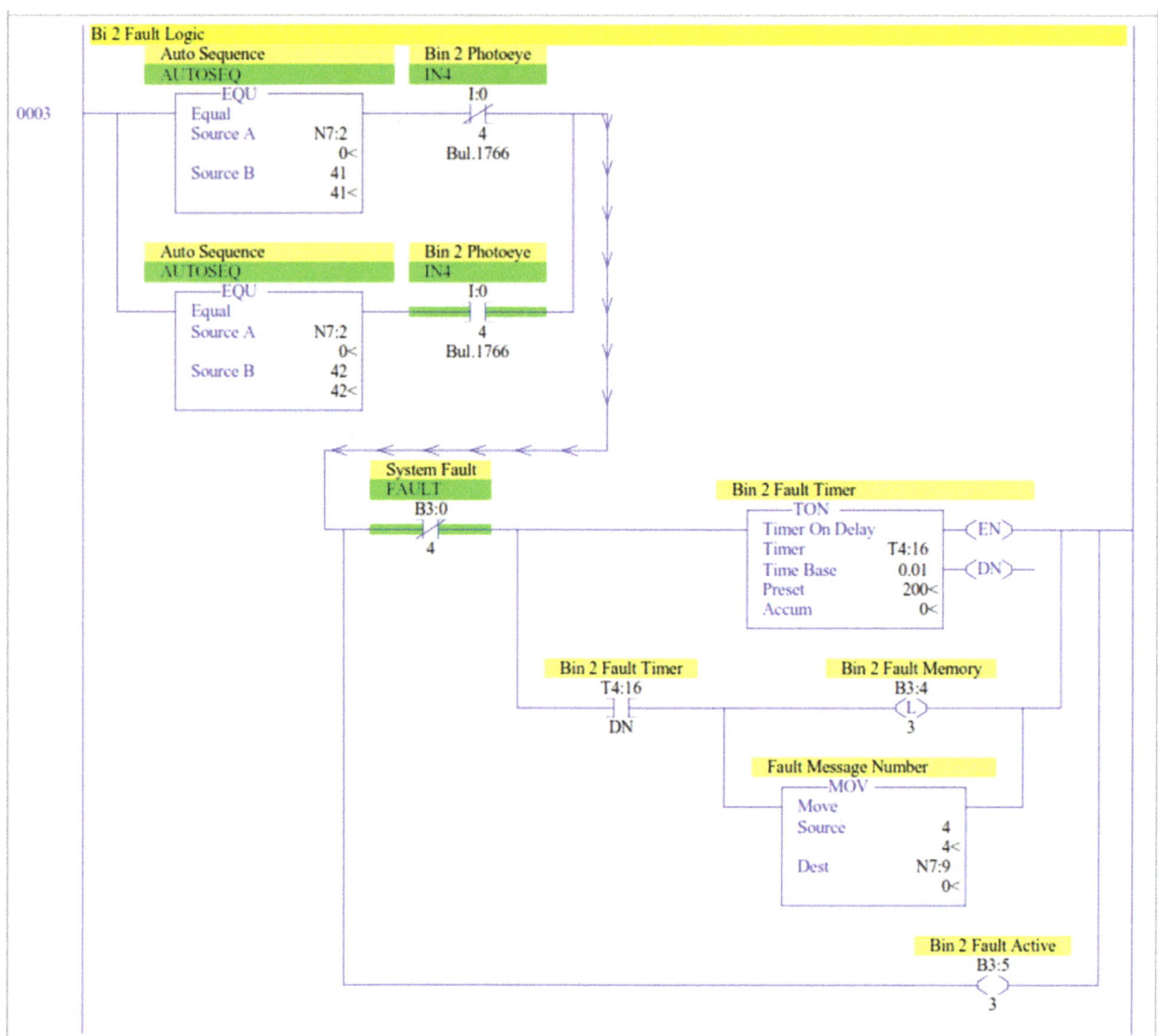

Ladder 7 Faults Continued

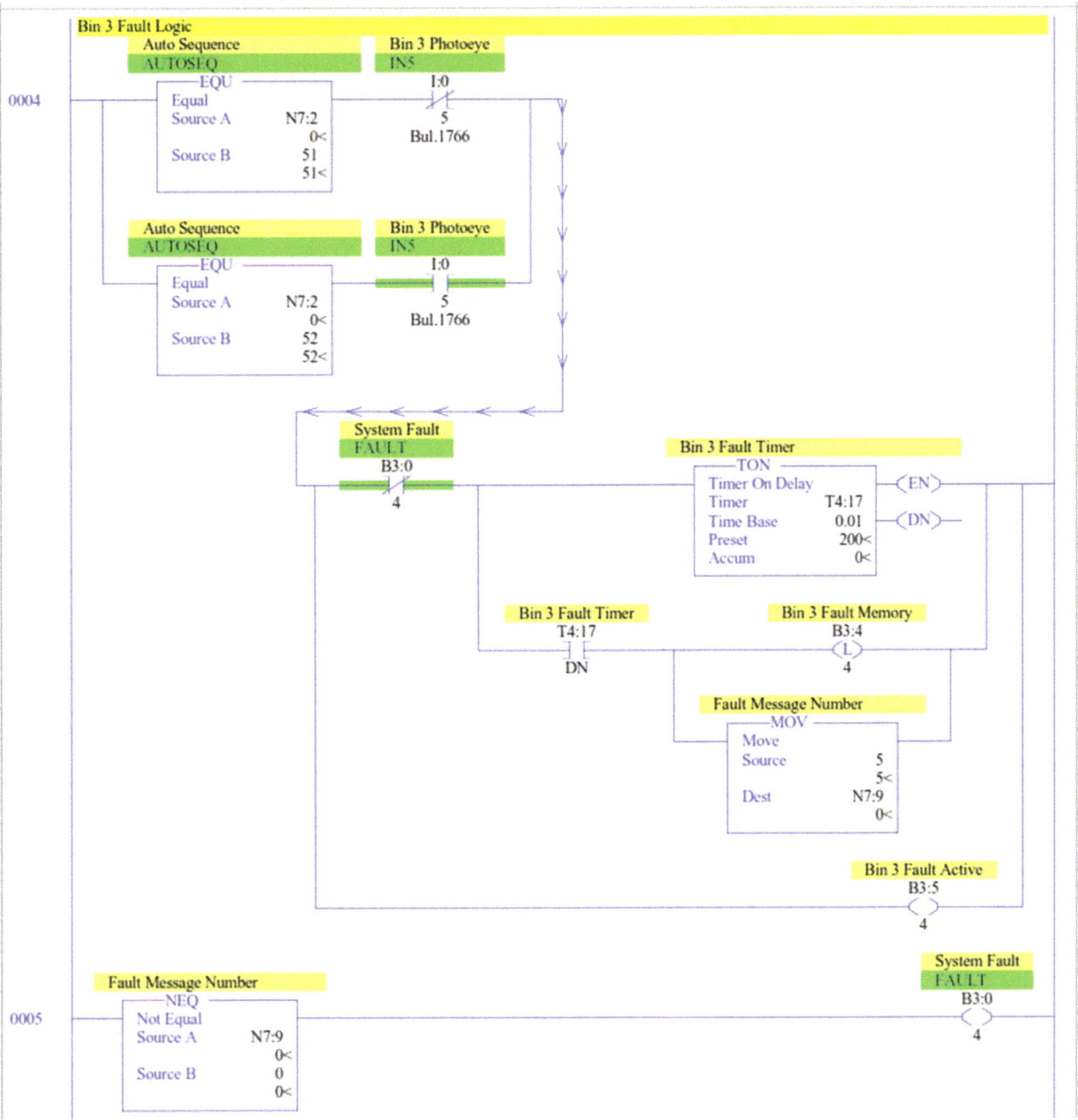

Ladder 7 Faults Continued

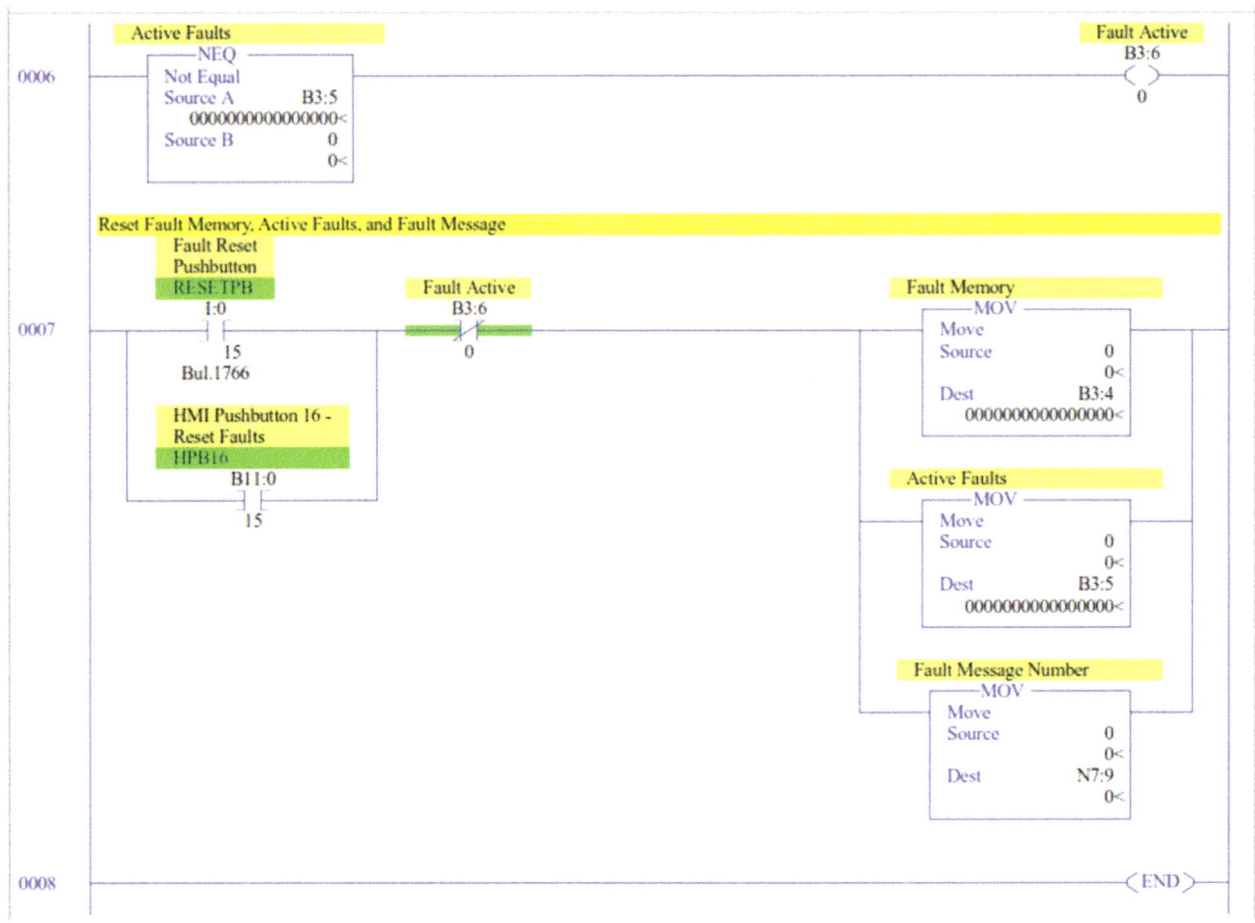

Ladder 8 – HMI

Ladder 8 HMI Continued

Ladder 9 – Production

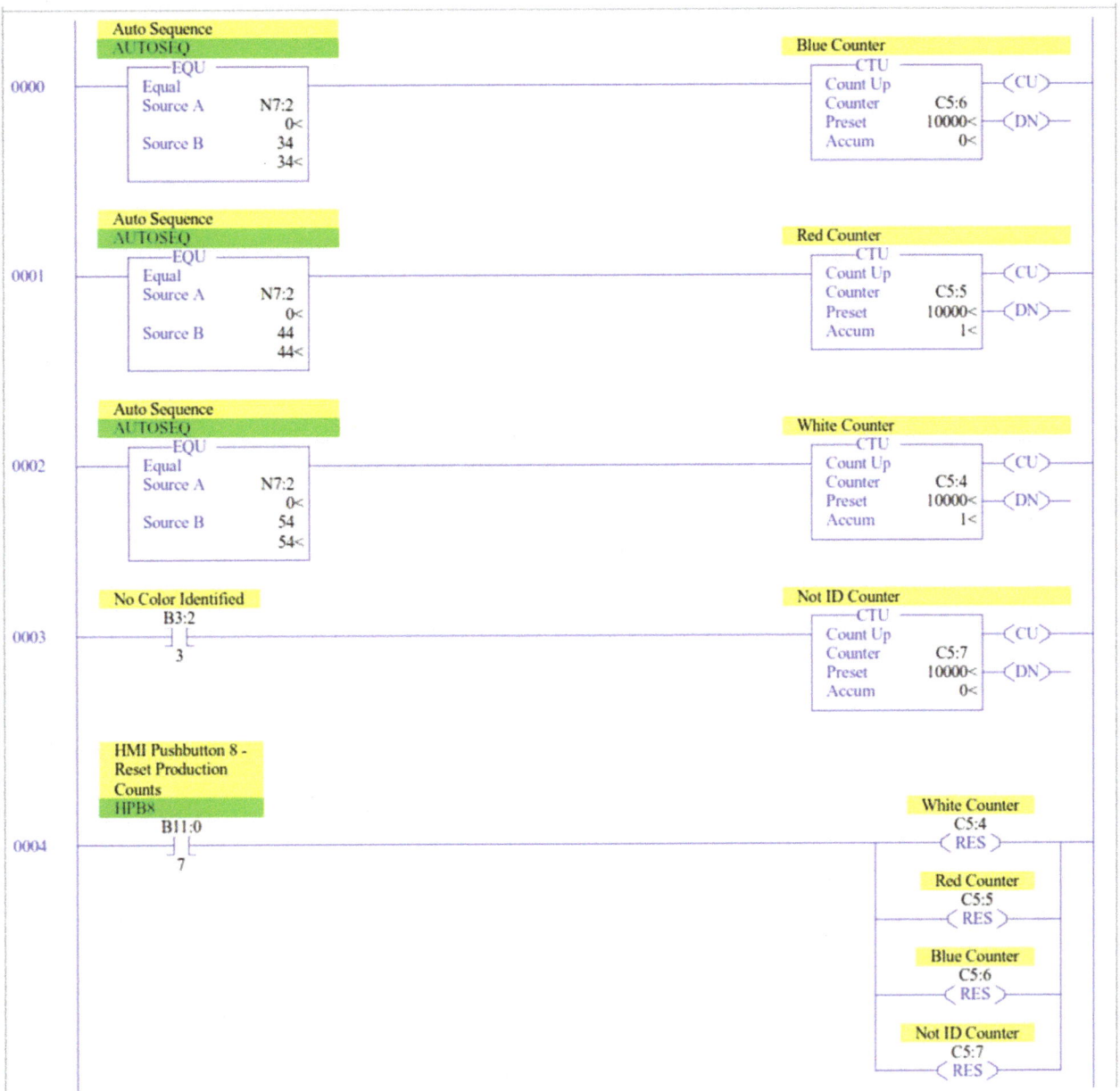

Ladder 9 Production Continued

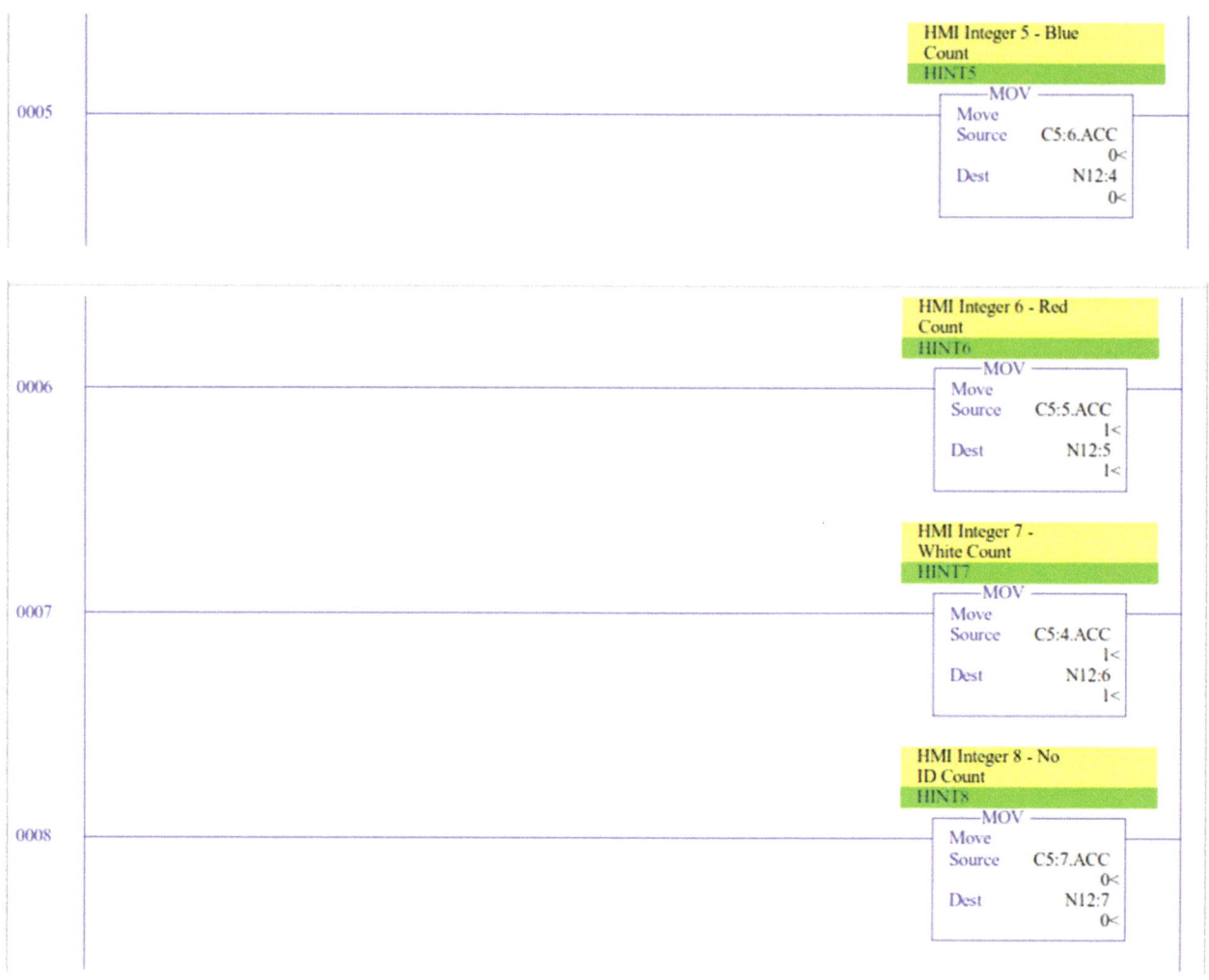

Ladder 9 Production Continued

Ladder 20 – Simulation

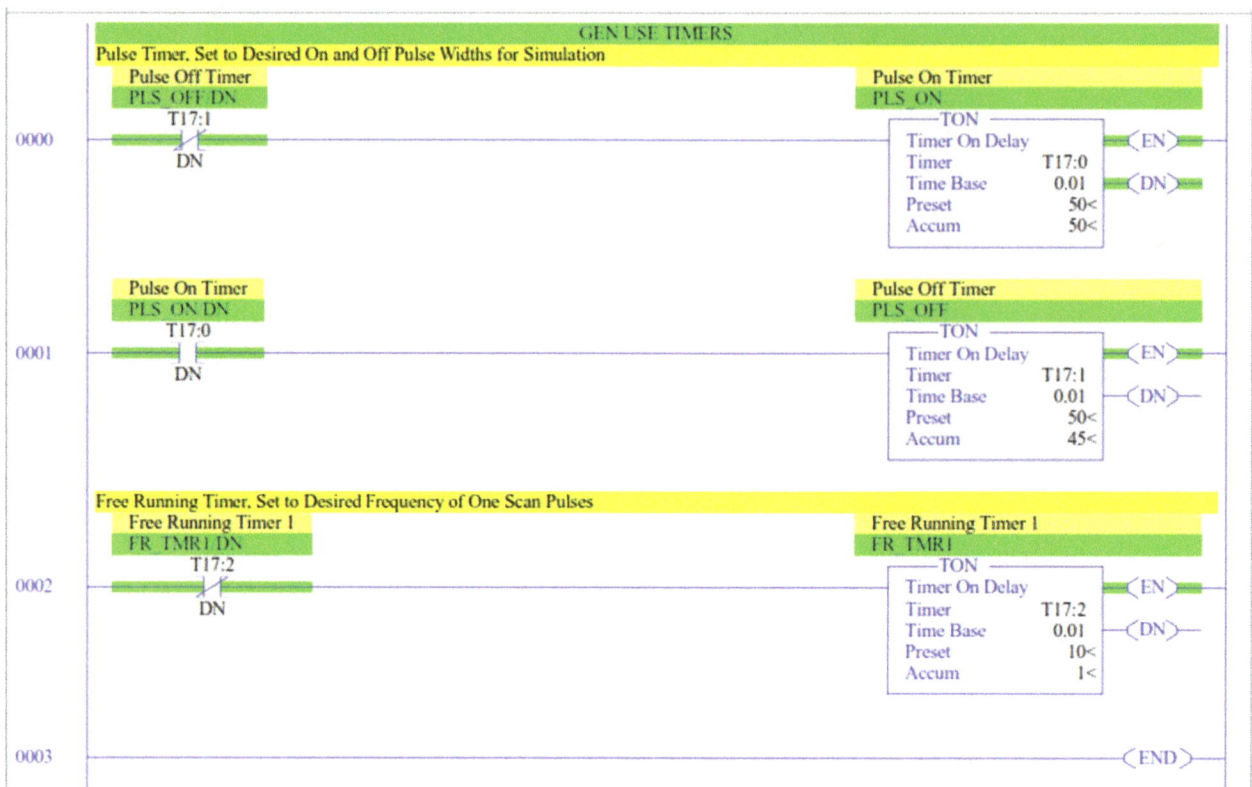

The simulation routine is not used in this lab. It will be used for future exercises with pre-programmed HMI screens and feedback elements from this routine to objects in the HMI.

About the Author

Frank Lamb is an industrial automation consultant and advanced PLC programming trainer with more than 30 years of experience in controls and machine automation.

From 1996 to 2006, Frank owned and operated Automation Consulting Services, Inc. (ACS), a panel building and machine integration company in Knoxville, TN. From 2006 to 2011, he worked as a senior-level project engineer for Wright Industries in Nashville, TN, where he led the design and implementation of large, complex systems and custom machines for multinational corporations and government agencies. In December 2011, Frank re-established Automation Consulting, LLC with a new vision: to use his experience in the field of industrial automation – from electrical, mechanical and controls engineering to project management, training, and machine documentation – to provide expert consulting and training services to manufacturers.

Frank is the president and owner of Automation Consulting, LLC in Nashville, TN and author of *Industrial Automation: Hands On*, published by McGraw-Hill Professional in 2013. He is a United States Air Force veteran, received his BSEE in Electrical and Computer Engineering from the University of Tennessee, and has a Green Belt in Lean Manufacturing/Six Sigma from Purdue University.